# Molecular Orbitals and Organic Chemical Reactions

# Molecular Orbitals and Organic Chemical Reactions

## Student Edition

**Ian Fleming**

Department of Chemistry,
University of Cambridge, UK

**WILEY**

A John Wiley and Sons, Ltd., Publication

This edition first published 2009
© 2009 John Wiley & Sons Ltd

Reprinted with corrections August 2011

*Registered office*
John Wiley & Sons Ltd, The Atrium, Southern Gate, Chichester, West Sussex, PO19 8SQ,
United Kingdom

For details of our global editorial offices, for customer services and for information about how to apply
for permission to reuse the copyright material in this book please see our website at www.wiley.com.

*Library of Congress Cataloging-in-Publication Data*

Fleming, Ian, 1935–
  Molecular orbitals and organic chemical reactions / Ian Fleming.—Student ed.
    p.  cm.
  Includes bibliographical references and index.
  ISBN 978-0-470-74660-8 (cloth)—ISBN 978-0-470-74659-2 (pbk.)   1. Molecular orbitals—
  Textbooks.  2. Physical organic chemistry—Textbooks.  I. Title.
  QD461.F533 2009
  547′.13—dc22
                                                                        2009028760

A catalogue record for this book is available from the British Library.

ISBN 978-0-470-74660-8 (H/B)
     978-0-470-74659-2 (P/B)

Set in 10/12pt Times by Integra Software Services Pvt. Ltd, Pondicherry, India

# Contents

# Preface

Molecular orbital theory is used by chemists to describe the arrangement of electrons in chemical structures. It provides a basis for explaining the ground-state shapes of molecules and their many other properties. As a theory of bonding it has largely replaced valence bond theory,[1] but organic chemists still implicitly use valence bond theory whenever they draw resonance structures. Unfortunately, misuse of valence bond theory is not uncommon as this approach remains in the hands largely of the less sophisticated. Organic chemists with a serious interest in understanding and explaining their work usually express their ideas in molecular orbital terms, so much so that it is now an essential component of every organic chemist's skills to have some acquaintance with molecular orbital theory. The problem is to find a level to suit everyone. At one extreme, a few organic chemists with high levels of mathematical skill are happy to use molecular orbital theory, and its computationally more amenable offshoot density functional theory, much as theoreticians do. At the other extreme are the many organic chemists with lower mathematical inclinations, who nevertheless want to understand their reactions at some kind of physical level. It is for these people that I have written this book. In between there are more and more experimental organic chemists carrying out calculations to support their observations, and these people need to know some of the physical basis for what their calculations are doing.[2]

I have presented molecular orbital theory in a much simplified, and entirely non-mathematical language, in order to make it accessible to every organic chemist, whether student or research worker, whether mathematically competent or not. I trust that every student who has the aptitude will look beyond this book for a better understanding than can be found here. To make it possible for every reader to go further into the subject, there is a larger version of this book,[3] with more discussion, with several more topics and with over 1800 references.

Molecular orbital theory is not only a theory of bonding, it is also a theory capable of giving some insight into the forces involved in the making and breaking of chemical bonds—the chemical reactions that are often the focus of an organic chemist's interest. Calculations on transition structures can be carried out with a bewildering array of techniques requiring more or less skill, more or fewer assumptions, and greater or smaller contributions from empirical input, but many of these fail to provide the organic chemist with insight. He or she wants to know what the physical forces are that give rise to the various kinds of selectivity that are so precious in learning how to control organic reactions. The most accessible theory to give this kind of insight is frontier orbital theory, which is based on the perturbation treatment of molecular orbital theory, introduced by Coulson and Longuet-Higgins,[4] and developed and named as frontier orbital theory by Fukui.[5] While earlier theories of reactivity concentrated on the product-like character of transition structures, perturbation theory concentrates instead on the other side of the reaction coordinate. It looks at how the interaction of the molecular orbitals of the starting materials influences the transition structure. Both influences are obviously important, and it is therefore helpful to know about both if we want a better understanding of what factors affect a transition structure, and hence affect chemical reactivity.

Frontier orbital theory is now widely used, with more or less appropriateness, especially by organic chemists, not least because of the success of the predecessor to this book, *Frontier Orbitals and Organic Chemical Reactions*,[6] which survived for more than thirty years as an introduction to the subject for many of the organic chemists trained in this period. However, there is a problem—computations show that the frontier orbitals do not make a significantly larger contribution than the sum of all the orbitals. As one theoretician put it to me: "It has no right to work as well as it does." The difficulty is that it works as an explanation in many situations where nothing else is immediately compelling. In writing this new book, I have therefore emphasised more the molecular orbital basis for understanding organic chemistry, about which there is less disquiet. Thus I have completely rewritten the earlier book, enlarging especially the chapters on molecular orbital theory itself. I have added a chapter on the effect of orbital interactions on the structures of organic molecules, a section on the theoretical basis for the principle of hard and soft acids and bases, and a chapter on the stereochemistry of the fundamental organic reactions. I have introduced correlation diagrams into the discussion of pericyclic chemistry, and a great deal more in that, the largest chapter. I have also added a number of topics, both omissions from the earlier book and new work that has taken place in the intervening years. I have used more words of caution in discussing frontier orbital theory itself, making it less polemical in furthering that subject, and hoping that it might lead people to be more cautious themselves before applying the ideas uncritically in their own work.

For all their faults and limitations, frontier orbital theory and the principle of hard and soft acids and bases remain the most accessible approaches to understanding many aspects of reactivity. Since they fill a gap between the chemist's experimental results and a state of the art theoretical description of his or her observations, they will continue to be used until something better comes along.

As in the earlier book, I begin by presenting ten experimental observations that chemists have wanted to explain. None of the questions raised by these observations has a simple answer without reference to the orbitals involved.

(i) Why does methyl tetrahydropyranyl ether largely adopt the conformation **P.1**, with the methoxy group axial, whereas methoxycyclohexane adopts largely the conformation **P.2** with the methoxy group equatorial?

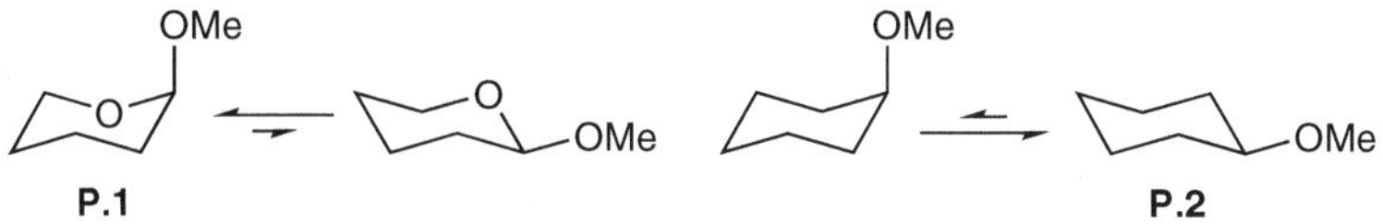

**P.1**                                                                      **P.2**

(ii) Reduction of butadiene **P.3** with sodium in liquid ammonia gives more *cis*-2-butene **P.4** than *trans*-2-butene **P.5**, even though the *trans* isomer is the more stable product.

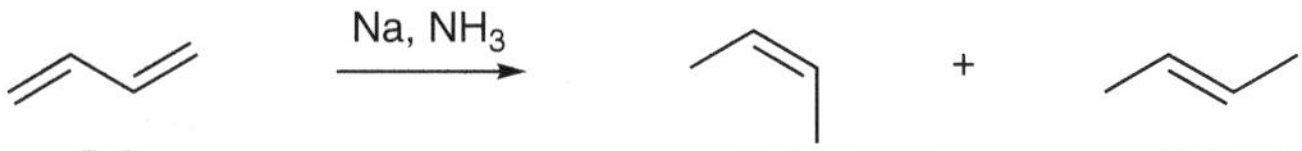

**P.3**                          **P.4**  60%          +          **P.5**  40%

(iii) Why do enolate ions react more rapidly with protons on oxygen **P.6**, but with primary alkyl halides on carbon **P.7**?

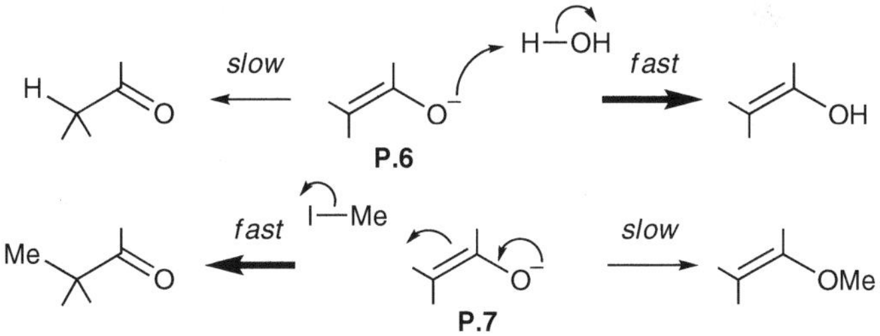

**P.6**

**P.7**

(iv) Hydroperoxide ion **P.8** is much less basic than hydroxide ion **P.9**. Why, then, is it so much more nucleophilic?

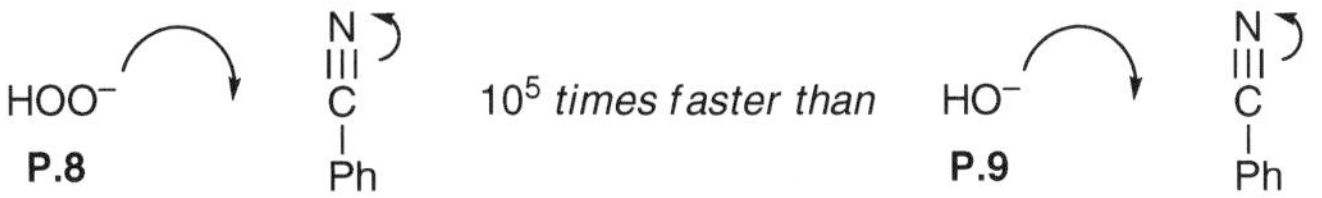

HOO⁻                                                    $10^5$ *times faster than*          HO⁻

**P.8**                    Ph                                                                          **P.9**                    Ph

(v) Why does butadiene **P.10** react with maleic anhydride **P.11**, but ethylene **P.12** does not?

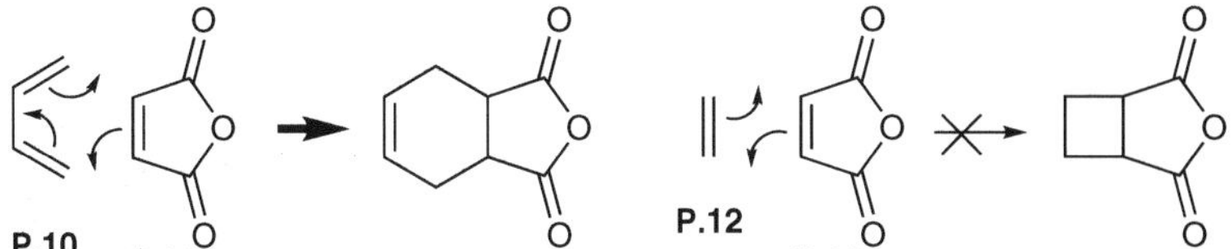

**P.10**        **P.11**                                              **P.12**        **P.11**

(vi) Why do Diels-Alder reactions of butadiene **P.10** go so much faster when there is an electron withdrawing group on the dienophile, as with maleic anhydride **P.11**, than they do with ethylene **P.12**?

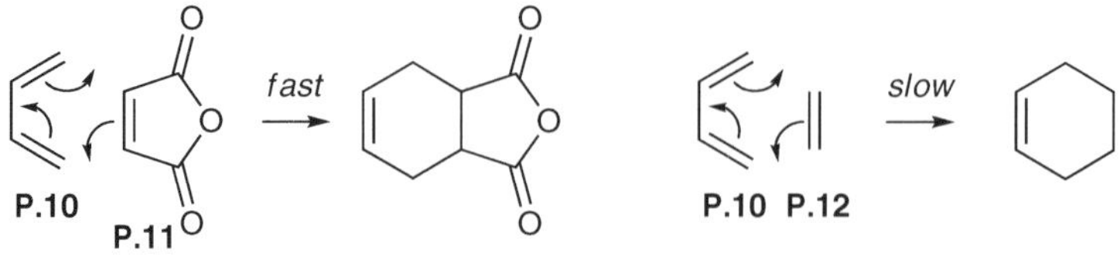

P.10
P.11
*fast*
P.10 P.12
*slow*

(vii) Why does diazomethane **P.14** add to methyl acrylate **P.15** to give the isomer **P.16** in which the nitrogen end of the dipole is bonded to the carbon atom bearing the methoxycarbonyl group, and not the other way round **P.13**?

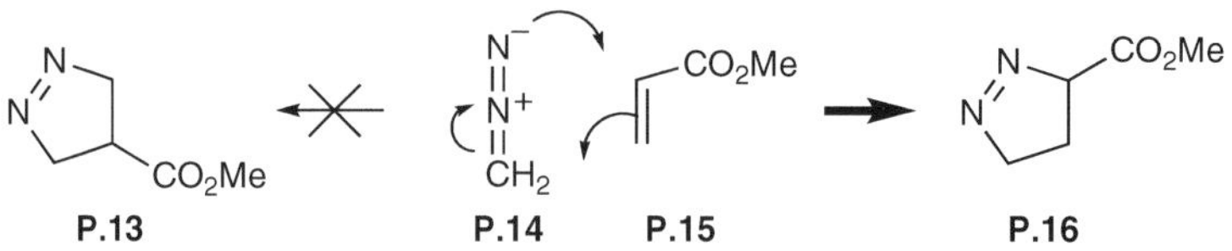

P.13
P.14
P.15
P.16

(viii) When methyl fumarate **P.17** and vinyl acetate **P.18** are co-polymerised with a radical initiator, why does the polymer **P.19** consist largely of *alternating* units?

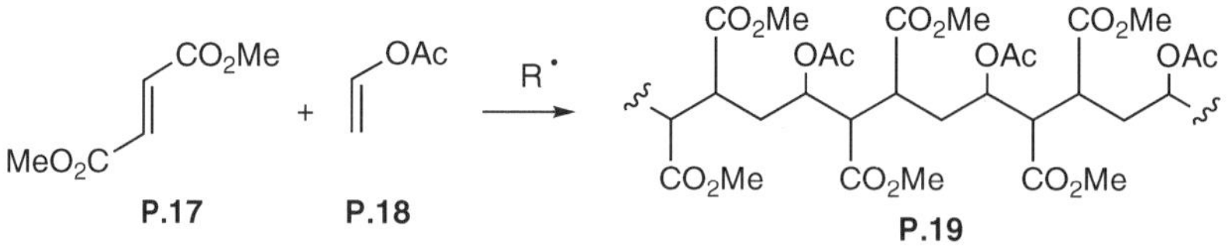

P.17
P.18
P.19

(ix) Why does the Paterno-Büchi reaction between acetone and acrylonitrile give only the isomer **P.20** in which the two 'electrophilic' carbon atoms become bonded?

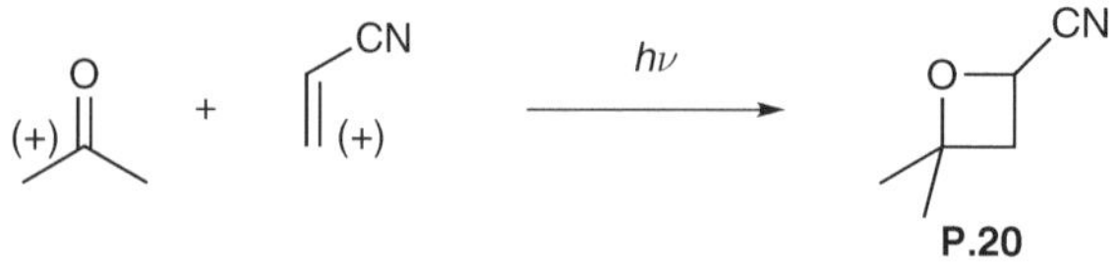

P.20

In the following chapters, each of these questions, and many others, receives a simple answer. Other books commend themselves to anyone able and willing to go further up the mathematical slopes towards a more acceptable level of explanation—a few introductory texts take the next step up[7], and several others take the journey further[8].

I have been greatly helped by a number of chemists: those who gave me advice for the earlier book, and who therefore made their mark on this: Dr. W. Carruthers, Professor R. F. Hudson, Professor A. R. Katritzky and Professor A. J. Stone. In addition, for this book, I am indebted to Dr. Jonathan Goodman for help with computer programmes, and to Professor A. D. Buckingham for much helpful correction. More than usually, I must absolve all of them of any errors left in this book.

# 1 Molecular Orbital Theory

## 1.1 The Atomic Orbitals of a Hydrogen Atom

The spatial distribution of the electron in a hydrogen atom is usually expressed as a *wave function* $\phi$, where $\phi^2 d\tau$ is the probability of finding the electron in the volume $d\tau$, and the integral of $\phi^2 d\tau$ over the whole of space is 1. The wave function is the underlying mathematical description, and it may be positive or negative. Only when squared does it correspond to anything with physical reality—the probability of finding an electron in any given space. Quantum theory gives us a number of permitted wave equations, but the one that matters here is the lowest in energy, in which the electron is in a 1s orbital. This is spherically symmetrical about the nucleus, with a maximum at the centre, and falling off rapidly, so that the probability of finding the electron within a sphere of radius 1.4 Å is 90% and within 2 Å better than 99%. This orbital is calculated to be 13.60 eV lower in energy than a completely separated electron and proton.

We need pictures to illustrate the electron distribution, and the most common is simply to draw a circle, Fig. 1.1a, which can be thought of as a section through a spherical contour, within which the electron would be found, say, 90% of the time. Fig. 1.1b is a section showing more contours and Fig. 1.1c is a section through a cloud, where one imagines blinking one's eyes a very large number of times, and plotting the points at which the electron was at each blink. This picture contributes to the language often used, in which the electron population in a given volume of space is referred to as the electron density. Taking advantage of the spherical symmetry, we can also plot the fraction of the electron population outside a radius $r$ against $r$, as in Fig. 1.2a, showing the rapid fall off of electron population with distance. The van der Waals radius at 1.2 Å has no theoretical significance—it is an empirical measurement from solid-state structures, being one-half of the distance apart of the hydrogen atoms in the C—H bonds of adjacent molecules. It is an average of several measurements. Yet another way to appreciate the electron distribution is to look at the radial density, where we plot the probability of finding the electron between one sphere of radius $r$ and another of radius $r + dr$. This has the form, Fig. 1.2b, with a maximum 0.529 Å from the nucleus, showing that, in spite of the wave function being at a maximum at the

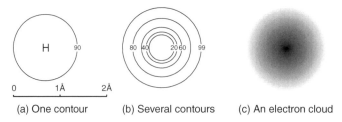

(a) One contour          (b) Several contours          (c) An electron cloud

**Fig. 1.1**    The 1s atomic orbital of a hydrogen atom

nucleus, the chance of finding an electron precisely there is very small. The distance
0.529 Å proves to be the same as the radius calculated for the orbit of an electron in the
early but untenable planetary model of a hydrogen atom. It is called the Bohr radius $a_0$,
and is often used as a unit of length in molecular orbital calculations.

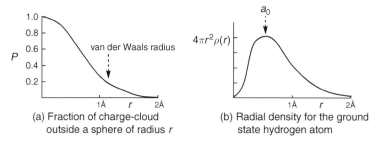

(a) Fraction of charge-cloud          (b) Radial density for the ground
outside a sphere of radius $r$          state hydrogen atom

**Fig. 1.2**    Radial probability plots for the 1s orbital of a hydrogen atom

## 1.2   Molecules made from Hydrogen Atoms

### 1.2.1   The H$_2$ Molecule

To understand the bonding in a hydrogen molecule, we have to see what happens
when the atoms are close enough for their atomic orbitals to interact. We need a
description of the electron distribution over the whole molecule. We accept that a first
approximation has the two atoms remaining more or less unchanged, so that the
description of the molecule will resemble the sum of the two isolated atoms. Thus we
combine the two atomic orbitals in a linear combination expressed in Equation 1.1,
where the function which describes the new electron distribution, the *molecular
orbital*, is called $\sigma$ and $\phi_1$ and $\phi_2$ are the atomic 1s wave functions on atoms 1 and 2.

$$\sigma = c_1\phi_1 + c_2\phi_2 \hspace{3cm} 1.1$$

The coefficients, $c_1$ and $c_2$, are a measure of the contribution which the atomic
orbital is making to the molecular orbital. They are of course equal in magnitude in
this case, since the two atoms are the same, but they may be positive or negative. To
obtain the electron distribution, we square the function in Equation 1.1, which is
written in two ways in Equation 1.2.

$$\sigma^2 = (c_1\phi_1 + c_2\phi_2)^2 = (c_1\phi_1)^2 + (c_2\phi_2)^2 + 2c_1\phi_1 c_2\phi_2 \qquad 1.2$$

Taking the expanded version, we can see that the molecular orbital $\sigma^2$ differs from the superposition of the two atomic orbitals $(c_1\phi_1)^2 + (c_2\phi_2)^2$ by the term $2c_1\phi_1 c_2\phi_2$. Thus we have two solutions (Fig. 1.3). In the first, both $c_1$ and $c_2$ are positive, with orbitals of the same sign placed next to each other; the electron population *between* the two atoms is increased (shaded area), and hence the negative charge which these electrons carry *attracts* the two positively charged nuclei. This results in a lowering in energy and is illustrated in Fig. 1.3, where the bold horizontal line next to the drawing of this orbital is placed low on the diagram. Alternatively, $c_1$ and $c_2$ are of opposite sign, and we represent the sign change by shading one of the orbitals, calling the plane which divides the function at the sign change a node. If there were any electrons in this orbital, the reduced electron population between the nuclei would lead to repulsion between them and a raised energy for this orbital. In summary, by making a bond between two hydrogen atoms, we create two new orbitals, $\sigma$ and $\sigma^*$, which we call the molecular orbitals; the former is *bonding* and the latter *antibonding* (an asterisk generally signifies an antibonding orbital). In the ground state of the molecule, the two electrons will be in the orbital labelled $\sigma$. There is, therefore, when we make a bond, a lowering of energy equal to twice the value of $E_\sigma$ in Fig. 1.3 (*twice* the value, because there are two electrons in the bonding orbital).

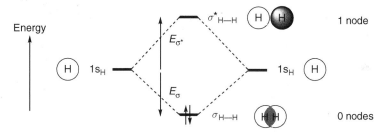

**Fig. 1.3** The molecular orbitals of the hydrogen molecule

The force holding the two atoms together is obviously dependent upon the extent of the overlap in the bonding orbital. If we bring the two 1s orbitals from a position where there is essentially no overlap at 2.5 Å through the bonding arrangement to superimposition, the extent of overlap steadily increases. The mathematical description of the overlap is an integral $S_{12}$ (Equation 1.3) called the *overlap integral*, which, for a pair of 1s orbitals rises from 0 at infinite separation to 1 at superimposition (Fig. 1.4).

$$S_{12} = \int \phi_1 \phi_2 d\tau \qquad 1.3$$

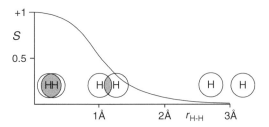

**Fig. 1.4** The overlap integral $S$ for two $1s_H$ orbitals as a function of internuclear distance

The mathematical description of the effect of overlap on the electronic energy is complex, but some of the terminology is worth recognising. The energy $E$ of an electron in a bonding molecular orbital is given by Equation 1.4 and for the antibonding molecular orbital is given by Equation 1.5:

$$E = \frac{\alpha + \beta}{1 + S} \qquad\qquad 1.4$$

$$E = \frac{\alpha - \beta}{1 - S} \qquad\qquad 1.5$$

in which the symbol $\alpha$ represents the energy of an electron in an isolated atomic orbital, and is called a *Coulomb integral*. The function represented by the symbol $\beta$ contributes to the energy of an electron in the field of both nuclei, and is called the *resonance* integral. It is roughly proportional to $S$, and so the overlap integral appears in the equations twice. It is important to realise that the use of the word resonance does not imply an oscillation, nor is it the same as the 'resonance' of valence bond theory. In both cases the word is used because the mathematical form of the function is similar to that for the mechanical coupling of oscillators. We also use the words *delocalised* and *delocalisation* to describe the electron distribution enshrined in the $\beta$ function—unlike the words resonating and resonance, these are not misleading, and are the better words to use.

The function $\beta$ is a negative number, lowering the value of $E$ in Equation 1.4 and raising it in Equation 1.5. In this book, $\beta$ will not be given a sign on the diagrams on which it is used, because the sign can be misleading. The symbol $\beta$ should be interpreted as $|\beta|$, the positive absolute value of $\beta$. Since the diagrams are always plotted with energy upwards and almost always with the $\alpha$ value visible, it should be obvious which $\beta$ values lead to a lowering of the energy below the $\alpha$ level, and which to raising the energy above it.

The overall effect on the energy of the hydrogen molecule relative to that of two separate hydrogen atoms as a function of the internuclear distance is given in Fig. 1.5. If the bonding orbital is filled, the energy derived from the electronic contribution (Equation 1.4) steadily falls as the two hydrogen atoms are moved from infinity towards one another (curve A). At the same time the nuclei repel each other ever more strongly, and the nuclear contribution to the energy goes steadily up (curve B).

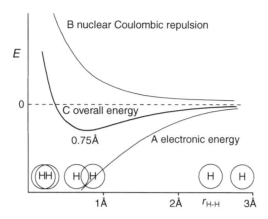

**Fig. 1.5**  Electronic attraction, nuclear repulsion and the overall effect as a function of internuclear distance for two $1s_H$ atoms

The sum of these two is the familiar Morse plot (curve C) for the relationship between internuclear distance and energy, with a minimum at the bond length. If we had filled the antibonding orbital instead, the resultant curve would be a steady increase in energy as the nuclei are pushed together. The characteristic of a bonding orbital is that the nuclei are held together, whereas the characteristic of an antibonding orbital, if it were to be filled, is that the nuclei would fly apart unless there are enough compensating filled bonding orbitals. In hydrogen, having both orbitals occupied is overall antibonding, and there is no possibility of compensating for a filled antibonding orbital.

We can see from the form of Equations 1.4 and 1.5 that the term $\alpha$ relates to the energy levels of the isolated atoms labelled $1s_H$ in Fig. 1.3, and the term $\beta$ to the drop in energy labelled $E_\sigma$ (and the rise labelled $E_{\sigma*}$). Equations 1.4 and 1.5 show that, since the denominator in the bonding combination is $1 + S$ and the denominator in the antibonding combination is $1 - S$, the bonding orbital is not as much lowered in energy as the antibonding is raised. In addition, putting *two* electrons into a bonding orbital does not achieve exactly twice the energy-lowering of putting *one* electron into it. We are *allowed* to put two electrons into the one orbital if they have opposite spins, but they still repel each other, because they have the same sign and have to share the same space; consequently, in forcing a second electron into the $\sigma$ orbital, we lose some of the bonding we might otherwise have gained. For this reason too, the value of $E_\sigma$ in Fig. 1.3 is smaller than that of $E_{\sigma*}$. This is why two helium atoms do not combine to form an He$_2$ molecule. There are four electrons in two helium atoms, two of which would go into the $\sigma$-bonding orbital and two into the $\sigma$*-antibonding orbital. Since $2E_{\sigma*}$ is greater than $2E_\sigma$, we would need extra energy to keep the two helium atoms together. Two electrons in the same orbital can keep out of each other's way, with one electron on one side of the orbital, while the other is on the other side most of the time, and so the energetic penalty for having a second electron in the orbital is not large. This synchronisation of the electrons' movements is referred to as *electron correlation*. The energy-raising effect of the repulsion of one electron by the other is automatically included in calculations based on Equations 1.4 and 1.5, but each electron is treated as having an average distribution with respect to the other. The effect of electron correlation is often not included, without much penalty in accuracy, but when it is included the calculation is described as being with *configuration interaction*, a bit of fine tuning added to a careful calculation.

The detailed form that $\alpha$ and $\beta$ take is where the mathematical complexity appears. They come from the Schrödinger equation, and they are integrals over all coordinates, represented here simply by $d\tau$, in the form of Equations 1.6 and 1.7:

$$\alpha = \int \phi_1 H \phi_1 d\tau \qquad\qquad 1.6$$

$$\beta = \int \phi_1 H \phi_2 d\tau \qquad\qquad 1.7$$

where $H$ is the energy operator known as a Hamiltonian. Even without going into this in more detail, it is clear how the term $\alpha$ relates to the atom, and the term $\beta$ to the interaction of one atom with another.

As with atomic orbitals, we need pictures to illustrate the electron distribution in the molecular orbitals. For most purposes, the conventional drawings of the bonding and antibonding orbitals in Fig. 1.3 are clear enough—we simply make mental reservations about what they represent. In order to be sure that we do understand enough detail, we can look at a slice through the two atoms showing the contours (Fig. 1.6). Here we see in the bonding orbital that the electron population close in to the nucleus is pulled in to the midpoint between the nuclei (Fig. 1.6a), but that further out the contours are an elliptical envelope with the nuclei as the foci. The antibonding orbital, however, still has some dense contours between the nuclei, but further out the electron population is pushed out on the back side of each nucleus. The node is halfway between the nuclei, with the change of sign in the wave function symbolised by the shaded contours on the one side. If there were electrons in this orbital, their distribution on the outside would pull the nuclei apart—the closer the atoms get, the more the electrons are pushed to the outside.

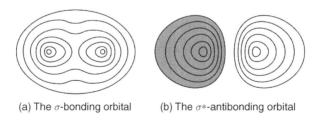

(a) The $\sigma$-bonding orbital        (b) The $\sigma*$-antibonding orbital

**Fig. 1.6**   Contours of the wave function of the molecular orbitals of $H_2$

The coefficients $c_1$ and $c_2$ in Equation 1.1 are a measure of the contribution which each atomic orbital is making to the molecular orbital. When there are electrons in the orbital, the squares of the $c$-values are a measure of the electron population in the neighbourhood of the atom in question. Thus *in each orbital* the sum of the squares of all the $c$-values must equal one, since only one electron in each spin state can be in the orbital. Since $|c_1|$ must equal $|c_2|$ in a homonuclear diatomic like $H_2$, we have defined what the values of $c_1$ and $c_2$ in the bonding orbital must be, namely $1/\sqrt{2} = 0.707$:

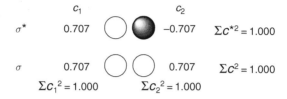

If all molecular orbitals were filled, then there would have to be one electron in each spin state on each atom, and this gives rise to a second criterion for $c$-values, namely that the sum of the squares of all the $c$-values *on any one atom* in *all* the molecular orbitals must also equal one. Thus the $\sigma^*$-antibonding orbital of hydrogen will have $c$-values of 0.707 and $-0.707$, because these values make the whole set fit both criteria.

### 1.2.2   The $H_3$ Molecule

We might ask whether we can join more than two hydrogen atoms together. We shall consider first the possibility of joining three atoms together in a triangular arrangement. With three atomic orbitals to combine, we can no longer simply draw an interaction diagram as we did in Fig. 1.3, where there were only two atomic orbitals. One way of dealing with the problem is first to take two of them together to form a hydrogen molecule, and then we combine the $\sigma$ and $\sigma^*$ orbitals, on the right of Fig. 1.7, with the 1s orbital of the third hydrogen atom on the left.

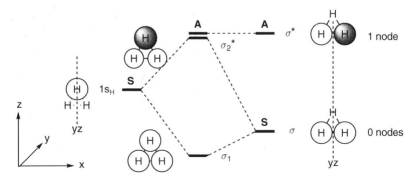

**Fig. 1.7**   Interacting orbitals for $H_3$

We now meet an important rule: we are only allowed to combine those orbitals that have the same symmetry with respect to all the symmetry elements present in the structure of the product and in the orbitals of the components we are combining. This problem did not arise in forming a bond between two identical hydrogen atoms, because they have inherently the same symmetry, but now we are combining different sets of orbitals with each other. The first task is to identify the symmetry elements, and classify the orbitals with respect to them. Because all the orbitals are s orbitals, there is a trivial symmetry plane in the plane of the page, which we shall label throughout this book as the xz plane. We can ignore it, and other similar symmetry elements, in this case. The only symmetry element that is not trivial is the plane in what we shall call the yz plane, running from top to bottom of the page and rising vertically from it. The $\sigma$ orbital and the 1s orbital are symmetric with respect to this plane, but the $\sigma^*$ orbital is antisymmetric, because the component atomic orbitals are out of phase. We therefore label the orbitals as S (symmetric) or A (antisymmetric).

The $\sigma$ orbital and the 1s orbital are both S and they can interact in the same way as we saw in Fig. 1.3, to create a new pair of molecular orbitals labelled $\sigma_1$ and $\sigma_2^*$. The former is lowered in energy, because all the s orbitals are of the same sign, and the latter is raised in energy, because there is a node between the top hydrogen atom and the two bottom ones. The latter orbital is antibonding overall, because, in the triangular arrangement we have chosen here, there are two antibonding interactions between hydrogen atoms and only one bonding interaction. As it happens, its energy is the same as that of the $\sigma^*$ orbital, but we cannot justify that now. In any case, the other orbital $\sigma^*$ remains unchanged in the $H_3$ molecule, because there is no orbital of the correct symmetry to interact with it.

Thus we have three molecular orbitals, just as we had three atomic orbitals to make them from. Whether we have a stable 'molecule' now depends upon how many electrons we have. If we have two in a protonated hydrogen molecule, $H_3^+$, they would both go into the $\sigma_1$ orbital, and the molecule would have a lower electronic energy than the separate proton and $H_2$ molecule. If we had three electrons $H_3 \cdot$ from combining three hydrogen atoms, we would also have a stable 'molecule', with *two* electrons in $\sigma_1$ and only *one* in $\sigma_2^*$, making the combination overall more bonding than antibonding. Only with four electrons in $H_3^-$ is the overall result antibonding, because the energy-raising interaction is, as usual, greater than the energy-lowering interaction. This device of building up the orbitals and only then feeding the electrons in is known as the *aufbau* method.

### 1.2.3    The $H_4$ 'Molecule'

There are several possible ways of arranging four hydrogen atoms, but we shall limit ourselves to tetrahedral, since we shall be using these orbitals later. This time, we combine them in pairs to create the orbitals for two hydrogen molecules, and then ask ourselves what happens to the energy when the two molecules are held within bonding distance.

With one hydrogen molecule aligned along the x axis, on the right in Fig. 1.8, and the other along the y axis, on the left of Fig. 1.8, the symmetry elements present are then the xz and yz planes. The bonding orbital $\sigma_x$ on the right is symmetric with respect to both planes, and is labelled SS. The antibonding orbital $\sigma_x^*$ is symmetric with respect to the xz plane but antisymmetric with respect to the yz plane, and is accordingly labelled SA. The bonding orbital $\sigma_y$ on the left is symmetric with respect to both planes, and is also labelled SS. The antibonding orbital $\sigma_y^*$ is antisymmetric with respect to the xz plane but symmetric with respect to the yz plane, and is labelled AS. The only orbitals with the same symmetry are therefore the two bonding orbitals, and they can interact to give a bonding combination $\sigma_1$ and an antibonding combination $\sigma_2^*$. As it happens, the latter has the same energy as the unchanged orbitals $\sigma_x^*$ and $\sigma_y^*$.

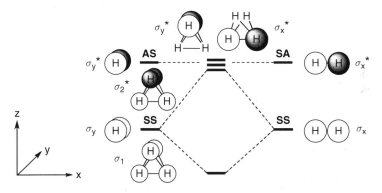

**Fig. 1.8**  The orbitals of tetrahedral $H_4$

We now have four molecular orbitals, $\sigma_1$, $\sigma_2^*$, $\sigma_x^*$ and $\sigma_y^*$, one lowered in energy and one raised relative to the energy of the orbitals of the pair of hydrogen molecules. If we have four electrons in the system, the net result is repulsion. Thus two $H_2$ molecules do not combine to form an $H_4$ molecule. This is true whatever geometry we use in the combination. It shows us why molecules exist—when two molecules approach each other, the interaction of their molecular orbitals usually leads to repulsion.

## 1.3 C—H and C—C Bonds

### 1.3.1 *The Atomic Orbitals of a Carbon Atom*

Carbon has s and p orbitals, but we can immediately discount the 1s orbital as contributing to bonding, because the two electrons in it are held so tightly in to the nucleus that there is no possibility of significant overlap with this orbital—the electrons simply shield the nucleus, effectively giving it less of a positive charge. We are left with four electrons in 2s and 2p orbitals to use for bonding. The 2s orbital is like the 1s orbital in being spherically symmetrical, but it has a

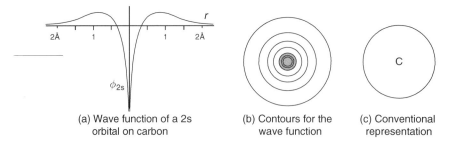

(a) Wave function of a 2s
orbital on carbon

(b) Contours for the
wave function

(c) Conventional
representation

**Fig. 1.9**   The 2s atomic orbital on carbon

spherical node, with a wave function like that shown in Fig. 1.9a, and a contour plot like that in Fig. 1.9b. The node is close to the nucleus, and overlap with the inner sphere is never important, making the 2s orbital effectively similar to a 1s orbital. Accordingly, a 2s orbital is usually drawn simply as a circle, as in Fig. 1.9c. The overlap integral $S$ of a 1s orbital on hydrogen with the outer part of the 2s orbital on carbon has a similar form to the overlap integral for two 1s orbitals (except that it does not rise as high, is at a maximum at greater atomic separation, and would not reach unity at superimposition). The 2s orbital on carbon, at −19.5 eV, is 5.9 eV lower in energy than the 1s orbital in hydrogen. The attractive force on the 2s electrons is high because the nucleus has six protons, even though this is offset by the greater average distance of the electrons from the nucleus and by the shielding from the other electrons. Slater's rules suggest that the two 1s electrons reduce the nuclear charge by 0.85 atomic charges each, and the other 2s and the two 2p electrons reduce it by $3 \times 0.35$ atomic charges, giving the nucleus an effective charge of 3.25.

The 2p orbitals on carbon also have one node each, but they have a completely different shape. They point mutually at right angles, one each along the three axes, x, y and z. A plot of the wave function for the $2p_x$ orbital along the x axis is shown in Fig. 1.10a, and a contour plot of a slice through the orbital is shown in Fig. 1.10b. Scale drawings of p orbitals based on the shapes defined by these functions would clutter up any attempt to analyse their contribution to bonding, and so it is conventional to draw much narrower lobes, as in Fig. 1.10c, and we make a mental

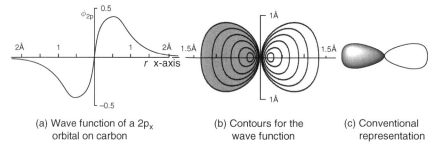

(a) Wave function of a $2p_x$
orbital on carbon

(b) Contours for the
wave function

(c) Conventional
representation

**Fig. 1.10**   The $2p_x$ orbital on carbon

reservation about their true size and shape. The 2p orbitals, at $-10.7$ eV, are higher in energy than the 2s, because they are held on average further from the nucleus. When wave functions for all three p orbitals, $p_x$, $p_y$ and $p_z$, are squared and added together, the overall electron probability has spherical symmetry, just like that in the corresponding s orbital, but concentrated further from the nucleus.

Bonds to carbon will be made by overlap of s orbitals with each other, of s orbitals with p orbitals, and, for carbon-carbon bonds, of p orbitals with each other. The overlap integral $S$ for a head on approach of an s orbital on hydrogen along the axis of a p orbital on carbon (Fig. 1.11a) grows as the orbitals begin to overlap (D), goes through a maximum when the nuclei are a little over 0.9 Å apart (C), falls fast as some of the s orbital overlaps with the back lobe of the p orbital (B), and goes to zero when the s orbital is centred on the carbon atom (A). In the last configuration, whatever bonding there would be from the overlap with the lobe of the same sign is exactly cancelled by overlap with the lobe (shaded) of opposite sign in the wave function. Of course this configuration is never reached, since the nuclei cannot coincide. The overlap integral for two p orbitals approaching head-on in the bonding mode (Fig. 1.11b), begins to grow when the nuclei approach (G), rises to a maximum when they are about 1.5 Å apart (F), falls to zero as overlap of the front lobes with each other is cancelled by overlap of the front lobes with the back lobes (E), and would fall eventually to $-1$ at superimposition. The signs of the wave functions for the individual s and p atomic orbitals can get confusing, which is why we adopt the convention of shaded and unshaded.

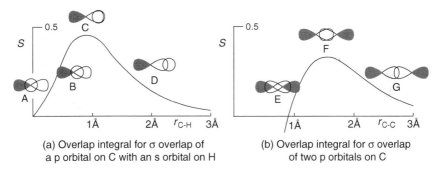

(a) Overlap integral for σ overlap of a p orbital on C with an s orbital on H

(b) Overlap integral for σ overlap of two p orbitals on C

**Fig. 1.11**   Overlap integrals for σ overlap with a p orbital on carbon

In both cases, s overlapping with p, and p overlapping with p, the overlap need not be perfectly head-on for some contribution to bonding to be still possible. For imperfectly aligned orbitals, the integral is inevitably less, because the build up of electron population between the nuclei, which is responsible for holding the nuclei together, is correspondingly less. The overlap integral for a 1s orbital on hydrogen and a 2p orbital on carbon is proportional to the cosine of the angle of approach $\theta$, where $\theta$ is 0° for head-on approach and 90° if the hydrogen atom is in the nodal plane of the p orbital.

### 1.3.2  Methane

In methane, there are eight valence electrons, four from the carbon and one each from the hydrogen atoms, for which we need four molecular orbitals. We can begin by combining two hydrogen molecules into a composite $H_4$ unit, and then combine the orbitals of that species (Fig. 1.8) with the orbitals of the carbon atom. It is not perhaps obvious where in space to put the four hydrogen atoms. They will repel each other, and the furthest apart they can get is a tetrahedral arrangement. In this arrangement, it is still possible to retain bonding interactions between the hydrogen atoms and the carbon atoms in all four orbitals, and the maximum amount of total bonding is obtained with this arrangement.

We begin by classifying the orbitals with respect to the two symmetry elements, the xz plane and the yz plane. The symmetries of the molecular orbitals of the $H_4$ 'molecule' taken from Fig. 1.8 are placed on the left in Fig. 1.12, but the energies of each are now close to the energy of an isolated 1s orbital on hydrogen, because the four hydrogen atoms are now further apart than we imagined them to be in Fig. 1.8. The s and p orbitals on the single carbon atom are shown on the right. There are two SS orbitals on each side, but the overlap integral for the interaction of the 2s orbital on carbon with the $\sigma_2{}^*$ orbital is zero—there is as much bonding with the lower lobes as there is antibonding with the upper lobes. This interaction leads nowhere. We therefore have four interactions, leading to four bonding molecular orbitals (shown in Fig. 1.12) and four antibonding (not shown). One is lower in energy than the other three, because it uses overlap from the 2s orbital on carbon, which is lower in energy than the 2p orbitals. The other three orbitals are actually equal in energy, just like the component orbitals on each side, and the four orbitals are all we need to accommodate the eight valence electrons.

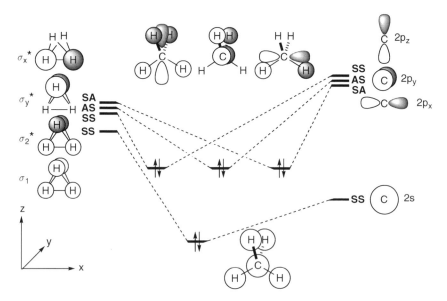

**Fig. 1.12**   The molecular orbitals of methane from the interaction of the orbitals of tetrahedral $H_4$ and a C atom

In this picture, the force holding any one of the hydrogen atoms bonded to the carbon is derived from more than one molecular orbital. The two hydrogen atoms drawn below the carbon atom in Fig. 1.12 have bonding from the low energy orbital made up of the overlap of all the s orbitals, and further bonding from the orbitals, drawn on the upper left and upper right, made up from overlap of the 1s orbital on the hydrogen with the $2p_z$ and $2p_x$ orbitals on carbon. These two hydrogen atoms are in the node of the $2p_y$ orbital, and there is no bonding to them from the molecular orbital in the centre of the top row. However, the hydrogens drawn above the carbon atom, one in front of the plane of the page and one behind, are bonded by contributions from the overlap of their 1s orbitals with the 2s, $2p_y$ and $2p_z$ orbitals of the carbon atom, but not with the $2p_x$ orbital.

Fig. 1.12 uses the conventional representations of the atomic orbitals, revealing which atomic orbitals contribute to each of the molecular orbitals, but they do not give an accurate picture of the resulting electron distribution. A better picture can be found in Jorgensen's and Salem's pioneering book, *The Organic Chemist's Book of Orbitals*, which is also available as a CD. There are also several computer programs which allow you easily to construct more realistic pictures. The pictures in Fig. 1.13 come from one of these, Jaguar, and show the four filled orbitals of methane. The wire mesh drawn to represent the outline of each molecular orbital shows one of the contours of the wave function, with the signs symbolised by lighter and heavier shading. It is easy to see what the component s and p orbitals must have been, and for comparison the four orbitals are laid out here in the same way as those in Fig. 1.12.

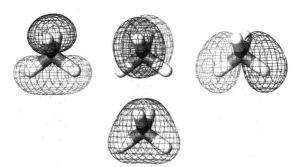

**Fig. 1.13**    One contour of the wave functions for the four filled molecular orbitals of methane

### 1.3.3 Methylene

Methylene, $CH_2$, is not a molecule that we can isolate, but it is a well known reactive intermediate with a bent H—C—H structure, and in that sense is a 'stable' molecule. Although more simple than methane, it brings us for the first time to another feature of orbital interactions which we need to understand. We take the orbitals of a hydrogen molecule from Fig. 1.3 and place them on the left of Fig. 1.14, except that again the atoms are further apart, so that the bonding and antibonding combination have relatively little difference in energy. On the right are the atomic orbitals of carbon. In this case we have three symmetry elements: (i) the xz plane, bisecting all three atoms; (ii) the yz plane, bisecting the carbon atom, and through which the hydrogen atoms reflect each other; and (iii) a two-fold rotation axis along the z coordinate, bisecting the H—C—H angle. The two orbitals, $\sigma_{HH}$ and $\sigma^*_{HH}$ in Fig. 1.14, are SSS and SAA with respect to these elements, and the atomic orbitals of carbon are SSS, SSS, ASA and SAA. Thus there are two orbitals on the right and one on the left with SSS symmetry, and the overlap integral is positive for the interactions of the $\sigma_{HH}$ and both the 2s and $2p_z$ orbitals, so that we cannot have as simple a way of creating a picture as we did with methane, where one of the possible interactions had a zero overlap integral.

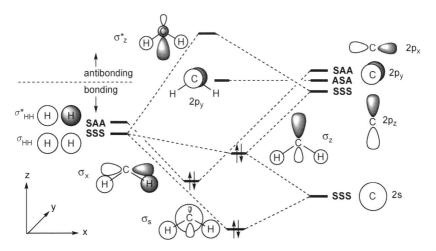

**Fig. 1.14**   The molecular orbitals of methylene from the interaction of the orbitals of $H_2$ and a carbon atom

In more detail, we have three molecular orbitals to create from three atomic orbitals, and the linear combination is Equation 1.8, like Equation 1.1 but with three terms:

$$\sigma = c_1\phi_1 + c_2\phi_2 + c_3\phi_3 \qquad\qquad 1.8$$

Because of symmetry, $|c_1|$ must equal $|c_3|$, but $|c_2|$ can be different. On account of the energy difference, it only makes a small contribution to the lowest-energy orbital in Fig. 1.14, where there is a small $p_z$ lobe, in phase, buried inside the s orbital $\sigma_s$. The second molecular orbital up in energy, the $\sigma_z$ orbital, is a mix of the $\sigma_{HH}$ orbital, the 2s orbital on carbon, out of phase, and the $2p_z$ orbital, out of phase, which has the effect of boosting the upper lobe, and reducing the lower lobe.

There is then a third orbital higher in energy, antibonding overall, with both the 2s and $2p_z$ orbitals out of phase with the $\sigma_{HH}$ orbital. Thus, we have created three molecular orbitals from three atomic orbitals. The other interaction, between the $\sigma^*_{HH}$ orbital and its SAA counterpart, the $2p_x$ orbital, gives a bonding combination $\sigma_x$ and an antibonding combination (not shown). Finally, the remaining p orbital $2p_y$ with no orbital of matching symmetry to interact with remains unchanged, and, as it happens, unoccupied.

We do not actually need to combine the orbitals of the two hydrogen atoms before we start. All we need to see is that the combinations of all the available s and p orbitals leading to the picture in Fig. 1.14 will account for the bent configuration which has the lowest energy. Methylene is a bent molecule, with a filled orbital of p character, labelled $\sigma_z$, in the same plane as the three atoms. The orbital $\sigma_s$ made up largely from the s orbitals is lowest in energy, both because the component atomic orbitals start off with lower energy, and because their combination is inherently head-on. An empty $p_y$ orbital is left unused, and this will be the lowest in energy of the unfilled orbitals—it is nonbonding and therefore lower in energy than the various antibonding orbitals created by the orbital interactions.

### 1.3.4 Hybridisation

One difficulty with these pictures, explaining the bonding in methane and in methylene, is that there is no single orbital which we can equate with the C—H bond. To get round this difficulty, chemists often use Pauling's idea of hybridisation; that is, they mix together the atomic orbitals of the carbon atom, adding the s and p orbitals together in various proportions, to produce a set of hybrids, *before* using them to make the molecular orbitals.

Thus one-half of the 2s orbital on carbon can be mixed with one-half of the $2p_x$ orbital on carbon, with its wave function in each of the two possible orientations, to create a degenerate pair of hybrid orbitals, called sp hybrids, leaving the $2p_y$ and $2p_z$ orbitals unused (Fig. 1.15, top). The 2s orbital on carbon can also be mixed with the $2p_x$ and $2p_z$ orbitals, taking one-third of the 2s orbital in each case successively with one-half of the $2p_x$ and one-sixth of the $2p_z$ in two combinations to create two hybrids, and with the remaining two-thirds of the $2p_z$ orbital to make the third hybrid. This trigonal set is called $sp^2$ (Fig. 1.15, centre); it leaves the $2p_y$ orbital unused at right angles to the plane of the page. For the $sp^3$ hybrids of tetrahedral carbon, the mixing is one-quarter of the 2s orbital with one-half of the $2p_x$ and one-quarter of the $2p_z$ orbital, in two combinations, to make one pair of hybrids, and one quarter of the 2s orbital with one-half of the $2p_y$ and one-quarter of the $2p_z$ orbital, also in two combinations, to make the other pair of hybrids (Fig. 1.15, bottom).

The conventional representations of hybrid orbitals used in Fig. 1.15 are just as misleading as the conventional representations of the p orbitals from which they are derived. A more accurate picture of the $sp^3$ hybrid is given by the contours of the wave function in Fig. 1.16. Because of the presence of the

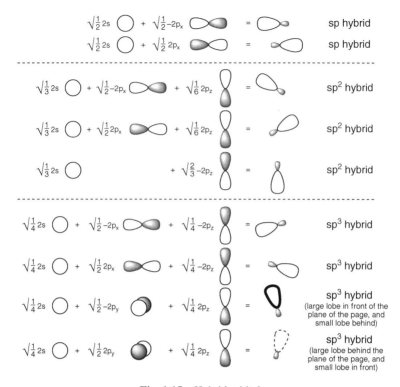

**Fig. 1.15**  Hybrid orbitals

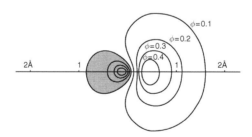

**Fig. 1.16**  A section through an sp$^3$ hybrid on carbon

inner sphere in the 2s orbital (Fig. 1.9a), the nucleus is actually inside the back lobe, and a small proportion of the front lobe reaches behind the nucleus. This follows from the way a hybrid is constructed by adding one-quarter of the wave function of the s orbital (Fig. 1.9a) and three-quarters in total of the wave functions of the p orbitals (Fig. 1.10a). We draw the conventional hybrids relatively thin, and make the mental reservation that they are fatter than they are usually drawn.

The interaction of the 1s orbital of a hydrogen atom with an sp$^3$ hybrid on carbon leads to a $\sigma_{CH}$-bonding orbital or a $\sigma^*_{CH}$-antibonding orbital (Fig. 1.17). Four of the bonding orbitals, with two electrons in each, point towards the corners of a regular tetrahedron, and give rise to the familiar picture for the bonds in methane shown in Fig. 1.18a.

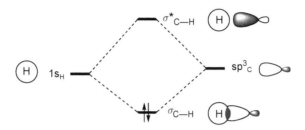

**Fig. 1.17**   Bonding and antibonding orbitals of a C—H bond

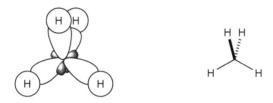

(a) The sp$^3$ hybrids on carbon overlapping          (b) Conventional bonds
     with the s orbitals of hydrogen

**Fig. 1.18**   Methane built up with sp$^3$-hybridised orbitals

This picture has the advantage over that in Fig. 1.12 that the C—H bonds do have a direct relationship with the lines drawn on the conventional structure (Fig. 1.18b). The two descriptions of the overall wave function for methane lead to identical electron distributions; hybridisation involves the same approximations, and the taking of s and p orbitals in various proportions and various combinations, as those used to arrive at Fig. 1.12.

In some ways Fig. 1.12 is a more realistic model. Measurements of ionisation potentials, for example, show that there are two different energy levels from which electrons may be removed; this is immediately easy to understand because it has filled orbitals of different energy, whereas the picture of four identical bonds from Fig. 1.17 hides this information. For other purposes, however, it is undoubtedly helpful to take advantage of the simple picture provided by the hybridisation model. It immediately reveals, for example, that all four bonds are equal. It can be used whenever it offers a similar simplification, but it is good practice to avoid it wherever possible. The common practice of referring to a molecule or an atom as 'rehybridising' is not good usage—the rehybridisation in question is in *our* picture, not in the molecule. It is likewise poor (but unfortunately common) practice to refer to atoms as being sp³, sp² or sp hybridised. Again the atoms themselves are not hybridised, it is *we* who have chosen to picture them that way. It is better in such circumstances to refer to the atoms as being tetrahedral, trigonal, or digonal.

### 1.3.5   C—C σ Bonds and π Bonds: Ethane

With a total of fourteen valence electrons to accommodate in molecular orbitals, ethane presents a more complicated picture, and we now meet a C—C bond. We will not go into the full picture—finding the symmetry elements and identifying which atomic orbitals mix to set up the molecular orbitals. It is easy enough to see the various combinations of the 1s orbitals on the hydrogen atoms and the 2s, $2p_x$, $2p_y$ and $2p_z$ orbitals on the two carbon atoms giving the set of seven bonding molecular orbitals in Fig. 1.19.

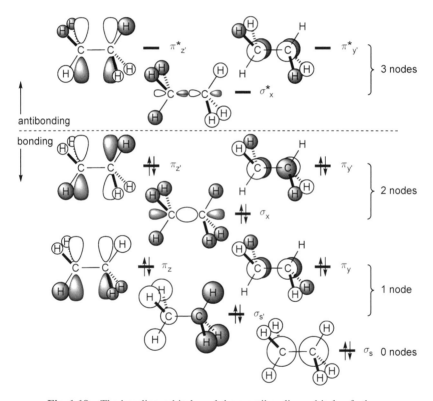

**Fig. 1.19**   The bonding orbitals and three antibonding orbitals of ethane

There is of course a corresponding picture using sp³ hybrids, but the following account shows how easy it is to avoid them. We shall concentrate for the moment on those orbitals which give rise to the force holding the two carbon atoms together; between them they make up the C—C bond. The molecular orbitals ($\sigma_s$ and $\sigma_s'$), made up largely from 2s orbitals on carbon, are like the orbitals in hydrogen, in that the region of overlap is directly on a line between the carbon nuclei; as before, they are called $\sigma$ orbitals. The bonding in the lower one is very strong, but it is somewhat offset by the antibonding (as far as the C—C bond is concerned) in the upper one. They are both strongly bonding with respect to the C—H bonds. There is actually a little of the $2p_x$ orbital mixed in with this orbital, just as we saw in Fig. 1.14 with a $2p_z$ orbital, but most of the $2p_x$ orbital contributes to the molecular orbital $\sigma_x$, which is also $\sigma$ in character, and very strong as far as the C—C bond is concerned. This orbital has a little of the 2s orbital mixed in, resulting in the asymmetric extension of the lobes between the two carbon nuclei and a reduction in size of the outer lobes. This time, its antibonding counterpart ($\sigma^*_x$) is not involved in the total bonding of ethane, nor is it bonding overall. It is in fact the lowest-energy antibonding orbital.

In the molecular orbitals using the $2p_y$ and $2p_z$ orbitals of carbon, the lobes of the atomic orbitals overlap sideways on. This is the distinctive feature of what is called $\pi$ bonding, although it may be unfamiliar to meet this type of bonding in ethane. Nevertheless, let us see where it takes us. The conventional way of drawing a p orbital (Fig. 1.10c) is designed to give uncluttered drawings, like those in Fig. 1.19. A better picture as we have already seen, and which we keep as a mental reservation when confronted with the conventional drawings, is the contour diagram (Fig. 1.10b). With these pictures in mind, the overlap sideways-on can be seen to lead to an enhanced electron population between the nuclei. However, since it is no longer directly on a line between the nuclei, it does not hold the carbon nuclei together as strongly as a $\sigma$-bonding orbital. The overlap integral $S$ for two p orbitals with a dihedral angle of zero has the form shown in Fig. 1.20, where it can be compared with the corresponding $\sigma$ overlap integral taken from Fig. 1.11b. Whereas the $\sigma$ overlap integral goes through a maximum at about 1.5 Å and then falls rapidly to a value of $-1$, the $\pi$ overlap integral rises more slowly but reaches unity at superimposition. Since C—C single bonds are typically about 1.54 Å long, the overlap integral at this distance for $\pi$ bonding is a little less than half that for $\sigma$ bonding. $\pi$ Bonds are therefore much weaker.

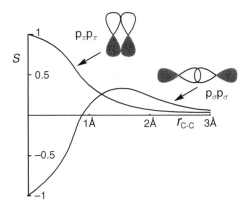

**Fig. 1.20**   Comparison of overlap integrals for $\pi$ and $\sigma$ bonding of p orbitals on C

Returning to the molecular orbitals in ethane made from the $2p_y$ and $2p_z$ orbitals, we see that they again fall in pairs, a bonding pair ($\pi_y$ and $\pi_z$) and (as far as C—C bonding is concerned, but not overall) an antibonding pair ($\pi_y'$ and $\pi_z'$). These orbitals have the wrong symmetry to have any of the 2s orbital mixed in with them. The electron population in the four orbitals ($\pi_y$, $\pi_z$, $\pi_y'$ and $\pi_z'$) is higher in the vicinity of the hydrogen atoms than in the vicinity of the carbon atoms, and these orbitals mainly contribute to C—H bonding. The amount both of bonding and antibonding that they contribute to the C—C bond is small, with the bonding and antibonding combinations more or less cancelling each other out. Thus the orbital ($\sigma_x$) is the most important single orbital making up the C—C bond. The other contribution to C—C bonding, which we cannot go into now, comes from the fact that $\sigma_s$ is more C—C bonding than $\sigma_s'$ is C—C antibonding.

Had we used the concept of hybridisation, the C—C bond would, of course, simply have been seen as coming from the bonding overlap of $sp^3$-hybridised orbitals on carbon with each other, and the overall picture for the C—C bond would have looked similar to $\sigma_x$ in Fig. 1.19, except that it would have had no contribution to bonding to the hydrogen atoms, and would have been labelled $\sigma$. For simplicity, we shall often discuss the orbitals of $\sigma$ bonds as though they could be localised into bonding and antibonding orbitals like $\sigma_x$ and $\sigma_x^*$. We shall not often need to refer to the full set of orbitals, except when they become important for one reason or another. Any property we may in future attribute to the bonding and antibonding orbitals of a $\sigma$ bond, as though there were just one such pair, can always be found in the full set of all the bonding orbitals, or they can be found in the interaction of appropriately hybridised orbitals.

### 1.3.6   C=C $\pi$ Bonds: Ethylene

The orbitals of ethylene are made up from the 1s orbitals of the four hydrogen atoms and the 2s, $2p_x$, $2p_y$ and $2p_z$ orbitals of the two carbon atoms (Fig. 1.21). One group,

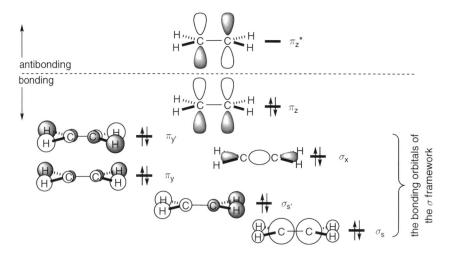

**Fig. 1.21**   The bonding orbitals and one antibonding orbital of ethylene

made up from the 1s orbitals on hydrogen and the 2s, $2p_x$ and $2p_y$ orbitals on carbon, is substantially $\sigma$ bonding, which causes the orbitals to be relatively low in energy. These orbitals make up what we call the $\sigma$-framework. Standing out, higher in energy than the $\sigma$-framework orbitals, is an orbital made up entirely from the $2p_z$ orbitals of the carbon atom overlapping in a $\pi$ bond. This time, the $\pi$ orbital is localised on the carbon atoms, free of any complications from involvement with bonding to the hydrogen atoms, and we can draw the interaction diagram in Fig. 1.22. The C—C bonding is greater in the $\sigma$ framework than the $\pi$ bonding, because of the larger overlap integral for $\sigma$ approach than for $\pi$ (Fig. 1.20).

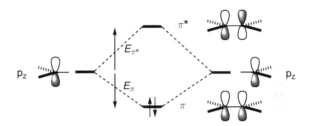

**Fig. 1.22**   A C=C $\pi$ bond

Another consequence of having an orbital localised on two atoms is that the equation for the linear combination of atomic orbitals contains only two terms (Equation 1.1), and the $c$-values are again 0.707 in the bonding orbital and 0.707 and $-0.707$ in the antibonding orbital. The energy of the p orbital on carbon is given the value $\alpha$, which is used as a reference point from which to measure rises and drops in energy, and will be especially useful when we come to deal with other elements. The value of $E_\pi$ in Fig. 1.22 is given the symbol $\beta$, and is also used as a reference with which to compare the degree of bonding in other $\pi$-bonding systems. To give a sense of scale, the value of $\beta$ for ethylene is approximately 140 kJ mol$^{-1}$ (= 1.45 eV = 33 kcal mol$^{-1}$). In other words, the total $\pi$ bonding in ethylene is 280 kJ mol$^{-1}$, since there are two electrons in the bonding orbital.

This separation of the $\sigma$ framework and the $\pi$ bond is the essence of *Hückel theory*. Because the $\pi$ bond in ethylene in this treatment is self-contained, we may treat the electrons in it in the same way as we do for the fundamental quantum mechanical picture of an electron in a box. We look at each molecular wave function as one of a series of sine waves, with the limits of the box *one bond length* out from the atoms at the end of the conjugated system, and then inscribe sine waves so that a node always comes at the edge of the box. With two orbitals to consider for the $\pi$ bond of ethylene, we only need the 180° sine curve for $\pi$ and the 360° sine curve for $\pi^*$. These curves can be inscribed over the orbitals as they are on the left of Fig. 1.23, and we can see on the right how the vertical lines above and below the atoms duplicate the pattern of the coefficients, with both $c_1$ and $c_2$ positive in the $\pi$ orbital, and $c_1$ positive and $c_2$ negative in $\pi^*$.

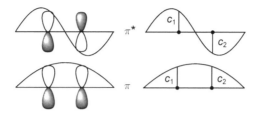

**Fig. 1.23**    The $\pi$ orbitals of ethylene and the electron in the box

A better picture, which we keep as a mental reservation when confronted with the conventional drawings, is the contour diagram. A better sense of the $\pi$ overlap from two p orbitals is given in Fig. 1.24, where we see more clearly from the contours on the left that in the bonding combination there is an enhanced electron population between the nuclei, but that it is no longer directly on a line between the nuclei. The wire-mesh diagrams illustrate the shapes of the $\pi$ and $\pi^*$ orbitals with some sense of their 3D character.

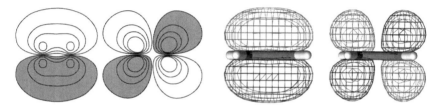

**Fig. 1.24**    A section through the contours of the $\pi$ and $\pi^*$ wave functions of ethylene, and wire-mesh outlines of one contour of each

## 1.4   Conjugation—Hückel Theory

The two p orbitals of ethylene are described as being conjugated with each other in making the $\pi$ bond. To make longer conjugated systems we add one p orbital at a time to the $\pi$ bond to make successively the allyl system, butadiene, the pentadienyl system and so on. We continue to separate completely the $\sigma$ framework (using the 2s, $2p_x$ and $2p_y$ orbitals on carbon with the 1s orbitals on hydrogen) from the $\pi$ system made up from the $2p_z$ orbitals.

### 1.4.1   The Allyl System

The members of the allyl system are reactive intermediates, and there are three of them: the allyl cation **1.1**, the allyl radical **1.2** and the allyl anion **1.3**. They have the same orbitals, but different numbers of electrons.

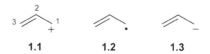

           **1.1**            **1.2**            **1.3**

The cation, radical and anion have the same $\sigma$ framework **1.4**, with fourteen bonding molecular orbitals filled with 28 electrons made by mixing the 1s orbitals of the five hydrogen atoms either with the 2s, $2p_x$ and $2p_y$ orbitals of the three carbon atoms or with the $sp^2$ hybrids. The allyl systems are bent not linear, but we shall treat the $\pi$ system as linear to simplify the discussion.

$$H_{\text{''''}}C-C_{\text{''''}}^{H}\underset{H}{\overset{}{\rightleftharpoons}}C-H$$

**1.4**

The $\pi$ system is made up from the three $p_z$ orbitals on the carbon atoms. The linear combination of these orbitals takes the form of Equation 1.9, with three terms, creating a pattern of three molecular orbitals, $\psi_1$, $\psi_2$ and $\psi_3{}^*$. In the allyl cation there are two electrons left to go into the $\pi$ system after filling the $\sigma$ framework (and in the radical, three, and in the anion, four).

$$\psi = c_1\phi_1 + c_2\phi_2 + c_3\phi_3 \qquad\qquad 1.9$$

We can derive a picture of these orbitals using the electron in the box, and recognising that we now have three orbitals and therefore three energy levels. If the lowest energy orbital is, as usual, to have no nodes (except the inevitable one in the plane of the molecule), and the next one up one node, we now need an orbital with two nodes. We therefore construct a diagram, Fig. 1.25, with one more turn of the sine curve, to include that for 540°, the next one up in energy that fulfils the criterion that there are nodes at the edges of the box, one bond length out, as well as the two inside.

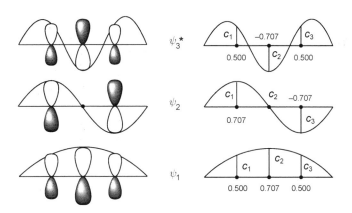

**Fig. 1.25**   The $\pi$ orbitals of the allyl system

The lowest-energy orbital, $\psi_1$, has bonding across the whole conjugated system, with the electrons concentrated in the middle. The next orbital up in energy $\psi_2$ must have a node in the middle of the conjugated system occupied by an *atom* and not by a *bond*. Having a node in the middle means having a zero coefficient $c_2$ on C-2, and hence the coefficients on C-1 and C-3 in this orbital must be $\pm 1/\sqrt{2}$, if, squared and summed, they are to equal one. The atomic orbitals in $\psi_2$ are far apart in space and their repulsive interaction does not significantly raise the energy of this molecular orbital relative to that of an isolated p orbital—whether filled or not, it does not contribute significantly to the overall bonding. If the sum of the squares of the three orbitals on C-2 is also to equal one, then the coefficients on C-2 in $\psi_1$ and $\psi_3^*$ must also be $\pm 1/\sqrt{2}$. Finally, since symmetry requires that the coefficients on C-1 and C-3 in $\psi_1$ and $\psi_3^*$ have the same absolute magnitude, and the sum of their squares must equal $1-(1/\sqrt{2})^2$, we can deduce the set of $c$-values in Fig. 1.25.

In this picture of the bonding, we get no immediate appreciation of the energies of these orbitals relative to those of ethylene. The nonbonding orbital $\psi_2$ is clearly on the $\alpha$ level, that of a p orbital on carbon, and $\psi_1$ is lowered by the extra $\pi$ bonding and $\psi_3^*$ is raised. To assess the energies, there is a simple geometrical device that works for linear conjugated systems. The conjugated system, *including the dummy atoms at the ends of the sine curves*, is inscribed vertically inside a circle of radius $2\beta$, following the convention that one $\pi$ bond in ethylene defines $\beta$. This is shown for ethylene and the allyl system in Fig. 1.26. The energies $E$ of the $\pi$ orbitals can then be calculated using Equation 1.10:

$$E = 2\beta\cos\frac{k\pi}{n+1} \qquad 1.10$$

where $k$ is the number of the atom along the sequence of $n$ atoms. This is simply an expression based on the trigonometry of Fig. 1.26. The $\pi$ orbital of ethylene, the placing of which defines the value of $\beta$, is on the first level ($k = 1$) of the sequence of two ($n = 2$) reading anticlockwise from the bottom. Thus the energies of the $\pi$ orbitals in the allyl system are $1.414\beta$ below the $\alpha$ level, and $1.414\beta$ above the $\alpha$ level.

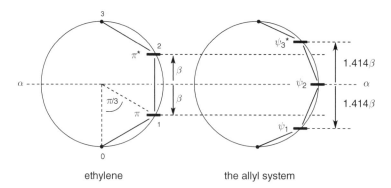

ethylene                    the allyl system

**Fig. 1.26**   Energies of $\pi$ molecular orbitals in ethylene and the allyl system

We can gain further insight by building the picture of the $\pi$ orbitals of the allyl system in another way. Instead of mixing together three p orbitals on carbon, we can combine two of them in a $\pi$ bond first, and then work out the consequences of having a third p orbital held within bonding distance of the C=C $\pi$ bond. We have to consider the effect of the p orbital, on the right of Fig. 1.27, on both the $\pi$ and $\pi^*$ orbitals of ethylene on the left. If we look *only* at the interaction of the p orbital with the $\pi$ orbital, we can expect to create two new orbitals in much the same way as we saw when the two $2p_z$ orbitals of carbon were allowed to interact in the formation of the $\pi$ bond of Fig. 1.22. One orbital $\psi_1$ will be lowered in energy and the other $\psi_x$ raised. Similarly if we look *only* at its interaction with the $\pi^*$ orbital, we can expect to create two new orbitals, one lowered in energy $\psi_y$ and one raised $\psi_3^*$. We cannot create four orbitals from three, because we cannot use the p orbital separately twice.

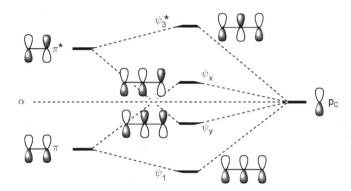

**Fig. 1.27**    A p orbital interacting independently with $\pi$ and $\pi^*$ orbitals. (No attempt is made to represent the relative sizes of the atomic orbitals)

We can see in Fig. 1.27 that the orbital $\psi_1$ has been created by mixing the p orbital with the $\pi$ orbital in a *bonding* sense, with the signs of the wave function of the two adjacent atomic orbitals matching. We can also see that the orbital $\psi_3^*$ has been created by mixing the p orbital with the $\pi^*$ orbital in an *antibonding* sense, with the signs of the wave functions unmatched. The third orbital that we are seeking, $\psi_2$ in Fig. 1.28, is a combination created by mixing the p orbital with the $\pi$ orbital in an *antibonding* sense *and* with the $\pi^*$ orbital in a *bonding* sense. We do not get two

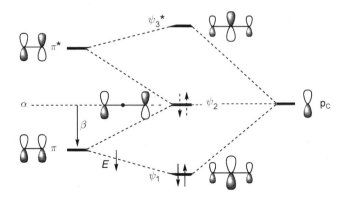

**Fig. 1.28**    The allyl system by interaction of a p orbital with $\pi$ and $\pi^*$ orbitals

orbitals, $\psi_x$ and $\psi_y$ in Fig. 1.27, but something in between, namely $\psi_2$ in Fig. 1.28. By adding $\psi_x$ and $\psi_y$ in this way, the atomic orbitals, drawn to the left of $\psi_x$ and $\psi_y$ in Fig. 1.27 cancel each other out on C-2 and reinforce each other on C-1 and C-3, thereby creating the molecular orbital $\psi_2$ in Fig. 1.28.

We have of course arrived at the same picture for the molecular orbitals as that created from mixing the three separate p orbitals in Fig. 1.25. The overall $\pi$ energy of the allyl cation, radical and anion has dropped relative to the energy of an isolated p orbital and ethylene by $2E$, which we know from Fig. 1.26 is $2 \times 0.414\beta$ or something of the order of 116 kJ mol$^{-1}$. (It is not uncommon to express these drops in energy as a 'gain' in energy—in this sense, the gain is understood to be to us, or to the outside world, and hence means a loss of energy in the system and stronger bonding.)

The electron population in any molecular orbital is derived from the square of the atomic orbital functions, so that the sine waves in Fig. 1.25 describing the coefficients are squared to describe the electron distribution. The $\pi$ electron population in the molecule as a whole is then obtained by adding up the electron populations, allowing for the number of electrons in each orbital, for all the *filled* $\pi$ molecular orbitals. Looking only at the $\pi$ system, we can see that the overall electron distribution for the cation is derived from the squares of the coefficients in $\psi_1$ alone. Roughly speaking, there is half an electron $(2 \times 0.5^2)$ on C-1 and C-3, and one electron $(2 \times 0.707^2)$ on C-2. This is illustrated graphically in Fig. 1.29a. Since the nucleus has a charge of $+1$, the excess charge on C-1 and C-3 is $+0.5$, in other words the electron deficiency in the cation is concentrated at the two ends.

(a) $\pi$ electron population in the allyl cation          (b) $\pi$ electron population in the allyl anion

**Fig. 1.29**   Total $\pi$ electron populations in the allyl cation and anion

For the anion, the electron population is derived by adding up the squares of the coefficients in both $\psi_1$ and $\psi_2$. Since there are two electrons in both orbitals, there are 1.5 electrons $(2 \times 0.5^2 + 2 \times 0.707^2)$ roughly centred on each of C-1 and C-3, and one electron $(2 \times 0.707^2)$ centred on C-2. This is illustrated graphically in Fig. 1.29b. Subtracting the charge of the nucleus then gives the excess charge as $-0.5$ on C-1 and C-3, in other words the electron excess in the anion is concentrated at the two ends.

One final detail with respect to this, the most important orbital, is that it is not quite perfectly nonbonding. Although the two atoms are separated they do interact slightly, as can be seen in $\psi_2$ in the wire-mesh drawing of the nonlinear allyl system in Fig. 1.30, where the perspective allows one to see that the right-hand lobes, which are somewhat closer to the viewer, are just perceptibly repelled by the left-hand lobes. This orbital does not therefore have exactly the same energy as an isolated p orbital—it is slightly higher in energy.

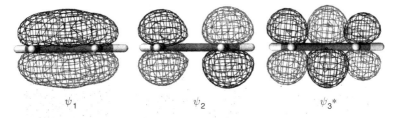

**Fig. 1.30**   The $\pi$ molecular orbitals of the allyl system

There is a problem with the conventional representation **1.1** of an allyl cation, which seems to imply that C-1 has the positive charge (an empty p orbital), and that C-2 and C-3 are in a double bond. But we could have drawn the cation **1.1**, redrawn as **1.1a**, equally well the other way round as **1.1b**, and the curly arrow symbolism shows how the two *drawings* are interconvertible. This device is at the heart of valence bond theory. For now we need only to recognise that these two drawings are representations of the *same* species—there is no reaction connecting them, although many people sooner or later fall into the trap of thinking that 'resonance' like **1.1a** → **1.1b** is a step in a reaction sequence. The double-headed arrow interconnecting them is a useful signal; this symbol should only be used to show the equivalence of 'resonance structures' and never to represent an equilibrium. There are corresponding pairs of drawings for the radical and for the anion.

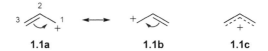

One way of avoiding these misleading structures is to draw the allyl cation as in **1.1c**, illustrating the delocalisation of the p orbitals with a dashed line, and placing the positive or negative charge in the middle. The trouble with these drawings is that they are hard to use clearly with curly arrows in mechanistic schemes, and they do not show that the positive charge is largely concentrated on C-1 and C-3. It is probably better in most situations to use one of the localised drawings for the cation, radical or anion, rather than the 'molecular orbital' version like **1.1c**, but always make the necessary mental reservation that each of the localised drawings implies the other, and that the molecular orbitals give a better picture.

As we shall see later, the most important orbitals with respect to reactivity are the highest occupied molecular orbital (HOMO) and the lowest unoccupied molecular orbital (LUMO). These are the *frontier orbitals*. For the allyl cation, the LUMO is $\psi_2$, and the drawings of the allyl cation **1.1a** and **1.1b** emphasise the electron distribution in the LUMO. Similarly, the corresponding drawings of the allyl anion emphasise $\psi_2$, the HOMO for that species. It is significant that it is the LUMO of the cation and the HOMO of the anion that will prove to be the more important frontier orbital in each case. Similarly in the allyl radical, the localised drawing illustrates the electron distribution in the singly occupied molecular orbital (SOMO), the most important orbital in that species.

### 1.4.2  Butadiene

The next step up in complexity comes with four p orbitals conjugated together, with butadiene **1.5** as the parent member. There is a $\sigma$ framework **1.6** with 36 electrons and four p orbitals to house the remaining four. Using the electron in the box with four p orbitals, we can construct Fig. 1.31, which shows the four wave functions, inside which the p orbitals are placed at the appropriate regular intervals. We get a new set of orbitals, $\psi_1$, $\psi_2$, $\psi_3{}^*$, and $\psi_4{}^*$, each described by Equation 1.11 with four terms.

**1.5**                    **1.6**

$$\psi = c_1\phi_1 + c_2\phi_2 + c_3\phi_3 + c_4\phi_4 \qquad\qquad 1.11$$

The lowest-energy orbital $\psi_1$ has all the $c$-values positive, and hence bonding is at its best. The next-highest energy level has one node, between C-2 and C-3; in

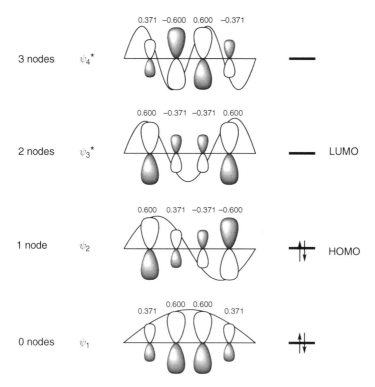

**Fig. 1.31**  $\pi$ Molecular orbitals of butadiene

other words, $c_1$ and $c_2$ are positive and $c_3$ and $c_4$ are negative. There is therefore bonding between C-1 and C-2 and between C-3 and C-4, but not between C-2 and C-3. With two bonding and one antibonding interaction this orbital is also overall bonding. Thus the lowest-energy orbital of butadiene, $\psi_1$, reasonably enough, has a high population of electrons in the middle, but in the next orbital up, $\psi_2$, because of the repulsion between the wave functions of opposite sign on C-2 and C-3, the electron population is concentrated at the ends of the conjugated system. Overall, summing the squares of the coefficients of the filled orbitals, $\psi_1$ and $\psi_2$, the $\pi$ electrons are evenly spread over all four carbon atoms of the conjugated system.

We can easily give numerical values to these coefficients for the linear conjugated system. The coefficients are proportional to the sine of the angle, as defined by the position of the atom within the sine curve. Equation 1.12 is the algebraic expression for this idea, illustrated in Fig. 1.31 with the atomic orbitals inscribed within the sine curves:

$$c_{jr} = \sqrt{\frac{2}{n+1}} \sin \frac{rj\pi}{n+1} \qquad\qquad 1.12$$

giving the coefficient $c_{jr}$ for atom $j$ in molecular orbital $r$ of a conjugated system of $n$ atoms (so that $j$ and $r = 1, 2, 3, \ldots, n$).

In butadiene, as in all alternant conjugated systems (those having no odd-membered rings), the $|c|$ values are reflected across a mirror plane placed horizontally, halfway between $\psi_2$ and $\psi_3^*$, and also across a mirror plane placed vertically, halfway between C-2 and C-3. It is only necessary therefore to calculate four of the 16 numbers in Fig. 1.31, and deduce the rest from the symmetry.

We can set up the conjugated system of butadiene by looking at the consequences of allowing two isolated $\pi$ bonds to interact, as they will if they are held within bonding distance. Let us first look at the consequence of allowing the orbitals close in energy to interact, which they will do strongly. (For a brief account of how the energy difference between interacting orbitals affects the extent of their interaction, see the discussion on page 47 of Equations 1.13 and 1.14.) We see that $\psi_1$ is derived by the interaction of $\pi$ with $\pi$ in a bonding sense, lowering the energy of $\psi_1$ below that of the $\pi$ orbital, and $\psi_2$ is derived from the interaction of $\pi$ with $\pi$ in an antibonding sense, raising the energy above that of the $\pi$ orbital. Similarly, the interaction of $\pi^*$ with $\pi^*$ in a bonding sense creates the orbital $\psi_3^*$ and in an antibonding sense the orbital $\psi_4^*$. Now we look at the consequence of the weaker interactions of $\pi$ with $\pi^*$. The interaction of $\pi$ with $\pi^*$ in a bonding sense lowers the energy of $\psi_1$ and $\psi_2$, and the interaction of $\pi$ with $\pi^*$ in an antibonding sense raises the energy of $\psi_3^*$ and $\psi_4^*$. Mixing these two sets together, and allowing for the greater contribution from the stronger interactions, we get the set of orbitals (Fig. 1.32), matching those we saw in Fig. 1.31. The net effect is to lower the energy of $\psi_1$ below the $\pi$ level, and to raise the energy of $\psi_2$ above the $\pi$ level, but without raising it up to the $\alpha$ level. Likewise $\psi_3^*$ is lowered in energy, but remains above the $\alpha$ level. Yet another way of looking at this

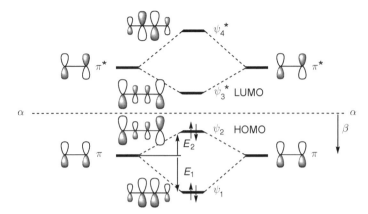

**Fig. 1.32**  Energies of the $\pi$ molecular orbitals of ethylene and butadiene by orbital interaction

system, is to say that the orbitals $\psi_1$ and $\psi_2$ and the orbitals $\psi_3{}^*$ and $\psi_4{}^*$ mutually repel each other.

We are now in a position to explain the well-known property that conjugated systems are often, but not always, lower in energy than unconjugated systems. It comes about because $\psi_1$ is lowered in energy more than $\psi_2$ is raised ($E_1$ in Fig. 1.32 is larger than $E_2$). The energy ($E_1$) given out in forming $\psi_1$ comes from the overlap between the atomic orbitals on C-2 and C-3; this overlap did not exist in the isolated $\pi$ bonds. It is particularly effective in lowering the energy of $\psi_1$, because the coefficients on C-2 and C-3 are large. By contrast, the increase in energy of $\psi_2$, caused by the repulsion between the orbitals on C-2 and C-3, is not as great, because the coefficients on these atoms are smaller in $\psi_2$. Thus the energy lost from the system in forming $\psi_1$ is greater than the energy needed to form $\psi_2$, and the overall $\pi$ energy of the ground state of the system ($\psi_1{}^2\psi_2{}^2$) is lower. We can of course see the same pattern, and attach some approximate numbers, using the geometrical analogy.

This is illustrated in Fig. 1.33, which shows that $\psi_2$ is raised above $\pi$ by $0.382\beta$ and $\psi_1$ is lowered below $\pi$ by $0.618\beta$. The overall lowering in energy for the extra

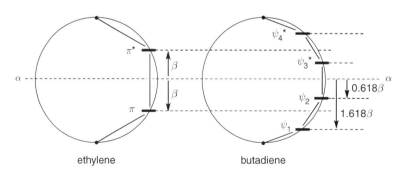

**Fig. 1.33**  Energies of the $\pi$ molecular orbitals of ethylene and butadiene by geometry

$\psi_1$      $\psi_2$      $\psi_3^*$      $\psi_4^*$

**Fig. 1.34**   The $\pi$ molecular orbitals of butadiene

conjugation is therefore $(2 \times 0.618 + 2 \times 1.618) - 4 = 0.472\beta$ or about 66 kJ mol$^{-1}$. Finally, it is instructive to look at the same $\pi$ orbitals in wire-mesh diagrams (Fig. 1.34) to reveal more accurately what the electron distribution in the $\pi$ molecular orbitals looks like.

### 1.4.3   Longer Conjugated Systems

The $\pi$ orbitals of longer linear conjugated systems are derived in essentially the same way. The energies and coefficients of the $\pi$ molecular orbitals for all six systems from an isolated p orbital up to hexatriene are summarised in Fig. 1.35. The viewpoint in this drawing is directly above the p orbitals, which appear therefore to be circular. This is a common simplification, rarely likely to lead to confusion between a p orbital and an s orbital, and we shall use it through much of this book.

The longer the conjugated system, the lower the energy of $\psi_1$, but each successive drop in energy is less than that of the system with one fewer atoms, with a limit at infinite length of $2\beta$. Among the even-atom species, the longer the conjugated system, the higher the energy of the HOMO, and the lower the energy

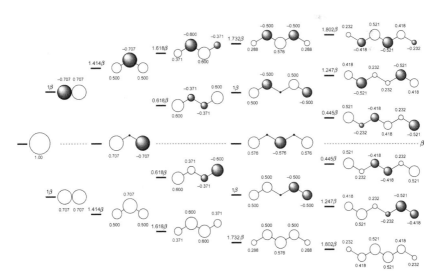

**Fig. 1.35**   The energies and coefficients of the $\pi$ molecular orbitals of the smaller conjugated systems

of the LUMO, with the energy gap becoming ever smaller. With a narrow HOMO—LUMO gap, polyenes require less energy in order to promote an electron from the HOMO to the LUMO, leading the absorption of UV and visible light to take place at ever longer wavelength.

At the extreme of an infinite polyene, however, simple Hückel theory reduces the HOMO—LUMO gap to zero, since the secants in diagrams like Fig. 1.33, would become infinitely small. Such a polyene would have equal bond lengths between each pair of carbon atoms, the gap between the HOMO and the LUMO would be zero.

A zero gap between the HOMO and the LUMO is characteristic of a metallic conductor, but long polyenes have alternating double and single bonds, and their interconversion, which is the equivalent of the movement of current along the chain, requires energy. The theoretical description of this modification to simple Hückel theory is known by physicists as a Peierls distortion. It has its counterpart for chemists in the Jahn-Teller distortion seen, for example, in cyclobutadiene, which distorts to have alternating double and single bonds, avoiding the degenerate orbitals and equal bond lengths of square cyclobutadiene. The simple Hückel picture is seriously wrong at the extreme of long conjugated systems. One way of appreciating what is happening is to think of the HOMO and the LUMO repelling each other more strongly when they are close in energy, just as the filled and unfilled orbitals of butadiene repel each other (Fig. 1.32), but more so.

## 1.5   Aromaticity

### 1.5.1   Aromatic Systems

One of the most striking properties of conjugated organic molecules is the special stability found in the group of molecules called aromatic, with benzene **1.7** as the longest established example. Hückel predicted that benzene was by no means alone, and that cyclic conjugated polyenes would have exceptionally low energy if the total number of $\pi$ electrons could be described as a number of the form $(4n + 2)$, where $n$ is an integer. Other $6\pi$-electron cyclic systems such as the cyclopentadienyl anion **1.8** and the cycloheptatrienyl cation **1.9** belong in this category, and the cyclopropenyl cation **1.10** ($n = 0$), [14]annulene **1.11** ($n = 3$), [18]annulene **1.12** ($n = 4$) and many other systems have been added over the years. Like conjugation in polyenes, however, aromaticity does not stretch to infinitely conjugated cyclic systems, even when they do have $(4n+2)$ electrons.

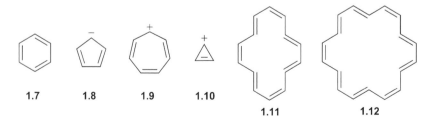

**1.7**          **1.8**          **1.9**          **1.10**

**1.11**                    **1.12**

Where does this special stability come from? We can approach this question in much the same way as we approached the derivation of the molecular orbitals of conjugated systems. We begin with a $\sigma$ framework containing the C—C and C—H $\sigma$ bonds. We must then deduce the nodal properties of the molecular orbitals created from six p orbitals in the ring. They are all shown both in elevation and in plan in Fig. 1.36. The lowest-energy orbital $\psi_1$ has no node as usual, but because the conjugated system goes round the ring instead of spilling out at the ends of the molecule, as it did with the linear conjugated systems, the coefficients on all six atoms are equal. The other special feature is that there are two orbitals having the same energy with one node $\psi_2$ and $\psi_3$, because they can be created in two symmetrical ways, one with the node horizontal $\psi_2$ and one with it vertical $\psi_3$. Similarly, there are two antibonding orbitals, $\psi_4^*$ and $\psi_5^*$, with the same energy having two nodes. Finally there is another antibonding orbital, $\psi_6^*$, with three nodes.

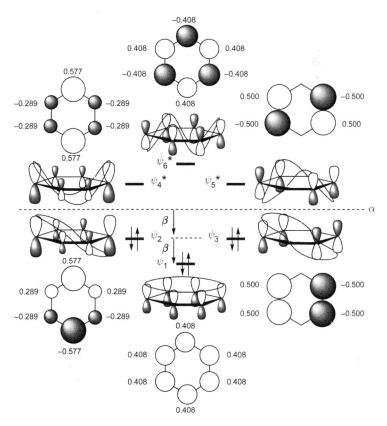

**Fig. 1.36**   The $\pi$ molecular orbitals of benzene

The energies of the molecular orbitals can be deduced by inscribing the conjugated system inside a circle of radius $2\beta$. There is no need for dummy atoms, since the sine curves go right round the ring, and the picture is therefore

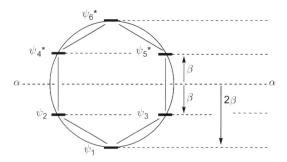

**Fig. 1.37** The energies of the $\pi$ molecular orbitals of benzene

that shown in Fig. 1.37. The lowering of energy for $\psi_2$ and for $\psi_3$ is equal to that of the $\pi$ bond in ethylene ($\beta$), and the fully bonding overlap of the six orbitals in $\psi_1$ gives rise to two $\pi$ bond's worth of bonding. The total of $\pi$ bonding is thus $2 \times 4\beta$, which is two more $\beta$ units than three *isolated* $\pi$ bonds. Benzene is also lowered in $\pi$ energy by one $\beta$ unit more than the $\pi$ energy for three *linearly conjugated* $\pi$ bonds: taking the numbers for hexatriene from Fig. 1.35, the total of $\pi$ bonding is $2 \times (1.802 + 1.247 + 0.445)\beta = 7\beta$.

One of the most striking artifacts of aromaticity, in addition to the lowering in $\pi$ energy, is the diamagnetic anisotropy, which is characteristic of these rings. Its most obvious manifestation is in the downfield shift experienced by protons on aromatic rings, and perhaps even more vividly by the upfield shift of protons on the inside of the large aromatic annulenes. The theory is beyond the scope of this book, but it is associated with the system of $\pi$ molecular orbitals, and can perhaps be most simply appreciated from the idea that the movement of electrons round aromatic rings is free, like that in a conducting wire, as epitomised by the equal C—C bond lengths.

We saw earlier that long polyenes do not approach a state with equal bond lengths as the number of conjugated double bonds increases. In the same way, the simple ($4n+2$) rule of aromaticity has been predicted to break down, with bond alternation setting in when $n$ reaches a large number. It is not yet clear what that number is as neither theory nor experiment has proved decisive. Early predictions that the largest possible aromatic system would be [22] or [26]annulene were too pessimistic, and aromaticity, using the ring-current criterion, probably peters out between [34] and [38]annulene.

### 1.5.2 Antiaromatic Systems

A molecule with $4n$ $\pi$ electrons in the ring, with the molecular orbitals made up from $4n$ p orbitals, does not show this extra stabilisation. Molecules in this class that have been made include cyclobutadiene **1.13** ($n = 1$), the cyclopentadienyl cation **1.14**, cyclooctatetraene **1.15** and pentalene **1.16** ($n = 2$), [12]annulene **1.17** ($n = 3$) and [16]annulene **1.18** ($n = 4$).

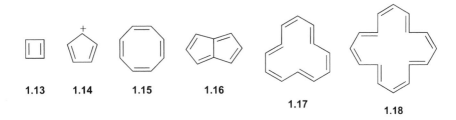

**1.13**  **1.14**  **1.15**  **1.16**

**1.17**  **1.18**

We can see this most easily by looking at the molecular orbitals of square cyclobutadiene in Fig. 1.38. As usual, the lowest-energy orbital $\psi_1$ has no nodes, and, as with benzene, and because of the symmetry, there are two exactly equivalent orbitals with one node, $\psi_2$ and $\psi_3$. The bonding in $\psi_1$ gives an energy lowering of $2\beta$ (this makes an assumption that the p orbitals are held at the same distance by the $\sigma$ framework). In contrast, the bonding interactions both in $\psi_2$ and $\psi_3$ are exactly matched by the antibonding interactions, and there is no lowering of the energy below the line ($\alpha$) representing the energy of a p atomic orbital on carbon. The molecular orbitals $\psi_2$ and $\psi_3$ are therefore nonbonding orbitals, and the net lowering in energy for the $\pi$ bonding in cyclobutadiene is only $2 \times 2\beta$. The energies of the four $\pi$ orbitals are similarly deduced from the model inscribing the conjugated system in a circle, with the point of the square at the bottom. The total $\pi$ stabilisation of $2 \times 2\beta$ is no better than having two isolated $\pi$ bonds. There is however less stabilisation than that found in a pair of *conjugated* double bonds—the overall $\pi$ bonding in butadiene, taking values from Fig. 1.33, is $4.5\beta$ and the overall $\pi$ bonding in cyclobutadiene is only $4\beta$.

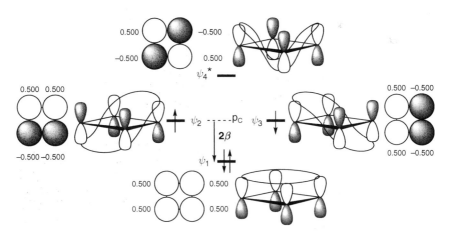

**Fig. 1.38** The $\pi$ molecular orbitals of cyclobutadiene

There is much evidence that cyclic conjugated systems of $4n$ electrons are significantly more reactive than the corresponding open-chain polyenes. There has been much speculation that they not only lack stabilisation but are actually destabilised. They have been called 'anti-aromatic' as distinct from nonaromatic.

Cyclobutadiene dimerises at extraordinarily low temperatures ($>35$ K). Cyclooctatetraene is not planar, and behaves like an alkene. When it is forced to be planar, as in pentalene, it becomes unstable to dimerisation at $0\,°C$. [12]Annulene and [16]annulene undergo electrocyclic reactions below $0\,°C$. We have seen that linear conjugation is more energy lowering than the cyclic conjugation of $4n$ electrons, which goes some way to setting the concept of anti-aromaticity on a physical basis. That $4n$ systems are unusually reactive is also explicable with an argument based on the frontier orbitals—the HOMO (at the nonbonding $\alpha$ level for cyclobutadiene) is unusually high in energy for a neutral molecule, significantly above the level of the HOMO of the linear conjugated hydrocarbon, and the LUMO is correspondingly low in energy.

The prediction from the argument in Fig. 1.38 is that square cyclobutadiene ought to be a diradical with one electron in each of $\psi_2$ and $\psi_3$, on the grounds that putting a second electron into an occupied orbital is not as energy lowering as putting the first electron into that orbital. This is not borne out by experiment, which has shown that cyclobutadiene is rectangular, with alternating double and single bonds, and shows no electron spin resonance (ESR) signal.

We can easily explain why the rectangular structure is lower in energy than the square. So far, we have made all $\pi$ bonds contribute equally one $\beta$-value to every $\pi$ bond. The difference in $\beta$-values, and hence in the strengths of $\pi$ bonds, is a function of how closely the p orbitals are held. In the rectangular structure of cyclobutadiene, in which we have moved the upper pair of carbon atoms away from the lower pair but moved the left-hand pair closer to the right-hand pair, the symmetry is lowered, and the molecular orbitals corresponding to $\psi_2$ and $\psi_3$ are no longer equal in energy (Fig. 1.39). The overall bonding in $\psi_1$ is more or less the same as in the square structure—C-1 and C-2 (and C-3 and C-4) move closer together in $\psi_1$, and the level of bonding is actually increased by about as much as the level of bonding is decreased in moving the other pairs apart. In the other filled orbital, $\psi_2$, the same distortion, separating the pair (C-1 from C-4 and C-2 from C-3) will reduce the amount of $\pi$ antibonding between them, and hence lower the energy. The corresponding argument on $\psi_3$ will lead to its being raised in energy and becoming an antibonding orbital. With one $\pi$ orbital raised in energy and the other lowered, the overall $\pi$ energy will be much the same, and the four electrons then go into the two bonding orbitals. This is known as a Jahn-Teller distortion, and can be expected to be a factor whenever a HOMO and a LUMO are close in energy, and can push each other apart.

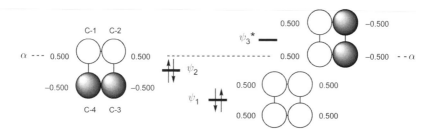

**Fig. 1.39**   The three lowest-energy $\pi$ molecular orbitals of rectangular cyclobutadiene

### 1.5.3 The Cyclopentadienyl Anion and Cation

The device of inscribing the pentagon in a circle sets up the molecular orbitals of the cyclopentadienyl anion and cation in Fig. 1.40. The total of $\pi$-bonding energy is $2 \times 3.236\beta$ for the anion, in which there are two electrons in $\psi_1$, two electrons in $\psi_2$, and two electrons in $\psi_3$. The anion is clearly aromatic, since the open-chain analogue, the pentadienyl anion has only $2 \times 2.732\beta$ worth of $\pi$ bonding (Fig. 1.35). The cyclopentadienyl anion **1.8**, a $4n+2$ system, is exceptionally stabilised, with the $pK_a$ of cyclopentadiene at 16 being strikingly low for a hydrocarbon. The cation **1.14**, however, has $\pi$-bonding energy of $2 \times 2.618\beta$, whereas its open-chain analogue, the pentadienyl cation, has $2 \times 2.732\beta$ worth of $\pi$ bonding. The cyclopentadienyl cation is not formed from its iodide by solvolysis under conditions where even the unconjugated cyclopentyl iodide ionises easily.

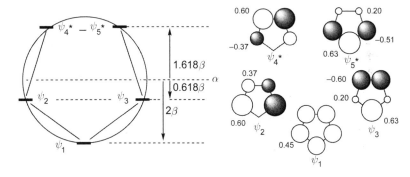

**Fig. 1.40** The energies and coefficients of the $\pi$ molecular orbitals of the cyclopentadienyl system

A striking difference between all the aromatic and all the anti-aromatic systems is the energy *difference* between the HOMO and the LUMO. The aromatic systems have a substantial gap between the frontier orbitals, and the anti-aromatic systems have, after the Jahn-Teller distortion, a small gap. The near degeneracy of the HOMO and the LUMO in the $4n$ annulenes allows a low-energy, one-electron transition between them with a magnetic moment perpendicular to the ring, whereas the aromatic systems do not. As a result, anti-aromatic rings with $4n$ electrons have a paramagnetic ring current—the protons at the perimeter of a $4n$ annulene come into resonance at high field, and protons on the inside of the ring at low field.

### 1.5.4 Homoaromaticity

The concept of aromaticity can also be extended to systems in which the conjugated system is interrupted, by a methylene group, or other insulating structural feature, provided that the overlap between the p orbitals of the conjugated systems can still take place through space. When such overlap has energy-lowering consequences, evident in the properties of the molecule, the phenomenon is called homoaromaticity. Examples are the homocyclopropenyl cation **1.19**, the trisho-mocyclopropenyl cation **1.20**, the bishomocyclopentadienyl anion **1.21** and the homocycloheptatrienyl cation **1.22**, each of which shows evidence of transannular overlap, illustrated, and emphasised with a bold line on the orbitals in the draw-ings **1.19b**, **1.20b**, **1.21b** and **1.22b**.

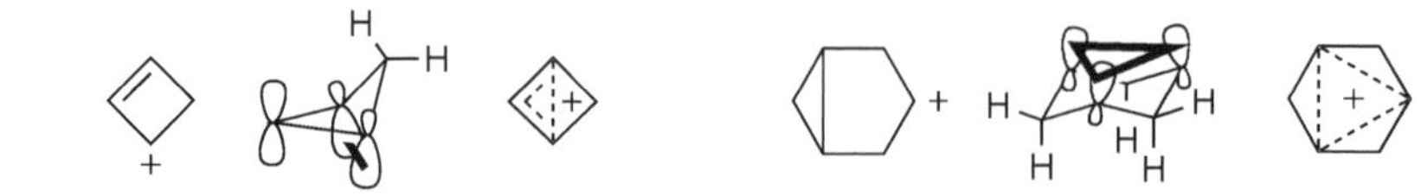

| 1.19a | 1.19b | 1.19c | 1.20a | 1.20b | 1.20c |

| 1.21a | 1.21b | 1.21c | 1.22a | 1.22b | 1.22c |

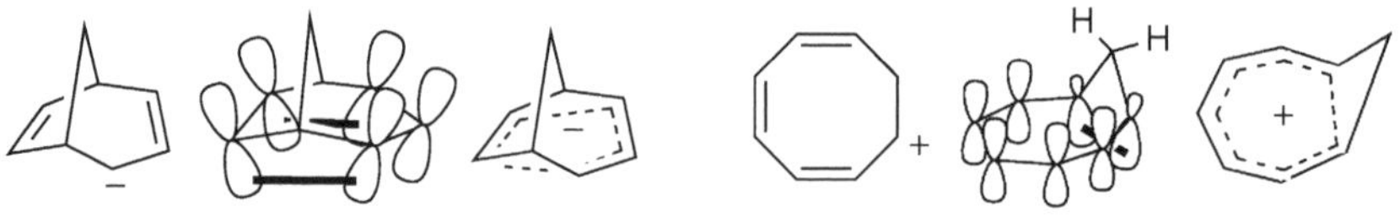

However, homoaromatic stabilisation appears to be absent in neutral systems. Homobenzene (cycloheptatriene) **1.23** and trishomobenzene (triquinacene) **1.26**, even though transannular overlap looks feasible, show no aromatic properties. In both cases, the conventional structures **1.23** and **1.24**, and **1.26** and **1.27** are lower in energy than the homoaromatic structures **1.25** and **1.28**, which appear to be close to the transition structures for the interconversion.

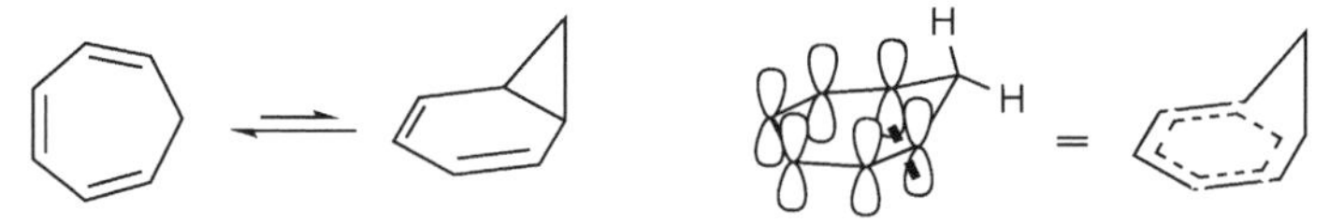

| 1.23 | 1.24 | 1.25a | 1.25b |

| 1.26 | 1.27 | 1.28a | 1.28b |

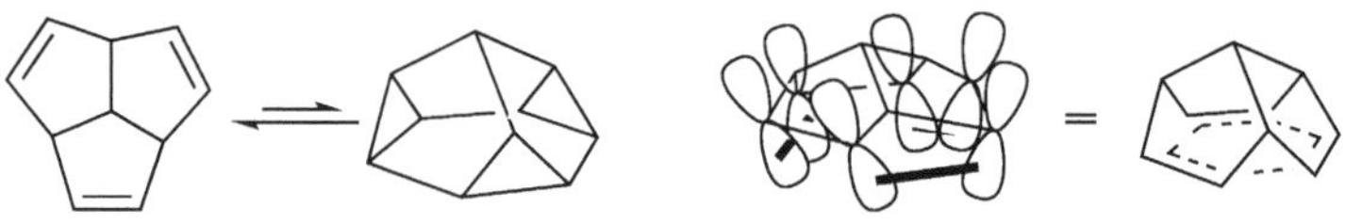

### 1.5.5  Spiro Conjugation

In addition to $\sigma$ and $\pi$ overlap, p orbitals can overlap in another way, even less effective in lowering the energy, but still detectable. If one conjugated system is held at right angles to another in a spiro structure **1.29**, the p orbitals of one can overlap with the p orbitals of the other, as symbolised by the bold lines on the front lobes. The overlap integral will be small, but overlap of molecular orbitals with the right symmetry can raise or lower the orbital energy in the usual way. The hydrocarbons **1.30** and **1.31** are representative examples.

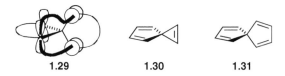

**1.29**          **1.30**          **1.31**

Take spiroheptatriene **1.30**, with the unperturbed orbitals of each component shown on the left and right in Fig. 1.41. The only orbitals that can interact are $\psi_2$ on the left and $\pi^*$ on right, all the others having the wrong symmetry. For example, the interaction of the top lobes of $\psi_1$ on the left and the upper p orbital of the $\pi$ orbital on the right, one in front and one behind, have one in phase and one out of phase, exactly cancelling each other out; similarly with the front $\pi$ lobes on the right and the upper and lower lobes of the front-right p orbital of $\psi_1$ on the left. The two orbitals that do interact create the usual pair of new orbitals, one raised and one lowered. Since there are only two electrons to go into the new orbitals, the overall energy of the conjugated system is lowered. The effect, $\Delta E_s$, is small, both because of the poor overlap, and because the two orbitals interacting are far apart in energy. Nevertheless, it is a general conclusion that if the total number of $\pi$ electrons is a $(4n+2)$ number, the spiro system is stabilised, leading to the concept of spiro-aromaticity.

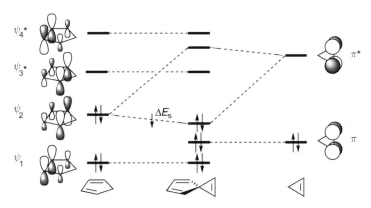

**Fig. 1.41**  $\pi$ Molecular orbitals of the 'aromatic' spiroheptatriene **1.30**

## 1.6   Strained σ Bonds—Cyclopropanes and Cyclobutanes

Another kind of imperfect sideways-on overlap, this time in $\sigma$ bonds, is found in strained molecules like cyclopropane and cyclobutane.

### 1.6.1   Cyclopropanes

There are several ways to describe the $\sigma$ bonds in cyclopropane. The most simple is to identify the C—H bonds as coming from the straightforward sp³ hybrids on the carbon atoms and the 1s orbitals on the hydrogen atoms **1.32** in the usual way, and the C—C bonds as coming from the remaining sp$^3$ hybrids imperfectly aligned **1.33**.

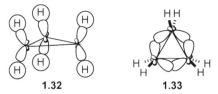

**1.32**                    **1.33**

In more detail, these orbitals ought to be mixed in bonding and antibonding combinations to create the full set of molecular orbitals, but even without doing so we can see that C—C bonding is somewhere between $\sigma$ bonding (head-on overlap) and $\pi$ bonding (sideways-on overlap). We can expect these bonds to have some of the character of each, which fits in with the general perception that cyclopropanes can be helpfully compared with alkenes in their reactivity and in their power to enter into conjugation. Thus cyclopropane is much less reactive than ethylene towards electrophiles, like bromine, but it does react, whereas ethane does not. Conjugation of a double bond or an aromatic ring with a cyclopropyl substituent is similar to conjugation with an alkene, but less effective.

Another way of understanding the C—C bonding, known as the Walsh description, emphasises the capacity of a cyclopropyl substituent to enter into $\pi$ bonding. In this picture, which is like the picture of the bonding in ethane without using hybridisation (Fig. 1.19), the six C—H bonds are largely made up from the s orbitals on hydrogen and the s, $p_x$ and $p_z$ orbitals on carbon, with the x, y and z axes redefined at each corner to be local x, y and z coordinates. The picture of C—H bonding can be simplified by choosing sp hybridisation from the combination of the 2s and $2p_x$ orbitals, and using the three sp hybrids with the large lobes pointing outside the ring and the three $p_z$ orbitals to make up the CH-bonding orbitals (Fig. 1.42). Some of these orbitals contribute to C—C bonding, notably the $\sigma_{CH}$, $\pi_{CC}$ orbital, but the major contributors are the overlap of the three sp hybrids with the large lobes pointing into the ring, which produce one bonding combination $\sigma_{CC}$, and the three $p_y$ orbitals, which combine to produce a pair of bonding orbitals $\pi_{CC}$, each with one node, and with coefficients to make the overall bonding between each of the C—C bonds equal.

The advantage of this picture is that it shows directly the high degree of $\pi$ bonding in the C—C bonds, and gives directly a high-energy filled p orbital, the $\pi_{CC}$ orbital at the top right, largely concentrated on C-1, and with the right symmetry for overlap with other conjugated systems [see (Section 2.2.1) page 72].

A remarkable property of cyclopropanes is that they are magnetically anisotropic, with the protons coming into resonance in their NMR spectra at unusually *high* field, typically 1 ppm *upfield* of the protons of an open-chain methylene group. For $^1$H NMR spectroscopy, this is quite a large effect, and it is also strikingly in the opposite direction from that expected by the usual analogy

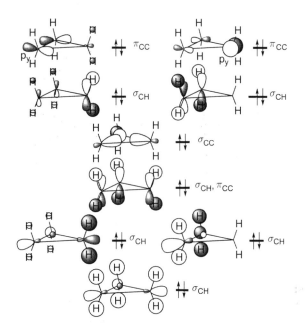

**Fig. 1.42** A simplified version of the occupied Walsh orbitals of cyclopropane

drawn between a cyclopropane and an alkene. The anisotropy is explained with the orbitals shown in Fig. 1.42: the pair of $\sigma_{CH}$ orbitals just below the highest occupied orbitals, together with the $\sigma_{CH},\pi_{CC}$ orbital, clearly have the same nodal pattern as the filled $\pi$ orbitals of benzene, and the pattern is repeated in the three filled orbitals of lowest energy. This pattern of orbitals is associated with the capacity to support a ring current, but, in contrast to benzene, the derived field places the protons in cyclopropanes in the region experiencing a reduced magnetic field **1.34**.

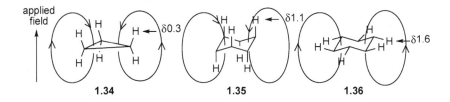

The same explanation, although we shall not show the molecular orbitals, has been advanced to account for the small difference in chemical shift between the axial and equatorial protons in cyclohexanes, detectable in the $^1$H NMR spectrum of cyclohexane by freezing out at −100 °C the otherwise rapid interconversion of the two chair conformations. The axial protons **1.35** come into resonance upfield at $\delta 1.1$ and the equatorial protons **1.36** downfield at $\delta 1.6$.

### 1.6.2   Cyclobutanes

The molecular orbitals of cyclobutanes, show many of the same features as cyclopropanes, only less so. Cyclobutanes also show enhanced reactivity over simple alkanes, but they are less reactive towards electrophiles, and cyclobutyl groups are less effective as stabilising substituents on electron-deficient centres than cyclopropyl groups. The most striking difference, however, is that the protons in cyclobutanes come into resonance in their $^1$H NMR spectra *downfield* of the protons from comparable methylene groups in open chains. The effect is not large, typically only about 0.5 ppm, with cyclobutane itself, for example, at δ1.96 in contrast to cyclohexane at δ1.44. The pattern of orbitals will be similar to those of cyclobutadiene (Figs 1.38 and 1.39), with the highest-energy orbitals degenerate and suffering Jahn-Teller distortion.

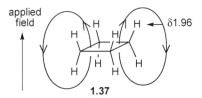

**1.37**

The ring current is therefore in the opposite direction, adding to the applied field at the centre of the ring, and the protons experience therefore an enhanced field **1.37**. The effect may be rather less in cyclobutanes than in cyclopropanes, because the cyclobutane ring is flexible, allowing the ring to buckle from the planar structure, and the C–H bonds thereby avoid the full eclipsing interactions inevitable in cyclopropanes, and compensated there by the aromaticity they create.

## 1.7   Heteronuclear Bonds, C–M, C–X and C=O

So far, we have been concentrating on symmetrical bonds between identical atoms (homonuclear bonds) and on bonds between carbon and hydrogen. The important interaction diagrams were constructed by combining atomic orbitals of more or less *equal* energy, and the coefficients, $c_1$ and $c_2$, in the molecular orbitals were therefore more or less equal in magnitude. It is true that C–H bonds, with and without hybridisation, involve the overlap of atomic orbitals of different elements, but the difference in the energy of the atomic orbitals of these two elements, did not make a significant difference in the earlier part of this chapter. The interaction of orbitals of different energy is inescapable when we come to consider molecules, like methyl chloride and methyllithium. As we have mentioned in passing, *orbitals of different energy interact to lower (and raise) the energy of the resultant molecular orbitals less than orbitals of comparable energy.*

### 1.7.1 Atomic orbital energies and electronegativity

One way of assessing the relative energies of the orbitals of different elements is to use one of the empirical scales of *electronegativity*. Pauling's is empirically derived from the differences in dissociation energy for the molecules XX, YY and XY. Several refinements of Pauling's scale have been made since it first appeared in 1932. A good recent one is Allen's, reproduced to scale in Fig. 1.43, along with values assigned by Mullay to the carbon atoms in methyl, vinyl and ethynyl groups. In spite of the widespread use of electronegativity as a unifying concept in organic chemistry, the electronegativity of an element is almost never included in the periodic table. Redressing this deficiency, Allen strikingly showed his electro-negativity scale as the third dimension of the periodic table, and his vivid picture is reproduced here as Fig. 1.44. The other way of assessing the relative energies of the orbitals of different elements is to accept the calculated values in Table 1.1, which are themselves supported by measurements of ionisation potentials (IPs).

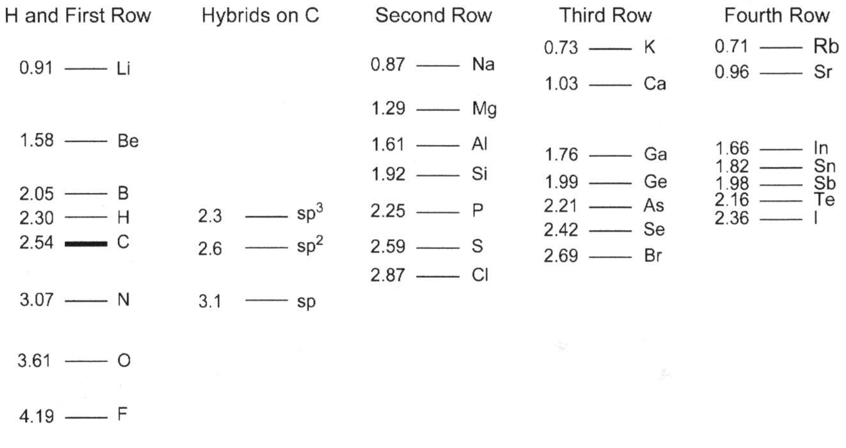

| H and First Row | Hybrids on C | Second Row | Third Row | Fourth Row |
|---|---|---|---|---|
| | | | 0.73 —— K | 0.71 —— Rb |
| 0.91 —— Li | | 0.87 —— Na | 1.03 —— Ca | 0.96 —— Sr |
| | | 1.29 —— Mg | | |
| 1.58 —— Be | | 1.61 —— Al | 1.76 —— Ga | 1.66 —— In |
| | | 1.92 —— Si | 1.99 —— Ge | 1.82 —— Sn |
| 2.05 —— B | 2.3 —— sp³ | 2.25 —— P | 2.21 —— As | 1.98 —— Sb |
| 2.30 —— H | 2.6 —— sp² | 2.59 —— S | 2.42 —— Se | 2.16 —— Te |
| 2.54 ▬▬ C | | 2.87 —— Cl | 2.69 —— Br | 2.36 —— I |
| 3.07 —— N | 3.1 —— sp | | | |
| 3.61 —— O | | | | |
| 4.19 —— F | | | | |

**Fig. 1.43** Allen electronegativity values and Pauling-based values for carbon hybrids

### 1.7.2 C–X σ Bonds

We are now ready to construct an interaction diagram for a bond made by the overlap of atomic orbitals with different energies. Let us take a C—Cl σ bond, in which the chlorine atom is the more electronegative element. Other things being equal, the energy of an electron in an atomic orbital on an electronegative element is lower than that of an electron on a less electronegative element.

As usual, we can tackle the problem with or without using the concept of hybridisation. The C—X bond in a molecule such as methyl chloride, like the C—C bond in ethane, has several orbitals contributing to the force which keeps the two atoms bonded to each other; but, just as we could abstract one pair of atomic orbitals of ethane and make a typical interaction diagram for it, so can we now take the corresponding pair of orbitals from the set making up a C—Cl σ bond.

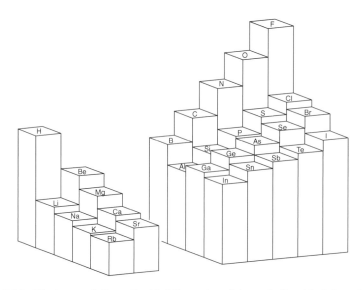

**Fig. 1.44** Electronegativity as the third dimension of the periodic table (adapted with permission from L. C. Allen, *J. Am. Chem. Soc.*, 1989, **111**, 9003. Copyright 1989 American Chemical Society)

**Table 1.1**    Valence atomic orbital energies for s, p and selected hybrid orbitals in eV(1 eV = 96.5 kJ mol$^{-1}$ = 23 kcal mol$^{-1}$)

|  | H |  |  |  |  |  |  |
|---|---|---|---|---|---|---|---|
| 1s | −13.6 |  |  |  |  |  |  |
|  | **Li** | **Be** | **B** | **C** | **N** | **O** | **F** |
| 2s | −5.4 | −9.4 | −14.7 | −19.4 | −25.6 | −32.4 | −40.1 |
| sp |  |  |  | −19.3 | −19.3 | −24.2 | −29.4 |
| sp$^2$ |  |  |  | −17.1 | −17.1 | −21.4 | −25.8 |
| sp$^3$ |  |  |  | −16.1 | −16.1 | −20.0 | −24.4 |
| 2p | −3.5 | −6.0 | −5.7 | −10.7 | −12.9 | −15.9 | −18.6 |
|  | **Na** | **Mg** | **Al** | **Si** | **P** | **S** | **Cl** |
| 3s | −5.2 | −7.6 | −11.3 | −15.0 | −18.4 | −20.9 | −25.3 |
| 3p |  | −6.0 | −7.8 | −9.8 | −11.7 | −13.7 |  |

In making a covalent bond between carbon and chlorine from the $2p_x$ orbital on carbon and the $3p_x$ orbital on chlorine, we have an interaction (Fig. 1.45) between orbitals of *unequal* energy (−10.7 eV for C and −13.7 eV for Cl, from Table 1.1). The interaction diagram could equally have been drawn using sp$^3$ hybrids on

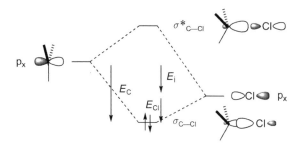

**Fig. 1.45**   A major part of the C—Cl σ bond

carbon and chlorine in place of the p orbitals. Because of the loss of symmetry, the chlorine atom has a larger share of the electron population. In other words, the coefficient on chlorine for the bonding orbital, $\sigma_{CCl}$ is larger than that on carbon. It follows from the requirement that the sum of the squares of all the $c$-values on any one atom in all the molecular orbitals must equal one, that the coefficients in the corresponding antibonding orbital, $\sigma^*_{CCl}$ must reverse this situation: the one on carbon will be larger than the one on chlorine.

What we have done in Fig. 1.45 is to take the lower-energy atomic orbital on the right and mix in with it, in a bonding sense, some of the character of the higher-energy orbital on the left. This creates the new bonding molecular orbital, which resembles the atomic orbital nearer to it in energy more than the one further away. We have also taken the higher-energy orbital and mixed in with it, in an anti-bonding way, some of the character of the lower-energy orbital. This produces the antibonding molecular orbital, which more resembles the atomic orbital nearer it in energy. When the coefficients are unequal, the overlap of a small lobe with a larger lobe does not lower the energy of the bonding molecular orbital as much as the overlap of two atomic orbitals of more equal size. The overlap integrals $S$ for forming bonds to the first-row elements N, O and F are essentially parallel to the overlap integral for the formation of a C—C bond (Figs 1.11b and 1.20), but displaced successively by about 0.2 Å to shorter internuclear distances for each element. This is because the orbitals of the first-row elements have similar shapes, but the electrons are held closer to the nucleus of the more electronegative elements.

Thus we can set up the filled orbitals for methyl chloride schematically in Fig. 1.46a, along with the lowest of the unoccupied orbitals. The major orbital contributing to C—Cl bonding is the one labelled $\sigma_{CCl}$. There is more bonding than antibonding from the overlap of the s orbitals, but probably nearly equal bonding and antibonding from the orbitals having π bonding between the carbon and the chlorine. The same degree of bonding can be arrived at by using the hybrid orbitals shown in Fig. 1.46b, where all of the C—Cl bonding comes from the sp³ hybrids.

We might be tempted at this stage to say that we have a weaker bond than we had for a C—C bond, but we must be careful in defining what we mean by a weaker bond. Tables of bond strengths give the C—Cl bond a strength of

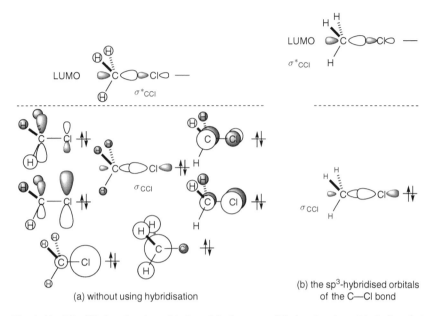

(a) without using hybridisation

(b) the sp³-hybridised orbitals of the C—Cl bond

**Fig. 1.46** The filled molecular orbitals and the lowest unfilled molecular orbital of methyl chloride

352 kJ mol⁻¹ (84 kcal mol⁻¹), whereas a typical C—C bond strength is a little lower at 347 kJ mol⁻¹ (83 kcal mol⁻¹). The C—Cl bond *is* strong, if we try to break it homolytically to get a pair of radicals, and the C—C bond *is* easier to break this way. This is what the numbers 352 and 347 kJ mol⁻¹ refer to.

Only part of the C—Cl bond strength represented by these numbers comes from the purely covalent bonding given by $2E_{Cl}$ in Fig. 1.45. The other contribution to C—Cl bonding comes from the electrostatic attraction between the high electron population on the chlorine atom and the relatively exposed carbon nucleus. We say that the bond is polarised, or that it has ionic character. Thus it is much easier to break a C—Cl bond heterolytically to the cation (on carbon) and the anion (on chlorine) than to cleave a C—C bond this way. The energy of an ionic bond is related to the value $E_i$ in Fig. 1.45, as we can see by taking it to an extreme: a 2s orbital on Li and a 2p orbital on F, with energies of –5.4 and –18.6 eV, respectively, will have negligible covalent overlap, and the molecule will consist almost entirely of isolated orbitals in which the higher-energy orbital has given up its electron to the lower-energy orbital. In other words, we shall have a pair of ions Li⁺ and F⁻. There will be no covalent bonding to speak of, and the drop in energy in going from the pair of radicals to the cation plus anion is now the difference in energy between the two orbitals, equivalent to $E_i$ in Fig. 1.45.

The important thing to remember is that *when two orbitals of unequal energy interact, the lowering in energy is less than when two orbitals of very similar energy interact.* Conversely, when it comes to transferring an electron, the ideal situation has the electron in a high-energy orbital and the 'hole' in a low-energy orbital. In a little more detail, the degree of energy lowering $E_{Cl}$ is a function of the overlap integral $S$ [see (Section 1.2.1) page 3]. When orbitals of significantly different energy interact, the energy lowering is roughly proportional to $S^2$ instead of $S$. The energy lowering is also inversely proportional to the difference in energy $E_i$. The equations for the energies of the lowered and raised orbitals in Fig. 1.45, take the form shown in Equations 1.13 and 1.14, where we see that the energy difference is in the denominator.

$$E_{\sigma_{CCl}} = E_{p_{Cl}} + \frac{(\beta_{CCl} - E_{p_{Cl}}S_{CCl})^2}{E_{p_{Cl}} - E_{p_C}} \qquad 1.13$$

$$E_{\sigma*_{CCl}} = E_{p_C} + \frac{(\beta_{CCl} - E_{p_C}S_{CCl})^2}{E_{p_C} - E_{p_{Cl}}} \qquad 1.14$$

A picture of the electron distribution in the $\sigma$ orbitals between carbon and chlorine is revealed in the wire-mesh diagrams for methyl chloride in Fig. 1.47, which shows one contour of the major orbital $\sigma_{CCl}$ contributing to C—Cl bonding together with the LUMO, $\sigma*_{CCl}$, which incidentally provides the cover illustration for this book. Comparing these with the schematic version in Fig. 1.46, we can see how the back lobe on carbon in $\sigma_{CCl}$ includes overlap with the s orbitals on the hydrogen atoms, and that the front lobe in $\sigma*_{CCl}$ wraps back a little behind the carbon atom to include some overlap to the s orbitals of the hydrogen atoms.

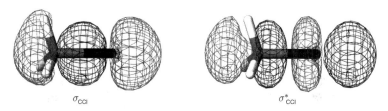

$$\sigma_{CCl} \qquad\qquad\qquad\qquad\qquad \sigma*_{CCl}$$

**Fig. 1.47**  The major C—Cl bonding orbital and the LUMO for methyl chloride

### 1.7.3  C—M σ Bonds

When the bond from carbon is to an electropositive element like lithium, carbon is the electronegative atom. The most strongly interacting orbitals are the 2s orbital on lithium and the $2p_x$ orbital on carbon, which has the form shown in Fig. 1.48, in which the lithium 2s orbital is at –5.4 eV and the carbon 2p orbital at –10.7 eV. The bonding orbital $\sigma_{LiC}$ is polarised towards carbon, and the antibonding $\sigma*_{LiC}$ towards lithium. Organic chemists often refer to organolithium compounds as anions, but it is as well to bear in mind that they are usually highly polarised covalent molecules. Furthermore, they are rarely monomeric, almost always existing as oligomers, in which the lithium is coordinated to more than one carbon atom, making the molecular orbital description below greatly oversimplified.

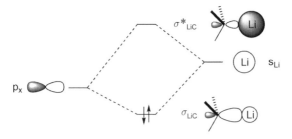

**Fig. 1.48**   A major contributory part of the Li—C $\sigma$ bond

The filled and one of the unfilled orbitals for monomeric methyllithium are shown in Fig. 1.49. The lowest energy orbital is made up largely from the 2s orbital on carbon and the 1s orbitals on hydrogen, with only a little mixing in of the 2s orbital of lithium and even less of the 2p. The next two up in energy are largely $\pi$ mixes of the $2p_z$ and $2p_y$ orbitals on carbon with a little of the $2p_z$ and $2p_y$ on lithium, and, as usual, the 1s orbitals on hydrogen. The $2p_z$ and $2p_y$ orbitals on lithium have the wrong symmetry to overlap with the 2s orbital on carbon. Then come the two orbitals we have seen in Fig. 1.48: the $2p_x$ orbital on carbon interacting productively with the 2s orbital on lithium, giving rise to the highest of the occupied orbitals $\sigma_{CLi}$, which has mixed in with it the usual 1s orbitals on hydrogen and a contribution from the $2p_x$ orbital on lithium, symbolised here by the displacement of the orbital on lithium towards the carbon. The next orbital up in energy, the lowest of the unfilled orbitals, is its counterpart $\sigma^*_{LiC}$, largely a mix of the 2s and the $2p_x$ orbital of lithium, symbolised again by the displacement of the orbital on lithium away from the carbon, with a little of the $2p_x$ orbital of carbon out of phase.

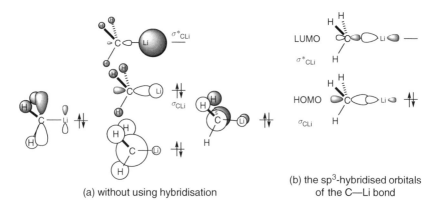

(a) without using hybridisation

(b) the $sp^3$-hybridised orbitals of the C—Li bond

**Fig. 1.49**   The filled and one of the unfilled molecular orbitals of methyllithium

The electron distribution in the $\sigma$ orbitals between carbon and lithium is revealed in Fig. 1.50, which shows one contour of the $\sigma_{CLi}$ and $\sigma^*_{CLi}$ orbitals of methyllithium, unrealistically monomeric and in the gas phase. Comparing Fig. 1.50 with Fig. 1.49, we can see better how the s and $p_x$ orbitals on lithium mix to boost the electron population between the nuclei in $\sigma_{CLi}$, and to minimise it in $\sigma^*_{CLi}$.

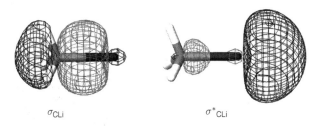

$\sigma_{CLi}$ $\qquad\qquad\qquad$ $\sigma^*_{CLi}$

**Fig. 1.50** The HOMO and LUMO for methyllithium in the vapour phase

### 1.7.4 C=O $\pi$ Bonds

The C=O $\pi$ bond is relatively straightforward, because the p orbitals in the $\pi$ system in Hückel theory are free from the complicating effect of having to mix in contributions from s orbitals. The $p_x$ orbital on oxygen is placed in Fig. 1.51 at a level somewhat more than $1\beta$ below that of the $p_x$ orbital on carbon. The energy of a p orbital on oxygen is $-15.9$ eV and that on carbon $-10.7$ eV (Table 1.1). As with $\pi$ bonds in general, the raising of the $\pi^*$ and lowering of the $\pi$ orbitals above and below the atomic p orbitals is less than for a C—O $\sigma$ bond, and less than for a $\pi$ bond between two carbon atoms. Both the $\pi_{C=O}$ and the $\pi^*_{C=O}$ orbitals are now lower in energy than the $\pi_{C=C}$ and $\pi^*_{C=C}$ orbitals of ethylene, which by definition are $1\beta$ above and $1\beta$ below the $\alpha$ level.

The contour plots in Fig. 1.52 show the electron distribution in more detail. To derive contours like these, we need values for $\alpha$ and $\beta$ for electronegative atoms in

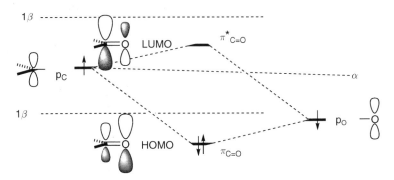

**Fig. 1.51** The interaction of p orbitals in the formation of the $\pi$ bond of a carbonyl group

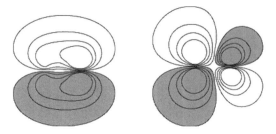

**Fig. 1.52**    Electron population contours for the $\pi$ and $\pi^*$ orbitals of formaldehyde

order to carry out Hückel calculations on the isolated $\pi$ system. There are no fundamentally sound values for $\alpha$ and $\beta$, which we need for systems with heteroatoms. Everything is approximate, and any values for energies and coefficients that come from simple calculations must be taken only as a guide. In simple Hückel theory, the value of $\alpha$ is adjusted for the element in question X from the reference value for carbon $\alpha_0$ by Equation 1.15. Likewise, the $\beta$ value for the C=C bond in ethylene $\beta_0$ is adjusted for C=X by Equation 1.16.

$$\alpha_X = \alpha_0 + h_X\beta_0 \qquad\qquad 1.15$$

$$\beta_{CX} = k_{CX}\beta_0 \qquad\qquad 1.16$$

The adjustment parameters $h$ and $k$ take into account the trends in electronegativity, but are not quantitatively related to the numbers in Fig. 1.43 and Table 1.1. The values of $h$ for some common elements and of $k$ for the corresponding C=X $\pi$ bonds in Table 1.2 have been recommended for use in Equations 1.15 and 1.16.

The polarisation of the carbonyl group is away from carbon towards oxygen in the bonding orbital, and the opposite in the antibonding orbital, as usual. The wire-mesh pictures in Fig. 1.53 show more realistically an outer contour of these two orbitals in formaldehyde itself. Note that in Figs 1.52 and 1.53, it *appears* from the shape of the outer contour that the electron population in the bonding orbital is very similar on oxygen to that on carbon.

**Table 1.2**    Parameters for simple Hückel calculations for bonds with heteroatoms

| Element | | $h$ | $k$ | Element | | $h$ | $k$ |
|---|---|---|---|---|---|---|---|
| **B** | Trigonal | −0.45 | 0.73 | | | | |
| **C** | Trigonal | 0 | 1 | **Si** | Trigonal | 0 | 0.75 |
| **N** | Trigonal | 0.51 | 1.02 | **P** | Trigonal | 0.19 | 0.77 |
| **N** | Tetrahedral | 1.37 | 0.89 | **P** | Tetrahedral | 0.75 | 0.76 |
| **O** | Trigonal | 0.97 | 1.06 | **S** | Trigonal | 0.46 | 0.81 |
| **O** | Tetrahedral | 2.09 | 0.66 | **S** | Tetrahedral | 1.11 | 0.69 |
| **F** | Tetrahedral | 2.71 | 0.52 | **Cl** | Tetrahedral | 1.48 | 0.62 |

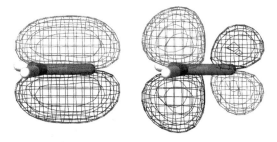

**Fig. 1.53** Wire-mesh plot of the $\pi$ and $\pi^*$ orbitals of formaldehyde

This is not the case, as shown by the extra contour around the oxygen atom in the plot in Fig. 1.52. The electron distribution around the oxygen atom is simply more compact, as a consequence of the higher nuclear charge on that atom.

### 1.7.5 Heterocyclic Aromatic Systems

The concept of aromaticity is not restricted to hydrocarbons. Heterocyclic systems, whether of the pyrrole type **1.38** with trigonal nitrogen in place of one of the C=C double bonds, or of the pyridine type **1.39** with a trigonal nitrogen in place of a carbon atom, are well known. The $\pi$ orbitals of pyrrole are like those of the cyclopentadienyl anion, and those of pyridine like benzene, but skewed by the presence of the electronegative heteroatom. The energies and coefficients of heteroatom-containing systems like these cannot be worked out with the simple

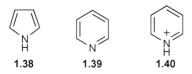

**1.38**     **1.39**     **1.40**

devices that work for linear and monocyclic conjugated hydrocarbons. The numbers in Fig. 1.54 are the results of simple Hückel calculations using parameters like those in Table 1.2 for equations like Equations 1.15 and 1.16. The overall $\pi$ energy is lowered by the cyclic conjugation. The lowest-energy orbital $\psi_1$ is polarised towards the electronegative atom, and the next orbital up in energy $\psi_2$ (and the highest unoccupied orbital) are polarised the other way. This polarisation is more pronounced in the pyridinium cation **1.40**, where the protonated nitrogen is effectively a more electronegative atom. In pyridine, the HOMO is actually localised as the nonbonding lone pair of electrons on nitrogen. The degeneracy of $\psi_2$ and $\psi_3$ is removed, but not by much. The orbitals with nodes through the heteroatoms are identical in energy and coefficients with those of the corresponding hydrocarbon, as in $\psi_3$ and $\psi_5^*$, both in pyrrole and in pyridine and its cation.

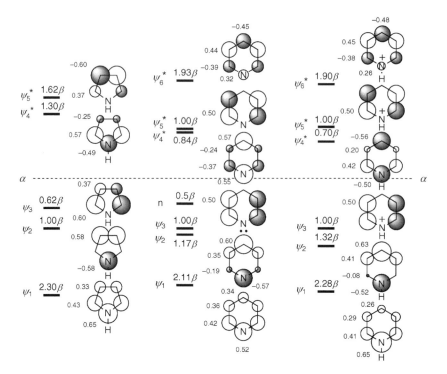

**Fig. 1.54**    $\pi$ Molecular orbitals of pyrrole, pyridine and the pyridinium ion. (Calculated using $h = 1$ and $k = 1$ for pyrrole, $h = 0.5$ and $k = 1$ for pyridine, and $h = 1$ and $k = 1$ for the pyridinium cation)

## 1.8   The Tau Bond Model

The Hückel version of molecular orbital theory, separating the $\sigma$ and $\pi$ systems, is not the only way of accounting for the bonding in alkenes. Pauling showed that it is possible to explain the electron distribution in alkenes and conjugated polyenes using only $sp^3$-hybridised carbon atoms. For ethylene, instead of having $sp^2$-hybridised carbons involved in full $\sigma$ bonding, and p orbitals involved in a pure $\pi$ bond, two $sp^3$ hybrids can overlap in something between $\sigma$ and $\pi$ bonding **1.41**. The overall distribution of electrons in this model is exactly the same as the combination of $\sigma$ and $\pi$ bonding in the conventional Hückel picture. In practice, this model, usually drawn with curved lines called $\tau$ bonds **1.42**, has found few adherents, and the insights it gives have not proved as useful as the Hückel model. It is useful, however, to recognise that it is perfectly legitimate, and that on occasion it might have some virtues in trying to explain aspects of stereochemistry.

**1.41**          $\equiv$          **1.42**

## 1.9 Spectroscopic Methods

A number of physical methods have found support in molecular orbital theory, or have provided evidence that the deductions of molecular orbital theory have some experimental basis. Electron affinities correlate moderately well with the calculated energies of the LUMO, ionisation potentials correlate moderately well with the calculated energies of the HOMO, and spectroscopic methods reveal features that support molecular orbital theory.

### 1.9.1 Ultraviolet Spectroscopy

When light of an appropriate energy interacts with an organic compound, an electron can be promoted from a low-lying orbital to a higher-energy orbital, with the lowest-energy transition being from the HOMO to the LUMO. Selection rules govern which transitions are allowed and which are forbidden. One rule states that electron spin may not change, and another that the orbitals should not be orthogonal. The remaining selection rule is based on the symmetries of the pair of orbitals involved. In most cases, the rules are too complicated to be made simple here. Group theory is exceptionally powerful in identifying which transitions are allowed, and it is one of the first applications of group theory that a chemist pursuing a more thorough understanding comes across. One case, however, is easy—that for molecules which only have a centre of symmetry, like s-*trans* butadiene. The allowed transitions for centro-symmetric molecules are between orbitals that are symmetric and antisymmetric with respect to the centre of symmetry. Thus the HOMO, $\psi_2$, is symmetric with respect to the centre of symmetry halfway between C-2 and C-3, and the LUMO, $\psi_3^*$, is antisymmetric (Fig. 1.35). Accordingly this transition is allowed and is strong, as is the corresponding transition for each of the longer linear polyenes.

Fig. 1.55 is a plot of the experimentally determined values of $\lambda_{max}$ for the longest wavelength absorption for a range of polyenes $R(CH=CH)_nR$, converted

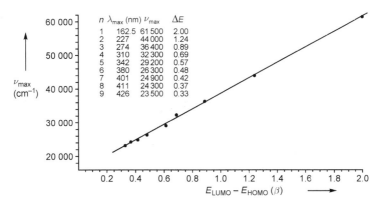

| $n$ | $\lambda_{max}$ (nm) | $\nu_{max}$ | $\Delta E$ |
|---|---|---|---|
| 1 | 162.5 | 61 500 | 2.00 |
| 2 | 227 | 44 000 | 1.24 |
| 3 | 274 | 36 400 | 0.89 |
| 4 | 310 | 32 300 | 0.69 |
| 5 | 342 | 29 200 | 0.57 |
| 6 | 380 | 26 300 | 0.48 |
| 7 | 401 | 24 900 | 0.42 |
| 8 | 411 | 24 300 | 0.37 |
| 9 | 426 | 23 500 | 0.33 |

**Fig. 1.55** Frequency of first $\pi \rightarrow \pi^*$ transitions of some representative polyenes $R(CH=CH)_nR$ plotted against $E_{LUMO} - E_{HOMO}$

to frequency units, against calculated values of $E_{LUMO} - E_{HOMO}$ in $\beta$ units. The correlation is astonishingly good—in view of the simplifications made in Hückel theory. Similarly impressive correlations can be made for aromatic and $\alpha,\beta$-unsaturated carbonyl systems. It is not however a good measure of *absolute* energies, and the energy of the $\pi \rightarrow \pi^*$ transition measured by UV cannot be used directly as a measure of the energy difference between the HOMO and the LUMO. This can be seen from that fact that the line in Fig. 1.55 does not go through the origin, as Hückel theory would predict, but intersects the ordinate at $15\,500$ cm$^{-1}$, corresponding to an energy of 185 kJ mol$^{-1}$ (44 kcal mol$^{-1}$).

### 1.9.2 Photoelectron Spectroscopy

Photoelectron spectroscopy (PES) measures the energies (Table 1.3) of filled orbitals, and overcomes the problem that UV spectroscopy does not give good absolute values for the energies of molecular orbitals.

**Table 1.3** Energies of HOMOs of some representative molecules as measured by PES (1 eV = 96.5 kJ mol$^{-1}$ = 23 kcal mol$^{-1}$)

| Entry | Molecule | Type of orbital | Energy (eV) |
|-------|----------|-----------------|-------------|
| 1 | :PH$_3$ | n | $-9.9$ |
| 2 | :SH$_2$ | n | $-10.48$ |
| 3 | :NH$_3$ | n | $-10.85$ |
| 4 | :OH$_2$ | n | $-12.6$ |
| 5 | :ClH | n | $-12.8$ |
| 6 | CH$_2$=CH$_2$ | $\pi$ | $-10.51$ |
| 7 | HC≡CH | $\pi$ | $-11.4$ |
| 8 | :O=CH$_2$ | n | $-10.88$ |
| 9 | | $\pi$ | $-14.09$ |
| 10 | CH$_2$=CH-CH=CH$_2$ | $\psi_2$ | $-9.1$ |
| 11 | | $\psi_1$ | $-11.4$ or $-12.2$ |
| 12 | HC≡C-C≡CH | $\psi_2$ | $-10.17$ |
| 13 | H$_2$NCH=O: | n | $-10.13$ |
| 14 | | $\pi$ | $-10.5$ |
| 15 | CH$_2$=CH-CH=O | n | $-10.1$ |
| 16 | | $\pi$ | $-10.9$ |
| 17 | Furan | $\pi$ | $-8.9$ |
| 18 | Benzene | $\pi$ | $-9.25$ |
| 19 | Pyridine | $\pi$ | $-9.3$ |
| 20 | | n | $-10.5$ |

Here we can see how the change from a simple double bond (entry 6) to a conjugated double bond (entry 10) *raises* the energy of the HOMO. Similarly, we can see how the change from a simple carbonyl group (entry 9) to an amide (entry 14) also raises the HOMO energy, just as it ought to, by analogy with the allyl anion, with which an amide is isoelectronic. The interaction between a C=C bond ($\pi$ energy –10.5 eV) and a C=O bond ($\pi$ energy –14.1 eV) gives rise to a HOMO of lower energy (–10.9 eV, entry 16) than when two C=C bonds are conjugated (–9.1 eV, entry 10). Finally, we can see that the more electronegative an atom, the lower is the energy of its HOMO (entries 1 to 5).

### 1.9.3 Nuclear Magnetic Resonance Spectroscopy

Chemical shift is substantially determined by the electron population surrounding the nucleus in question and shielding it from the applied field. Chemical shifts are therefore used to probe the *total* electron population. The chemical shift range with protons is so small that aromatic ring currents and other anisotropic influences make such measurements using proton spectra unreliable, but the agreement between calculated and observed $^{13}C$ chemical shifts is good.

Coupling constants $J$ measure the efficiency with which spin information from one nucleus is transmitted to another. This is not usually mediated through space, but by interaction with the electrons in the intervening orbitals. The interaction between the nuclear magnetic moment and the electrons is only with s orbitals (only s orbitals have an electron population at the nucleus), but p and $\pi$ orbitals can transmit spin information when the s orbital carrying that information is not orthogonal to a p orbital or $\pi$ system. Transmission of information about the magnetic orientation of one nucleus to another is therefore dependent upon how well the intervening electrons interact with the nuclear spin and then with each other. In a crude approximation, the number of intervening orbital interactions affects both the sign and the magnitude of the coupling constant. Although the sign of the coupling constant does not affect the appearance of the $^1H$-NMR spectrum, it does change the way in which structural variations affect the magnitude of the coupling constant. In general, although not always, one-bond couplings $^1J$ and three-bond couplings $^3J$ are positive in sign, and two- and four-bond couplings $^2J$ and $^4J$ are negative in sign. The full treatment of this topic is beyond this book, but manifestations of orbital interactions can be seen in several well known features of NMR spectra.

Thus, the $^1J$ values for $^1H$—$^{13}C$ coupling are positive, and correlated with the degree of s character at carbon **1.43–1.45**, whereas the $^2J$ coupling between geminal protons is negative. Although negative, it is larger in absolute magnitude when both C—H bonds are conjugated to the same $\pi$ bond **1.47** than when they are not **1.46**. Vicinal hydrogen atoms, have positive $^3J$ coupling constants, large in absolute magnitude when the C—H bonds are antiperiplanar **1.48** or syncoplanar **1.50**, because they are conjugated with each other, but virtually zero when the C—H bonds are orthogonal **1.49** (the Karplus equation). Coupling constants are usually larger when the intervening bond is a $\pi$ bond, with the *trans* and *cis* $^3J$ coupling in alkenes typically 15 and 10 Hz for the same 180° and 0° dihedral angles. Longer-range coupling is most noticeable when one or more of the intervening bonds is a $\pi$ bond, most strikingly demonstrated by $^5J$ values as high as 8–10 Hz in 1,4-cyclohexadienes **1.51**. When there are no $\pi$ bonds, the strongest long-range coupling is found when the intervening $\sigma$ bonds are oriented and held rigidly for efficient conjugation with $^4J$ W-coupling **1.52** and **1.53**.

### 1.9.4 Electron Spin Resonance Spectroscopy

A final technique which both confirms some of our deductions and provides useful quantitative data for frontier orbital analysis is ESR spectroscopy. This technique detects the odd electron in radicals; the interaction of the spin of the electron with the magnetic nuclei ($^1$H, $^{13}$C, etc.) gives rise to splitting of the resonance signal, and the degree of splitting is proportional to the electron population at the nucleus. Since we already know that the coefficients of the atomic orbitals, $c$, are directly related to the electron population, we can expect there to be a simple relationship between these coefficients and the observed coupling constants. The nucleus most often used is $^1$H, and the atomic orbital whose coefficient is measured in this way is that on the carbon atom to which the hydrogen atom in question is bonded.

The McConnell equation (Equation 1.17) shows the relationship of the observed coupling constant ($a_H$) to the unpaired spin population on the adjacent carbon atom ($\rho_C$).

$$a_H = Q_{CH}^H \rho_C \qquad\qquad 1.17$$

The constant Q is different from one situation to another, but when an electron in a $p_z$ orbital on a trigonal carbon atom couples to an adjacent hydrogen, it is about –24G. Applied to aromatic hydrocarbons, where it is particularly easy to generate radical anions and cations, there proves to be a good correlation between coupling constants and the calculated coefficients. The standard ways of generating radicals for ESR measurements involve adding an electron to a molecule or taking one away. In the former case the odd electron is fed into what was the LUMO, and in the latter case the odd electron is left in the HOMO. Since these are the orbitals which appear to be the most important in determining chemical reactivity, it is particularly fortunate that ESR spectroscopy should occasionally give us access to their coefficients.

## 1.10 Exercises

1. In the discussion about the $H_3$ molecule, we could have combined the three atoms in a straight line. Show that there would be less bonding both in $\sigma_1$ and $\sigma_2{}^*$ and less antibonding in the $\sigma^*$ orbital.

2. Using Equation 1.12, or by drawing the geometrical equivalent, calculate the coefficients for the frontier orbitals of the heptatrienyl cation.

3. Construct a diagram equivalent to that in Fig. 1.26 to calculate the energies of the frontier orbitals of the heptatrienyl cation.

4. Spiro-aromaticity (Section 1.5.5) is barely detectable. Explain why the spiro cation **1.54**, might have more $\pi$ stabilisation by spiro conjugation than spiro-heptatriene **1.30**.

| | | |
|---|---|---|
| **1.30** | **1.54** | **1.31** |

5. Using a diagram adapted from that in Fig. 1.41, show that the spirononat-etraene **1.31** is spiro-antiaromatic.

# 2 The Structures of Organic Molecules

This chapter uses the language of molecular orbital theory developed in Chapter 1 to explain some of the better known structural features of organic molecules. It is concerned with the ground state and with thermodynamic properties, not with how molecules behave in chemical reactions. It is important to realise that conjugation, for example, may, and usually does, make a molecule *thermodynamically more stable* than an unconjugated one, but it does not follow that conjugated systems are less reactive. Indeed, they are often more reactive or, we might say, *kinetically less stable*. Organic chemists use 'stable' and 'stability' without always identifying which meaning they are assuming. In this chapter we shall look at thermodynamic stability, and reserve reactivity for later chapters.

## 2.1 The Effects of π Conjugation

Much of organic chemistry is explained by making analogies from one compound or reaction to another. We use substituent effects, manifest in one compound or reaction, to inform us about the effects of the same substituents in other compounds or reactions. Substituent effects may be derived by calculation or from experimental measurements like heats of combustion or hydrogenation. But calculations alone do not make immediate *chemical* sense, and an experimental measurement still needs an explanation. The discussion in the following pages shows that we can work out, crudely but usefully, the effects of substituents on π systems in an easy, nonmathematical way, both on the overall energy, and on the energy and polarisation of the frontier orbitals.

### 2.1.1 A Notation for Substituents

There are three common types of substituents, each of which modifies the orbitals of conjugated systems in a different way (Fig. 2.1). They are: (a) simple conjugated systems, like vinyl or phenyl, which we shall designate with the letter C; (b) π-bonded systems which are also electron withdrawing, like formyl, acetyl,

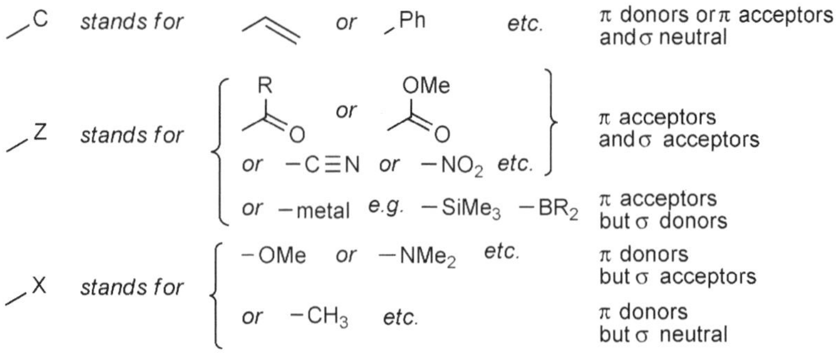

**Fig. 2.1** Definitions and character of C-, Z- and X-substituents

cyano, nitro, and carboxy, which we shall designate with the letter Z; and (c) heteroatoms which carry a lone pair of electrons, which we shall designate with the letter X.

The C-substituents can be electron donating or electron withdrawing, depending upon what they are bonded to. They are called $\pi$ donors or $\pi$ acceptors, as appropriate, and their effect on the $\sigma$ framework is small. The Z-substituents withdraw electrons from double bonds that they are conjugated to, and, since most of them have electronegative heteroatoms, they are also weakly electron withdrawing by an inductive effect within the $\sigma$ framework. Such substituents are therefore strong $\pi$ acceptors and usually weak, but occasionally, strong $\sigma$ acceptors, especially for substituents like nitro, where an electronegative heteroatom is the point of attachment. Another group of $\pi$ electron-withdrawing substituents is slightly different. Metals, and metalloids like the silyl group, are $\pi$ acceptors [see (Section 2.2.3.2) page 77] but, because metals are more electropositive than carbon, they are $\sigma$ donors. These substituents do not have a separate symbol, but their effect on the $\pi$ system is more often than not what we shall be interested in, and they are included among the group labelled Z above. The X-substituents are electronegative heteroatoms with a lone pair of electrons, and they are therefore $\pi$ donors and $\sigma$ acceptors. We include simple alkyl groups in the category of X-substituents, because overlap of the C—H (or C—C) bonds [hyperconjugation, see (Section 2.2) page 69] supplies electrons to the $\pi$ system. The effect is usually in the same direction as that of a lone pair but smaller, but alkyl groups are largely neutral with respect to the $\sigma$ framework.

### 2.1.2 The Effect of Substituents on the Stability of Alkenes

**2.1.2.1 C-Substituents.** A double bond, lowers the $\pi$ energy when it is conjugated to another double bond (see pages 28–30). The phenyl group is similar— the filled $\pi$ molecular orbitals in styrene come at $2.14\beta$, $1.41\beta$, $1.00\beta$, $0.66\beta$ below the $\alpha$ level, and the LUMO at $0.66\beta$ above the $\alpha$ level. The total $\pi$ stabilisation is $2 \times 5.21\beta$, whereas the total $\pi$ stabilisation for the separate components benzene and ethylene is $2 \times 5.0\beta$. A C-substituent raises the energy of the HOMO (from $1\beta$ below the $\alpha$ level in ethylene to $0.62\beta$ below it in butadiene and $0.66\beta$ in styrene), and it lowers the energy of the LUMO (from $1\beta$ in ethylene to $0.62\beta$ in butadiene and $0.66\beta$ in styrene). Similarly with the coefficients—the terminal carbon atom, both in the HOMO and in the LUMO has a larger coefficient than the internal atom, 0.6 and $\pm0.37$ in butadiene and 0.6 and $\pm0.39$ in styrene.

### 2.1.2.2 Z-Substituents.

The simplest Z-substituent is the formyl group in acrolein **2.1**. A simple Hückel calculation gives Fig. 2.2 in which we are looking at the p orbitals from above. This figure gives us what we want, but no insight.

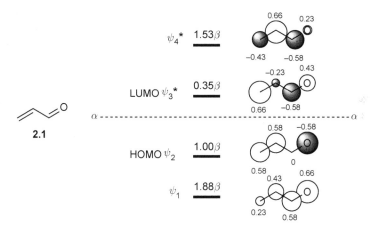

**Fig. 2.2** The $\pi$ molecular orbitals of acrolein. (These energies and coefficients were calculated using $h = 1$ and $k = 1$)

To estimate the $\pi$ energies, we go to extremes: first we ignore the fact that one of the atoms is an oxygen atom and not a carbon atom, which gives us the orbitals of butadiene **2.3**. Then we go to the other extreme and treat the carbonyl group as a carbocation **2.2**. Normally we do not draw it this way, because such good stabilisation is better expressed by drawing the molecule (as in **2.1**) with a full $\pi$ bond. The truth is somewhere in between. Organic chemists make the mental reservation about drawings like **2.1** and **2.2** that the butadiene-like system, implied by **2.1**, is only one extreme approximation of the true orbital picture for acrolein, and the other extreme approximation is an allyl cation, substituted by a noninteracting oxy-anion. The truth is somewhere in between—an allyl cation substituted by a strongly interacting oxy-anion.

The energies for the molecular orbitals for these two extremes are shown in Fig. 2.3. The true orbital energy for the orbitals of acrolein must be in between those of the corresponding orbitals of the allyl cation and butadiene. We can perhaps expect the true structure to be more like the butadiene system than the allyl cation system (for the same reason that we prefer to draw it as **2.1** rather than **2.2**). We can see that the effect of having a Z-substituent conjugated with the double bond of

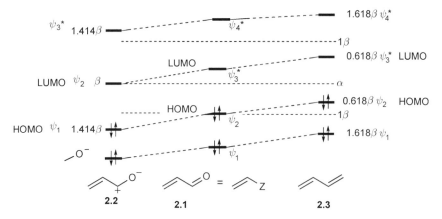

**Fig. 2.3**   The energies of the $\pi$ orbitals of acrolein **2.1** as a weighted sum of the $\pi$ orbitals of an allyl cation **2.2** and butadiene **2.3**

ethylene is to lower the $\pi$ energy of the system, with $\psi_1$ and $\psi_2$ together having more $\pi$ bonding than the separate orbitals of ethylene and a carbonyl group. The energy of the HOMO of acrolein, $\psi_2$, is, however, little changed from that of the $\pi$ orbital of ethylene at the $\beta$ level. Also, because it is butadiene-like, the HOMO and the LUMO will be closer in energy than they are in ethylene—the LUMO will have been lowered in energy relative to ethylene's and the HOMO will be very similar in energy. We have superimposed the orbitals of an allyl cation on those of butadiene, and, with suitable weighting, added them together.

We can use the same ideas to deduce the pattern of the coefficients. We have again contributions from the allyl-cation-like nature of acrolein and from its butadiene-like nature in Fig. 2.4. The coefficients for acrolein on the right can

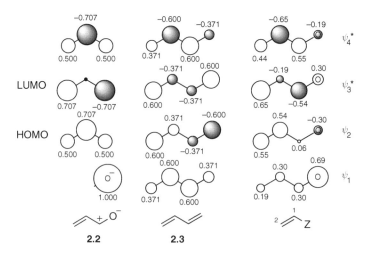

**Fig. 2.4**   Crude estimates of the coefficients of the $\pi$ orbitals of a Z-substituted alkene as an unweighted average of the coefficients of an allyl cation **2.2** and butadiene **2.3**

then be expected to be somewhere in between the corresponding coefficients in the two components on the left and in the middle. The average of the two components is given on the right in Fig. 2.4, these arbitrarily representing a simple unweighted sum. These numbers are not coefficients, because they have not been arrived at with legitimate algebra, and, squared and summed, they do not, of course, add up either horizontally or vertically to one. They are however similar in their general pattern to those obtained by calculation in Fig. 2.2.

To take the LUMO of a Z-substituted alkene ($\psi_3^*$) as an example, the carbon atom C-1 with the Z-substituent on it has a zero coefficient on the corresponding atom in the allyl cation and a small coefficient in butadiene ($-0.371$). The coefficient on C-1 in the LUMO of a Z-substituted alkene is therefore likely to be small ($-0.19$ in Fig. 2.4, and $-0.23$ in Fig. 2.2). In contrast, the carbon atom C-2 has large coefficients both in the allyl cation (0.707) and in butadiene (0.60). The coefficient on C-2 in the LUMO of a Z-substituted alkene is therefore large (0.65 in Fig. 2.4, and 0.66 in Fig. 2.2).

Turning now to the HOMO of acrolein ($\psi_2$) and C-1, the allyl cation has a large coefficient (0.707) on the central atom, but butadiene has a small coefficient on the corresponding atom (0.371). The two effects act in opposite directions—the conjugation causing a reduction in the coefficient on the carbon atom carrying the formyl group, and the allyl-cation-like character causing an increase in this coefficient. The result is a medium-sized coefficient (0.54 in Fig. 2.4, and 0.58 in Fig. 2.2). For C-2, it is the allyl cation that has the smaller coefficient (0.500) and the butadiene the larger (0.600). The combination is again a medium-sized coefficient (0.55 in Fig. 2.4 and 0.58 in Fig. 2.2). We have already said that acrolein is probably better represented by the drawing **2.1** than by the drawing **2.2**, from which we may guess that it is the butadiene-like character which makes the greater contribution to the HOMO, in which case acrolein will have its HOMO coefficients polarised in the same way as those of butadiene, but less so (as they are in Fig. 2.4).

### 2.1.2.3 X-Substituents.

In an X-substituted alkene like methyl vinyl ether **2.4**, we have a lone pair of electrons brought into conjugation with the double bond. We can deduce the pattern of molecular orbitals by an interaction diagram, Fig. 2.5, resembling that for the allyl anion in Fig. 1.28. The difference is that the lone pair on oxygen, being on an electronegative element, is lower in energy than the p orbital on carbon. This lowers the energy of all the orbitals $\psi_1$–$\psi_3^*$ relative to their counterparts in the allyl system. However the orbital $\psi_1$ is created by the interaction of the lone-pair orbital on the oxygen atom, labelled n, in a bonding sense with both $\pi$ and $\pi^*$, strongly with the former and weakly with the latter, because of the greater separation of energy of the interacting orbitals. In contrast, $\psi_2$ is derived by the weak interaction of n with $\pi^*$ in a bonding sense, and strongly with $\pi$ in an antibonding sense. As a result $\psi_1$ is lowered in energy more than $\psi_2$ is raised, and the overall energy is lowered relative to the energy of the separate orbitals of the $\pi$ bond and the lone pair. As usual, conjugation has lowered the overall energy. The net $\pi$ stabilisation has been measured crudely by comparing the heats of hydrogenation of ethylene and ethyl vinyl ether as

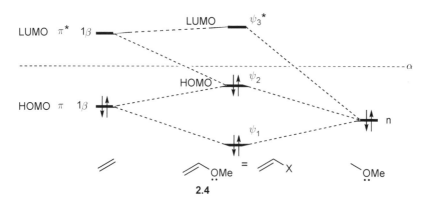

**Fig. 2.5**   Energies of the $\pi$ orbitals of an X-substituted alkene

25 kJ mol$^{-1}$ (6 kcal mol$^{-1}$). We should also note that both the HOMO and the LUMO of an X-substituted alkene are raised in energy relative to the HOMO and LUMO of ethylene, with the HOMO raised more than the LUMO.

In order to deduce the coefficients for an X-substituted alkene, we adopt the idea that at one extreme, the lone pair on the oxygen atom is fully and equally involved in the overlap with the $\pi$ bond, so that the orbitals will be those of an allyl anion **2.5**. At the other extreme, to make allowance for the fact that the lone pair on an electronegative atom like oxygen is not as effective a donor as a filled p orbital on carbon, it is an alkene with no participation from the lone pair on the oxygen atom. Thus we add a bit of allyl anion-like character, on the left in Fig. 2.6, to the unperturbed alkene, in the centre of Fig. 2.6. The average of the two components is given on the right in Fig. 2.6, these representing a simple unweighted sum. As with the Z-substituted alkene, these numbers are not coefficients, but they match the *pattern* of large, medium and small coefficients

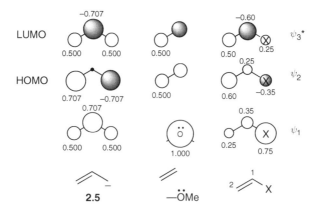

**Fig. 2.6**   Crude estimates of the coefficients of the $\pi$ orbitals of an X-substituted alkene as an unweighted sum of the coefficients of an allyl anion **2.5** and an alkene

obtained from a simple Hückel calculation. The lowest-energy orbital $\psi_1$ has a large contribution from the lone pair added to the lowest-energy orbital of the allyl anion, creating an orbital strongly polarised towards the X-substituent. For the HOMO, the unperturbed alkene has (necessarily) equal coefficients on each atom, and the allyl anion has a zero coefficient on the atom bearing the X-substituent. The result of mixing these two is $\psi_2$, a strongly polarised orbital as far as the coefficients on C-1 and C-2 are concerned. For the LUMO, the unperturbed alkene again has equal coefficients, but the allyl anion has a larger coefficient on the carbon atom carrying the X-substituent than on the other one. The result is $\psi_3{}^*$, an orbital mildly polarised in the opposite direction.

### 2.1.3   The Effect of Substituents on the Stability of Carbocations

**2.1.3.1   C- and X-substituents.**   A molecule having an empty p orbital on carbon, and therefore carrying a positive charge, will be lowered overall in energy by $\pi$ conjugation with a C-substituent. We have seen this already, from the opposite direction, when we moved from the orbitals of an alkene to those of an allyl cation in Fig. 1.28. Similarly, the effect of an X-substituent is even more stabilising, as we saw in considering the orbitals of a carbonyl group in Fig. 1.51, which could equally well have been drawn with two electrons in the $p_O$ orbital and none in the $p_C$. The weakest kind of X-substituent is an alkyl group to which we shall return while discussing hyperconjugation [see (Section 2.2) page 69] .

**2.1.3.2   Z-substituents.**   The effect of a Z-substituent is not so straightforward. Fig. 2.7 shows the interaction between the orbitals of a carbonyl group and an empty p orbital on carbon. The set of $\pi$ orbitals in the middle is essentially the same as the set of orbitals in the middle of Fig. 2.5, but with two fewer electrons in the $\pi$ system. We deduce that there is an overall lowering of $\pi$ energy, because $\psi_1$ is lower in energy than the $\pi_{C=O}$ orbital as a result of the interaction with the empty p orbital, $p_C$. However, this lowering is not large, because this interaction is

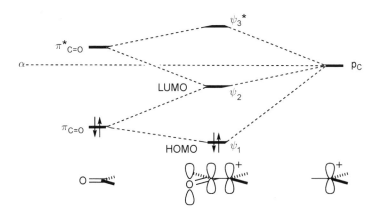

**Fig. 2.7**   The $\pi$ orbitals of a carbocation conjugated to a Z-substituent

between an orbital at the $\alpha$ level and a $\pi$ orbital, $\pi_{C=O}$, low in energy (Fig. 1.51). The overall lowering in $\pi$ energy is not therefore as great as the corresponding lowering in energy in $\psi_1$ of the allyl cation ($E$ in Fig. 1.28). We might notice at this stage that $\psi_2$ is lowered in energy, whereas it was not lowered at all in the allyl cation. This orbital is made up by interaction of the p orbital with the $\pi$ orbital of the carbonyl group in an antibonding sense and with the $\pi^*$ orbital in a bonding sense, as with the allyl cation, but, since both the $\pi$ and $\pi^*$ orbitals are lower in energy in a carbonyl group than in an alkene, the antibonding contribution to $\psi_2$ is weaker and the bonding contribution stronger.

It is well known, however, that a carbonyl group does not appear to be a stabilising influence on a carbocation, and yet we have just deduced that it is stabilised in the $\pi$ system. The most obvious factor that we have left out in the argument above is the Coulombic effect of the partially ionic character of both the $\sigma$ and the $\pi$ bond of a carbonyl group. The polarisation of both bonds towards the oxygen atom (Fig. 1.51) places a significant positive charge on the carbonyl carbon atom, immediately adjacent to the full positive charge on the nucleus of the carbon atom carrying the empty p orbital. This is energy-raising, because the now relatively exposed nuclei repel each other. We thus have a small energy-lowering contribution from the $\pi$ overlap, but an energy-raising contribution from an adverse Coulombic effect. Evidently the latter wins. For the first time, we see that conjugation cannot always be relied upon to lower the overall energy.

### 2.1.4 The Effect of Substituents on the Stability of Carbanions

The orbitals for the interactions of C-, Z- and X-substituents with a filled p orbital on carbon are the same as those we have just used for their interaction with an empty p orbital, but with two more electrons to feed into the $\pi$ orbitals. The interaction of a C-substituent with a filled p orbital gives us the allyl anion orbitals, and these are just as $\pi$ stabilised as the allyl cation (Fig. 1.28).

Even better, conjugation of a filled p orbital with a Z-substituent gives us the same orbitals as in Fig. 2.7, but now $\psi_2$ is filled, and, since it is lowered in energy by the interaction more than $\psi_2$ of the allyl anion, the overall $\pi$ energy is lower still. This is the $\pi$ system for an enolate ion. Furthermore, the extra electrons mean that a partial positive charge is no longer adjacent to an unshielded nucleus, and the Coulombic repulsion is no longer unusually large.

This time it is the X-substituent that might be destabilising rather than stabilising. The interaction of a lone pair of electrons on an oxygen atom, as a model for an X-substituent, and a filled p orbital on carbon create the $\pi$ orbitals of the carbonyl group (Fig. 1.51), but with two electrons in $\pi^*_{CO}$. Since this interaction is straightforwardly the interaction of atomic orbitals, the overall effect is a rise in energy, because $\pi^*_{CO}$ is raised more in energy than $\pi_{CO}$ is lowered. In practice, although this effect in the $\pi$ system must be present, electronegative elements usually stabilise an adjacent 'anionic' carbon. The reason is twofold. In the first place, there is a Coulombic effect working in the $\sigma$ framework against the effect in the $\pi$ system. In the second place, we do not usually have an anion—what we have is a carbon-metal (C—M) bond. A C—M bond is polarised, with the filled $\sigma$ orbital having a large coefficient on the electronegative element, which in this case is the carbon atom (Fig. 1.48). Thus, the C—M bond has many of the properties of a genuine carbanion, the repulsive interaction of the lone pair on the X-substituent and the $\sigma_{CM}$ orbital will be energy-raising, just as it would be for a carbanion. But the presence of a metal changes the story, because it has empty orbitals that can accept coordination from the lone pairs of the electronegative heteroatom. This coordination may be directly within the molecule, but is more often present in an aggregate, and it is always powerfully energy-lowering, making any effect on the $\pi$ overlap less important.

The one X-substituent that probably does destabilise an anion is an alkyl group. An alkyl group, although classified as an X-substituent, is not a $\sigma$ acceptor, nor does it have coordination sites. Its destabilising effect is by $\sigma$ conjugation [see (Section 2.2.1) page 69]. In contrast, sulfur- and phosphorus-based groups like phenylthio or diphenylphosphinyl are X-substituents that are well known to be anion-stabilising. This phenomenon, and the even less expected effect of an electropositive element like silicon stabilising an adjacent anion, has most simply been accounted for by invoking overlap of an empty d orbital on sulfur, phosphorus or silicon with the filled p orbital of the anion. This is unmistakably stabilising, as usual with the overlap of a filled with an unfilled orbital, but the contribution it makes is unlikely to be significant, because the 3d orbitals on these second-row elements and a 2p orbital on carbon are too far apart in energy and too ill-matched in size to have a significant interaction. Anion stabilisation by sulfur, phosphorus and silicon appears to be best accounted for by $\sigma$ conjugation [see (Section 2.2.3.2) page 77].

### 2.1.5    The Effect of Substituents on the Stability of Radicals

**2.1.5.1    C-, Z- and X-Substituents.**    All three kinds of substituent stabilise radicals. A C-substituent gives the orbitals of the allyl radical, which is just as stabilised as it was for the cation and anion (Fig. 1.28). A Z-substituent gives us the same orbitals as those in Fig. 2.7, but with one electron in $\psi_2$, leading to an overall drop in $\pi$ energy and a reduction in the amount of Coulombic repulsion that destabilised cations. Finally an X-substituent gives us the orbitals of the carbonyl group (Fig. 1.51) but with one electron in $\pi^*_{CO}$. With two drops in energy from the doubly filled orbital $\pi_{CO}$ matched by only one rise in energy from the singly occupied $\pi^*_{CO}$, the overall effect is a drop in energy.

**2.1.5.2    Captodative Stabilisation.**[9]    A particularly telling case is a radical that has both an X- and a Z-substituent, either directly attached to a radical centre or conjugated to it through a $\pi$ system, as in the long-lived radicals **2.6–2.10**. Radicals like this are called captodative or merostabilised, and have been claimed to be more stabilised than the sum of each component would suggest. Since both types of substituent can stabilise a radical, it is reasonable that both together can continue to stabilise a radical. We can see how this might be in Fig. 2.8. The interaction of an X-substituent and a Z-substituent creates the set of orbitals in the centre. There is a rise in energy in creating $\psi_3$, but there is only one electron in this

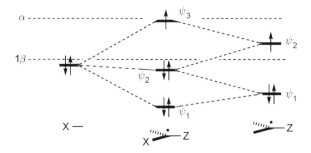

**Fig. 2.8**    The effect of bringing an X-substituent into conjugation with a
Z-substituted radical

orbital. There is a small drop in energy in creating $\psi_2$ and a more significant drop in energy in creating $\psi_1$, both of which have two electrons in them. Overall the energy may have dropped and the radical as a whole is lower in $\pi$ energy than the separate components.

| 2.6 | 2.7 | 2.8 | 2.9 | 2.10 |

However, it is not obvious whether captodative substitution is better at lowering the overall energy than having two Z-substituents or two X-substituents. It is clear experimentally that, if there is a specific captodative effect, it is small, never more than about 25 kJ mol$^{-1}$ (6 kcal mol$^{-1}$). Experimental evidence that seems to imply special stabilisation to captodative radicals is the ease of the reversible C—C fragmentation of the diaminosuccinate **2.11**, in which the rate implies that the captodative radical **2.12** is some 17 kJ mol$^{-1}$ (4 kcal mol$^{-1}$) lower in energy than might be expected by adding the stabilising effects of each of the substituents. More evidence comes from measurements of the rate of rotation about the C-2 to C-3 bond of a range of allyl radicals **2.13**. At the point of highest energy in the rotation, the radical will lose its allylic character, and be stabilised only by the substituents $R^1$ and $R^2$. The captodative radical with $R^1 = $ OMe and $R^2 = $ CN has the lowest activation energy, some 12 kJ mol$^{-1}$ (2.9 kcal mol$^{-1}$) lower than the sum of the substituent effects would have suggested, but for the radicals with

| 2.11 | 2.12 | 2.13a | 2.13b |

$R^1 = R^2 = OMe$ and $R^1 = R^2 = CN$ the activation energies are some 24 kJ mol$^{-1}$ (5.7 kcal mol$^{-1}$) higher in energy. It seems that neither two X- nor two Z-substituents have twice the stabilising effect of one, but one of each does have an additive effect. In this formulation at least, the captodative effect does appear to be real.

### 2.1.6  Energy-Raising Conjugation

Not all conjugation is energy-lowering, especially when a filled p orbital is conjugated with an X-substituent, in which the repulsive effect of two filled atomic orbitals inherently destabilises the $\pi$ system. Examples of the repulsive interaction of two filled p orbitals where there are no mitigating factors are the conformations adopted by hydrogen peroxide **2.14** and hydrazine **2.15**. The overlap is avoided by twisting about the X—X bond, so that the two lone pairs are as little in conjugation as possible.

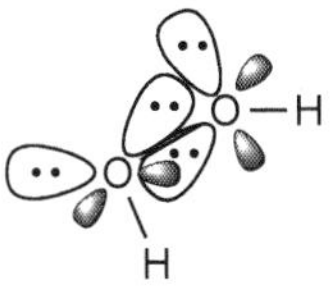

**2.14** 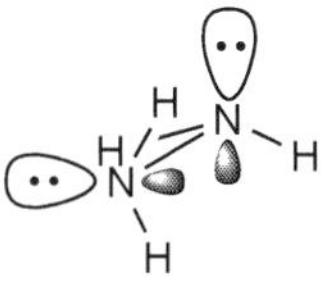                              **2.15**

Another energy-raising conjugation can be found when two carbonyl groups are adjacent. There may be some $\pi$ stabilisation, but Coulombic repulsion between the two carbon atoms, both of which bear a partial positive charge, is substantially destabilising. Evidence for destabilisation comes from the extent to which $\alpha$-diketones like 1,2-cyclohexanedione **2.16** enolise, and evidence for the $\pi$ stabilisation can be found in such molecules as glyoxal **2.17**, where the carbonyl groups stay in conjugation rather than twisting. Twisting would do nothing to relieve the Coulombic repulsion between the charges on the carbon atoms, but it would remove the $\pi$ conjugation. The s-*trans* conformation is favoured, because the relatively large partial negative charges on the oxygen atoms repel each other. With two more electrons—the enediolate ion **2.18** has $\pi$ molecular orbitals like those of butadiene, but with two electrons in $\psi_3$*. It is a conjugated system higher in energy than the sum of the separated components.

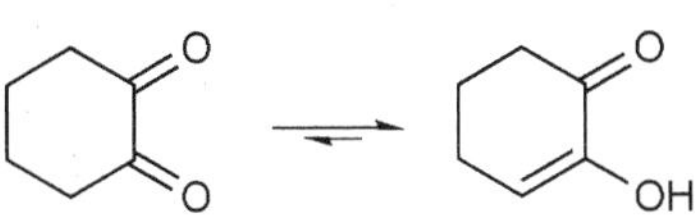

**2.16** 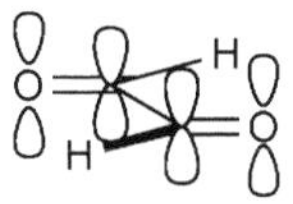                    **2.17** 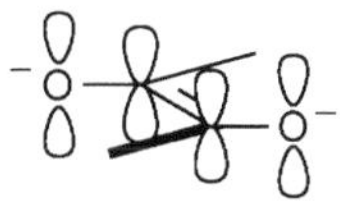                    **2.18**

## 2.2  σ Conjugation—Hyperconjugation

Conjugation has been discussed so far only as something taking place between p orbitals in a $\pi$ system. However, it is possible to consider the conjugation of

$\sigma$ bonds constructed from hybridised orbitals with p orbitals. The major interactions will be between the C—H orbitals $\sigma$ and $\sigma^*$ of a C—H or C—C bond with a p orbital. The overlap of $\sigma$ bonds with p orbitals is called hyperconjugation, a serious misnomer, because hyperconjugation, far from being especially strong, as the prefix hyper implies, is a feeble level of conjugation compared with the kind of $\pi$-conjugation that we have seen so far. Another term that is sometimes used is $\sigma$ conjugation, on the grounds that it is conjugation of a $\sigma$ bond with something else, but this is not satisfactory either, since the overlap is $\pi$ in nature not $\sigma$. Yet another term that is used is vertical stabilisation, which is not a misnomer, but is not usefully specific about its nature.

### 2.2.1   C—H and C—C Hyperconjugation

**2.2.1.1   Stabilisation of Alkyl Cations.**   Hyperconjugation is most evident in the stabilisation given to an empty p orbital on carbon by a neighbouring alkyl group, explaining how alkyl substituents stabilise carbocations. Fig. 2.9 shows the interaction of the $\sigma$ orbitals of the C—H bond on the left with the empty p orbital on the right. The interaction in Fig. 2.9 is similar to that shown in Fig. 1.28 for the allyl cation, except that it is a $\sigma$ bond instead of a $\pi$ bond interacting with the empty p orbital. Because the $\sigma_{CH}$ orbital in Fig. 2.9 is lower in energy than the $\pi$ orbital in Fig. 1.28, the hyperconjugative interaction with the empty p orbital is less effective, and the overall drop in energy $2E$ is less than it was for simple $\pi$-conjugation.

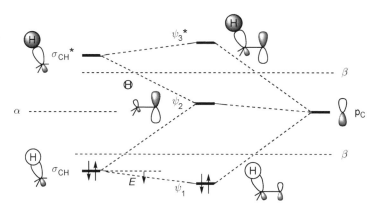

**Fig. 2.9**   Interaction of the orbitals of a $\sigma$ C—H bond with an empty p orbital on carbon

As usual, hybridisation, although a convenient device, is unnecessary—the energy lowering could equally well have been explained using the $p_z$ orbital on carbon, with the most significant interaction illustrated on the left in Fig. 2.10. Indeed, this provides a more simple way to appreciate that the lowest-energy conformation of the cation is not overwhelmingly that in which one of the $\sigma$ bonds is aligned to overlap with the empty p orbital. Because the two $\pi$-type orbitals, $\pi_z$ and $\pi_y$ have the same energy, the interactions in the two conformations shown in

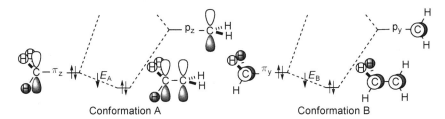

Conformation A                                          Conformation B

**Fig. 2.10**   Orbital interactions stabilising two conformations of the ethyl cation

Fig. 2.10 are, to a first approximation, equal ($E_A = E_B$). We can expect that the barrier to rotation about the C—C bond of the ethyl cation will be small. Although intuitively reasonable, it is not so easy to set up an interaction diagram using hybridisation to show that the energy-lowering effect of *imperfectly* lined up overlap of *two* C—H σ orbitals with the empty p orbital is the same as *perfectly* lined up overlap of *one*.

The overlap and its consequences, as illustrated in Fig. 2.9, could equally well have been drawn with C—C bonds in place of the C—H bonds. The energies of C—C and C—H orbitals are similar to each other, and the value of $E$ will be similar. Alkyl groups in general are effectively π electron donors, in much the same way as, but to a lesser extent than, a double bond or a lone pair and we classify alkyl groups as X-substituents.

One case where C—C bonds are exceptionally effective in hyperconjugation is in the stabilisation provided by a cyclopropyl substituent to an empty p orbital. The cyclopropylmethyl cation is actually better stabilised than an allyl cation, as judged by the more rapid ionisation of cyclopropylmethyl chloride **2.19** than of crotyl chloride **2.20**. In this case, hyperconjugation appears, unusually, to be better than π conjugation.

$$\text{(reaction scheme)}$$

**2.19**   H₂O, EtOH, 50°, $k_1$ (rel) 41   →   (cyclopropylmethyl OEt)   (crotyl)   **2.20**   H₂O, EtOH, 50°, $k_1$ (rel) 1   →   (crotyl OEt)

It can be explained using the Walsh orbitals of a cyclopropane (Fig. 1.42), where one of the degenerate pair of highest occupied orbitals $\pi_{CC}$ has a large $p_y$ coefficient on carbon which can orient itself in such a way as to stabilise an empty p orbital on a neighbouring atom **2.21a**, seen from a different perspective in **2.21b**. This is like conjugation with a full p orbital, and is more effective in lowering the π energy than conjugation with a π bond is in the allyl cation (Fig. 1.28). The other high-energy filled orbital in the Walsh diagram has the wrong symmetry for overlap with the neighbouring p orbital, and has no effect on its energy one way or the other. This picture is supported experimentally by the preferred conformation in many systems matching that in **2.21**, as can be seen in the two main conformations adopted by cyclopropane carboxaldehyde **2.22a** and **2.22b**, if the carbonyl group is thought of as a highly stabilised carbocation.

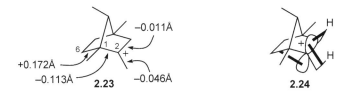

**2.21a**          **2.21b**          **2.22a**          **2.22b**

*2.2.1.2 Bridging in Carbocations.* As usually defined, hyperconjugation implies no change in the shape of the molecule caused by the extra overlap. However, the extra bonding in $\psi_1$ between the $\sigma$ C—H bond and the p orbital ought to have the effect of shortening the C—C bond and lengthening the C—H bond (or C—C bond if that is involved), and there is experimental evidence from X-ray crystal structures that this does indeed happen. Thus the bicyclo[2.2.1]heptyl cation **2.23** shows shortening of the three C—C bonds to the cationic centre relative to a typical bond between a tetrahedral and a trigonal carbon (1.522 Å), and lengthening of the bond between C-1 and C-6 relative to a typical bond between two tetrahedral carbons (1.538 Å). This shows the effects expected from the hyperconjugative interactions shown with bold lines on the drawing **2.24**.

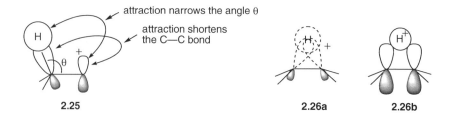

Hyperconjugative overlap will also reduce the H—C—C angle $\theta$, because there is now extra bonding between the hydrogen atom and the empty p orbital **2.25**. There is even the possibility that the lowest energy structure has the hydrogen atom sitting halfway between the two carbon atoms **2.26**. The bonding in this structure can be represented with hybridisation as two half filled orbitals made up from sp$^3$ hybrids and the 1s orbital of hydrogen **2.26a**, or without hybridisation as largely made up by the interaction of the empty 1s orbital of an isolated proton with both lobes of the $\pi$ bond of ethylene **2.26b**. The bonding, however it is described, is the same, and similar in nature to that of other two-electron, two-bond bridged systems, such as those in diborane. This structure may be the minimum in the energy profile, as it is in diborane, or it may be a maximum, in which case it is the transition structure for the [1,2]-shift of the hydrogen or carbon atom from one carbon to the next.

Although tertiary cations like **2.23** are well established not to have bridged structures, it is not easy to discover whether a localised structure with hyperconjugation, with the minimum movement of the atoms, or the fully bridged structure is the lower in energy for primary and secondary cations. In the 1960s, a large amount of effort went into trying to solve experimentally and computationally the problem of the nonclassical ion, as it was called. Both experiment and calculations gave conflicting or ambiguous answers, one of many problems being that calculations on ions in the gas phase, however high in level, inherently favour bridged structures, because bridged structures spread the charge more effectively. The present state of opinion probably favours structures like **2.25** *without bridging* for almost every alkyl cation except the most simple, the ethyl cation itself, which is only found in the gas phase. A bridged structure like **2.26** is therefore a low-energy transition structure for a 1,2-hydride shift, and, with carbon in the bridge, the transition structure for the Wagner-Meerwein type of cationic rearrangement.

### 2.2.1.3 Stabilisation of a π Bond.

Hyperconjugation has also been used to explain another well-known thermodynamic property—that having alkyl substituents stabilises alkenes. An alkene like 2-methyl-1-butene **2.27** undergoes easy protonation in acid to give the t-amyl cation **2.28**, which can lose a proton again to give 2-methyl-2-butene **2.29**. The ease of the reaction is explained by the hyperconjugative stabilisation given to the intermediate tertiary cation **2.28**. What is not so obvious is why the more-substituted alkene **2.29** is lower in energy then the less-substituted alkene **2.27**, which it certainly is, because the equilibrium lies well to the right. Heats of hydrogenation of alkenes provide quantitative evidence of the greater thermodynamic stability of the more substituted alkenes, with the attachment of one or more alkyl group more or less additively increasing the heat of hydrogenation of an alkene by about 10 kJ mol$^{-1}$ (2.4 kcal mol$^{-1}$).

One factor is the hyperconjugative stabilisation of the C=C π bond by the alkyl groups. Fig. 2.11 shows the interaction of the orbitals of a σ bond with the orbitals of a π bond, which is similar to the interaction of two π bonds in butadiene (Fig. 1.32). Although the σ-bonding orbital and the π*-antibonding orbital are further apart than $\psi_2$ and $\psi_3{}^*$ in butadiene, they are just close enough to mix in a bonding sense effectively to lower the energies of $\psi_1$ and $\psi_2$ in Fig. 2.11, making the drop in energy $E_1$ a little greater than the rise in energy $E_2$.

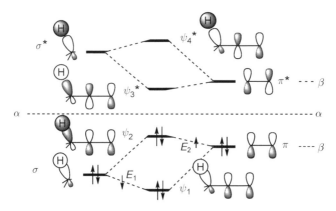

**Fig. 2.11**  Hyperconjugative stabilisation of a C=C $\pi$ bond

### 2.2.2  C–M Hyperconjugation

In Fig. 2.9, the stabilising effect of the hyperconjugation was small, because the energy gap between the $\sigma$-bonding orbital and the empty p orbital on carbon was large. A $\sigma$ bond closer in energy to the empty p orbital should have a larger interaction and be more stabilising. This is the case when the $\sigma$ bond is between a metal and carbon, because a metal is inherently more electropositive than carbon. Fig. 2.12 shows the energies of the bonding and antibonding orbitals to carbon both to a generic electropositive element M and to a generic electronegative element X. The $\sigma_{CM}$-bonding and antibonding orbitals will be higher in energy than the corresponding $\sigma_{CH}$ or $\sigma_{CC}$ orbitals, and the $\sigma_{CX}$ orbitals will be lower in energy than the $\sigma_{CH}$ or $\sigma_{CC}$ orbitals.

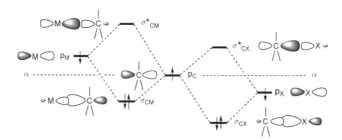

**Fig. 2.12**  $\sigma$-Bonding and antibonding orbitals from carbon to an electropositive element M and to an electronegative element X

Transferring the energy levels for a C—M bond on the left in Fig. 2.12 to an interaction diagram leads to Fig. 2.13 as a description of a $\beta$-metalloethyl cation **2.30**. With the $\sigma_{CM}$-bonding orbital higher in energy than the bonding $\sigma_{CH}$ orbital,

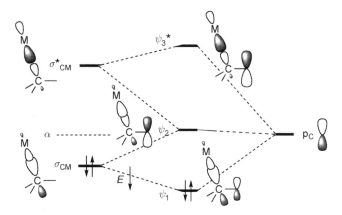

**Fig. 2.13** Interaction of the orbitals of a carbon-metal bond with an empty p orbital on carbon

the interaction with the empty p orbital on carbon will be stronger than it was for C—H, and the drop in energy $E$ will be greater. Such cations are well stabilised by hyperconjugation.

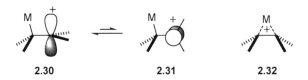

The alternative conformation **2.31**, with the empty p orbital at right angles to the M—C bond, is not stabilised any better than it is by an alkyl group, because the M—C bond is in the node of the empty p orbital and there will be no interaction between them. Thus metal-stabilised cations can be expected to adopt and retain the configuration **2.30**. The stabilisation seen in Fig. 2.13 is enhanced by the polarisation of the M—C bond. The coefficients in the $\sigma_{CM}$ orbital are large on the carbon and small on the metal, just as the coefficients of the C—X-bonding orbital are large on the X atom and small on the carbon. The bonding interaction of the $\sigma_{CM}$ orbital with the empty p orbital will therefore be greater than it was for the corresponding overlap of the $\sigma_{CH}$ orbital, where the coefficient on the carbon atom is smaller. Thus we have a more favourable energy match and a more favourable coefficient for the overlap of the M—C bond than for the H—C bond.

The degree of this stabilisation is of course dependent upon what the metal is. In practice, cations with this general structure have been investigated using barely metallic metals, like silicon. Even a trimethylsilyl group as the atom M in a cation of general structure **2.30** is lost too easily for the cation itself to be studied directly with any ease. It is clear from much evidence that silyl groups are substantially stabilising of $\beta$ cations. The Si—C bond is oriented in the plane of the empty p orbital, and rotation about the C—C bond is dramatically slowed down so that cations like **2.30** are configurationally stable during most reactions. The question of bridging also arises here, since the lowest energy structure may be the bridged cation **2.32**. Calculations in simple systems indicate that only the least substituted β-silyl cation, the trimethylsilylethyl cation itself, might be bridged. The more substantially metallic elements, however, very probably have bridged structures—coordination of a metal cation to the centre of a π bond is a familiar theme in organometallic chemistry, and the structure is **2.32**.

### 2.2.3 Negative Hyperconjugation

#### 2.2.3.1 Negative Hyperconjugation with a Cation.
If instead the carbon is bonded to an electronegative element like fluorine, the interaction diagram corresponding to Fig. 2.13 changes. The orbitals of the X—C bond, taken from

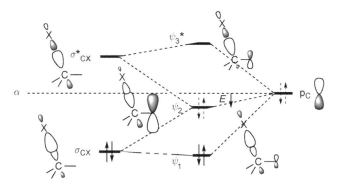

**Fig. 2.14**   Interaction of the orbitals of a bond between carbon and an electronegative element X with a p orbital on carbon

Fig. 2.12 and moved into Fig. 2.14, are lower in energy than the corresponding C—H orbitals. The interaction of $\sigma_{CX}$ with the p orbital will now have little energy-lowering effect on $\psi_1$, because the orbitals are so far apart in energy. There is therefore little π stabilisation afforded to a cation in the conformation **2.33**, and in addition there will be the usual strong inductive electron withdrawal destabilising it in the σ framework. The alternative conformation **2.34** possesses the greater degree of hyperconjugative stabilisation, as long as the other substituents on the carbon atom are not as electronegative as X, and will be preferred, but the inductive withdrawal will still make it a relatively high-energy cation.

2.33            2.34

### 2.2.3.2 Negative Hyperconjugation with an Anion.

However, if it is a carbanion that is conjugated to the X—C bond, the p orbital is filled. The orbital $\psi_2$ in Fig. 2.14 is lowered in energy significantly by an amount $E$ as a consequence of the orbital $\sigma^*_{CX}$ being so much closer in energy to the p orbital than either of the orbitals $\sigma^*_{CM}$ in Fig. 2.13 or $\sigma^*_{CH}$ in Fig. 2.9. Since this orbital is filled, there is a drop in overall energy $2E$, which the cation does not benefit from. As a consequence of the hyperconjugation, the conformation **2.35** is now stabilised more than the alternative **2.36**. Furthermore, the large coefficient on carbon in the $\sigma^*_{CX}$ orbital makes its overlap with the filled p orbital even more bonding than without the electronegative element X, and the small coefficient on carbon in the $\sigma_{CX}$ orbital makes its overlap with the filled p orbital even less antibonding, both factors further contributing to $E$, the lowering in energy of the $\psi_2$ orbital. This type of hyperconjugation is sometimes called 'negative' hyperconjugation, because it is conjugation with a negative charge, but it is another misnomer, since energy-lowering is usually regarded as a positive outcome.

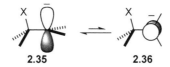

**2.35**          **2.36**

In practice, this phenomenon is not usually seen with carbanions themselves. Even if it were, simple carbanions would not be trigonal as they are shown in Fig. 2.14 and in **2.35**. Organic chemists use the word carbanion loosely, but most often they refer either to compounds with trigonal carbons carrying substantial excess negative charge, as with enolate ions [see (Section 2.1.4) page 66], or to compounds with C—M bonds. In enolate ions, the orbital $p_C$ used in Fig. 2.14 would correspond to the p orbital on the terminal carbon in $\psi_2$ of an X-substituted alkene [see (Section 2.1.2.3) page 63], which has a large coefficient on C-2 (Fig. 2.6). In compounds containing a C—M bond, the orbital $p_C$ used in Fig. 2.14 would correspond to the orbital $\sigma_{CM}$ in Fig. 2.13, which also has a large coefficient on carbon. The origins of the $\pi$ stabilisation would be seen to be similar to those identified above, but made a little more complicated by having to bring in more orbitals. Fig. 2.14 is therefore the paradigm for the general case. The well-known electron-withdrawing power of a trifluoromethyl group is explained by negative hyperconjugation substantially supporting the inductive effect.

The same explanation applies to the well-known stabilisation of carbanions by a neighbouring sulfur, phosphorus or silicon group. The main stabilisation comes from overlap of the filled orbital on carbon with the $\sigma^*_{YR}$ orbital **2.37**, and is at a maximum when the orbitals are anti-periplanar, accounting for the exceptional ease with which the anion **2.38** can be prepared by removing the bridgehead proton. The interaction diagram is essentially the same as that in Fig. 2.14, except that the energy of the $\sigma^*_{YR}$ orbital, when Y is one of Si, P or S, is lower than it is when Y is the corresponding first-row element. The bonding interaction between a hydrogen atom or a first-row atom and a second-row atom is inherently less energy-lowering. The sulfur and the phosphorus have the added advantage of being (mild) $\sigma$-withdrawing groups, but silicon has the advantage of having the Si—R bonds polarised from silicon towards the R group, if R is hydrogen or a carbon group, making them electronegative elements in this context. This explanation has largely replaced that using overlap with empty d orbitals.

**2.37** Y = Si, P or S    **2.38**          **2.39**

A lone pair on an electronegative element can take the place of the carbanion in the arguments above, and overlap with an appropriately electron-withdrawing $\sigma$ bond can be $\pi$-stabilising. Trisilylamine **2.39**, unlike trimethylamine, is planar, with a trigonal nitrogen atom, probably largely as a result of the overlap of the nitrogen lone pair with the antibonding Si—H orbitals, which are polarised from silicon towards the hydrogen.

**2.2.3.3    The Anomeric Effect.**[10]    A lone pair conjugated to a C—X bond, in which X is an electronegative element, is a special category of negative hyperconjugation. The best-known illustration of this *anomeric effect*, as it is called, is in the equilibrium for the glucosides **2.40** and **2.41**, where the diastereoisomer with the axial methoxy group **2.41** is favoured, in spite of the usual equatorial preference for substituents in six-membered rings.

**2.40**          **2.41**          **2.42**

The generally accepted explanation for this phenomenon is associated with negative hyperconjugation, similar to the stabilisation of a carbanion discussed in the preceding section, but with the lone pair on the ring oxygen atom taking the place of the filled p orbital on carbon. Lone pairs are often given the letter n as a distinctive label. The anomeric effect is a consequence of the overlap of the nonbonding lone pair $n_O$ with the low-lying $\sigma^*$ orbitals of the exocyclic C—O bond **2.42**, superimposed, of course, on all the usual interactions of filled orbitals with filled orbitals. Only when the exocyclic C—O bond is axial are its orbitals able to overlap well with the axial sp$^3$ hybrid lone pair on the ring oxygen. Alternatively, without using hybridisation, it is the nonbonding p$_z$ lone pair that overlaps better with an axial C—O bond.

At the same time, the methyl group on the exocyclic oxygen adopts a conformation in which it sits gauche to the ring oxygen atom, as a consequence of the lone pair on the exocyclic oxygen atom being conjugated anti-periplanar with the endocyclic C—O bond **2.43**. This is perhaps a little clearer in the Newman projection from above **2.44**. The preference for the gauche orientation is called the exo anomeric effect. The exo anomeric effect operates even with those tetrahydropyrans that have equatorial substituents at the anomeric centre—although the endocyclic oxygen cannot indulge in an anomeric interaction, the exocyclic oxygen can **2.45** (= **2.46**).

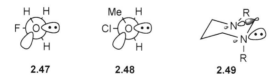

| 2.43 | 2.44 | 2.45 | 2.46 |

The anomeric effect can be seen to affect the conformation of many systems with the features RO—C—X, most of which adopt a conformation with the R group gauche to the X group rather than anti, as one might have expected. This is the generalised anomeric effect, and it has many manifestations, such as the preferred conformations for fluoromethanol **2.47** and methoxymethyl chloride **2.48**. Nor, of course, is it confined to oxygen lone pairs. The preferred conformation **2.49** for $N,N$-dialkyl-1,3-diazacyclohexanes has one of the alkyl groups axial in order that the lone pair on that nitrogen can be conjugated with the C—N bond. The optimum anomeric effect in this system would have both alkyl groups axial, so that both lone pairs could be involved in an anomeric interaction, but this conformation would have a 1,3-diaxial interaction between the alkyl groups, and this steric repulsion, not surprisingly, overrides the anomeric effect.

|  |  |  |
| :---: | :---: | :---: |
| 2.47 | 2.48 | 2.49 |

Bond lengths are also affected. When the two heteroatoms are different **2.50**, with one lone pair on a less electronegative atom like oxygen and the other on a more electronegative element like halogen, bond shortening is more noticeable in the O—C bond, and the C—X bond is increased in length. The anomeric effect between $n_O$ and $\sigma^*_{CX}$ increases the $\pi$ bonding in the C—O bond but, because it mixes in an antibonding orbital between the C atom and the halogen, that bond is weakened and made longer. The anomeric effect of $n_X$ with $\sigma^*_{CO}$ is lower, because $\sigma^*_{CX}$ is lower in energy than $\sigma^*_{CO}$ and $n_O$ is higher in energy than $n_X$, making the energy match better between $n_O$ and $\sigma^*_{CX}$. Thus the consequence of a lop-sided anomeric effect is overall to weaken the C—X bond—as the electron population is increased on the carbon atom, the X atom moves away. This is dependent upon the geometry, as seen in the structure of *cis*-2,3-dichlorodioxan **2.51**. The equatorial C—Cl bond is the same length as that in methyl chloride, because it is oriented at an angle giving little conjugation with the lone pairs on the neighbouring O-1. In contrast, the axial C—Cl bond is lined up for an anomeric effect with the axial lone pair on O-4, and it is longer. At the same time, the bond between O-4 and C-3 is shortened, whereas the bond between O-1 and C-2 is close to that for a normal ether link.

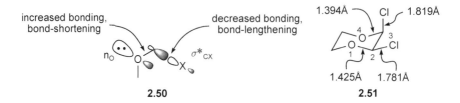

**2.50**                    **2.51**

### 2.2.3.4   Syn-coplanar and Anti-Periplanar Overlap.

In the discussion about the anomeric effect, the lone pair has been oriented, without comment, anti to the C—X bond. The lone pair and the C—X bond are able to overlap in this orientation **2.52** because they are coplanar, but at first sight they could equally easily have overlapped had they been syn **2.53**. Undoubtedly, coplanarity is the single most important constraint for good overlap, but what about the choice between syn and anti? One answer is that the syn arrangement **2.53** carries with it at least one eclipsing interaction with the substituent R, whereas the anti arrangement **2.52** has all substituents and lone pairs staggered. The eclipsed arrangement is not even a minimum, but a transition structure for rotation about the O—C bond.

**2.52**                    **2.53**

This simple difference alone seems to account for why anti arrangements, both in anomeric effects and in $\beta$-eliminations (to be discussed in Chapter 4) are so common, but it is not the whole story, because there are systems where this factor is not present, and yet there is still a preference for anti anomeric effects [and anti eliminations, see (Section 5.1.2.1) page 156].

A tempting way to explain the inherent preference for anti over syn arrangements is to picture the antibonding hybridised orbitals with the large lobes outside the bond instead of between the atoms. Thus we might redraw the $\sigma^*_{CX}$ orbital in **2.52** as **2.54**. This seems to make sense—the atomic orbitals of opposite sign will repel each other. Many organic chemists succumb to this temptation, for we immediately see that there is better overlap with the $n_O$ orbital in the anti arrangement—the large lobes are close. In the corresponding syn arrangement, the large lobes are on opposite sides and the overlap is 'obviously' less.

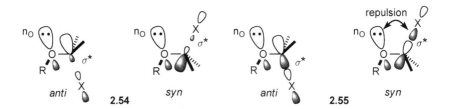

Unfortunately it is illegitimate. When we mix two atomic orbitals, the bonding orbital with an attendant drop in energy is paired with an antibonding orbital with its corresponding rise in energy. If we establish $\sigma$ overlap for two p orbitals in the usual way, the antibonding orbital $\sigma^*$ must match it. Although the lobes may be compressed by the repulsion, one cannot arbitrarily move them in and out, however commonly you may come across this device in your reading. A better picture can be seen in the wire-mesh drawing for methyl chloride in Fig. 1.47, where the $\sigma^*_{CCl}$ orbital shows that both the inside lobe and the outside are large, and not like those in the drawing **2.54**. We are left therefore with the problem of accounting for the preference for anti overlap. Perhaps the simplest explanation is a more careful use of pictures like those drawn in **2.55** with a somewhat more realistic hybrid orbital. The anti arrangement still has good bonding overlap, but in the syn arrangement, there are both attractions and repulsions between the $n_O$ orbital and $\sigma^*_{CX}$ orbital. Furthermore, and perhaps more important still, the anti arrangement keeps the centres of negative charge as far apart as they can be.

## 2.3 The Configurations and Conformations of Molecules

Defining the terms *configuration* and *conformation* poses a problem, because there is no sharp boundary between them. Conformational changes are usually those that can take place rapidly at room temperature or below, making the isolation of separate conformers difficult, and configurational changes have energy barriers high enough to make it easy to isolate configurational isomers. But conformational barriers can rise above those that can be crossed at room temperature, and configurational barriers, like double bond geometries, can become so low that they are easily crossed, but the ambiguity is usually not a problem.

Conjugation, whether it is in the $\pi$ system or in the $\sigma$ system, is one of the factors responsible both for the configurations and the conformations that molecules adopt. The energy-lowering induced by $\pi$ conjugation usually has the effect of making the planar arrangement with the maximum $\pi$ overlap the lowest in energy, and imparting a barrier to rotation about any single bonds separating the elements of conjugation. At one extreme, strong $\pi$ overlap fixes the *configuration* of benzene as a perfectly flat hexagonal ring. At the other extreme, weak $\pi$ overlap in the anomeric effect controls the *conformation* of methoxymethyl chloride **2.48**, but with a low barrier to rotation.

### 2.3.1 Restricted Rotation in π-Conjugated Systems

***2.3.1.1 One π Bond.*** It hardly needs saying that a π bond is not usually free to rotate. The π energy $2E_\pi$ we saw in Fig. 1.22 ($\sim$280 kJ mol$^{-1}$) is high enough to stop rotation about the π bond. An experimental value for the activation barrier for the thermal isomerisation of *cis*-2-butene **2.60** is 259 kJ mol$^{-1}$ (62 kcal mol$^{-1}$). For rotation about a π bond to become easier in the ground state either the transition structures like the diradical **2.57** or the zwitterion **2.58** must be stabilised or the planar structure **2.56** must be destabilised. Phenyl groups stabilise radical centres, and the barrier to rotation in stilbenes **2.61** is lower than that in 2-butene. Steric interaction between the *cis*-vicinal substituents, raising the energy of the planar structure, contributes to lowering the barrier to rotation, as in the bifluorenylidene **2.62**, which benefits from both effects. Alternatively, the substituents A and B may stabilise a cationic centre on one side and the substituents C and D an anionic centre on the other **2.58**. Alkenes having donor

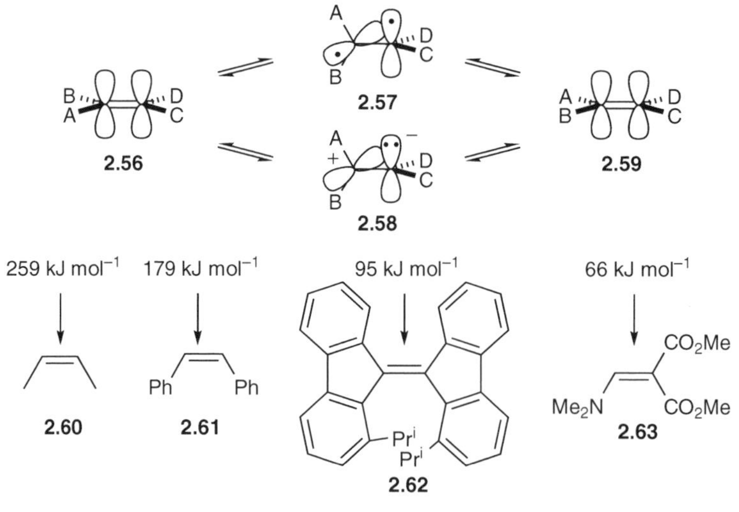

substituents (X) at one end and acceptors (Z) at the other are called 'push-pull' alkenes, and the barriers to rotation are indeed lowered, with the enamine system of the alkene **2.63** having a strikingly low barrier.

Photochemical excitation, however, takes one electron from the π orbital and promotes it to the π*. The π energy now is $(E_\pi - E_{\pi*})$, which is negative, removing the energetic benefit of conjugation altogether, and making the conformation **2.57**, with the two p orbitals orthogonal, the lowest in energy. Initially, the excited state must be in the high energy, planar conformation **2.56**, but if the photochemically excited molecule has a long enough lifetime the conformation will change to that with the lower energy **2.57**. Later, when the electron in the $\pi^*$ orbital returns to the $\pi$ orbital, the molecule will return to the planar arrangement **2.56** or **2.59**.

### 2.3.1.2 Allyl and Related Systems.

The allyl conjugated system is also more or less configurationally stable, whether it is the cation, the radical or the anion. The drawing of a $\sigma$ bond in the localised structure **2.64a** disguises the $\pi$ bonding still present between C-1 and C-2. The alternative and exactly equivalent drawing **2.64b** reveals that this is not the case, and C-1 and C-2 are just as strongly $\pi$-bonded as C-2 and C-3.

2.64a                                        2.64b

The molecular orbitals of the allyl system (Fig. 1.28) demonstrate the configurational stability more powerfully—the lowest filled orbital, $\psi_1$, has $\pi$ bonding across the whole conjugated system, and the nonbonding $\psi_2$, makes no contribution to $\pi$ bonding whether it is empty or filled. The total $\pi$-bonding energy for all three allyl systems (Fig. 1.26) is $2 \times 1.414\beta$. If rotation were to take place about the bond between C-1 and C-2, the transition structure would have a full $\pi$ bond between C-2 and C-3 and an orthogonal p orbital on C-1. The difference in $\pi$ energy between the conjugated allyl system ($2 \times 1.414\beta$) and this transition structure with a full $\pi$ bond ($2\beta$) is therefore $2 \times 0.414\beta$, or about 116 kJ mol$^{-1}$ (28 kcal mol$^{-1}$), making the $\pi$ bond strength between C-1 and C-2 nearly half that of a simple $\pi$ bond, quite large enough to restrict rotation under normal conditions. Higher levels of calculation confirm that a substantial barrier is present, but reveal that the cation, radical and anion are not in detail the same— the unsubstituted cation is calculated to have a rotation barrier in the gas phase of 140 kJ mol$^{-1}$ (33.5 kcal mol$^{-1}$), the radical a barrier of 63 kJ mol$^{-1}$ (15 kcal mol$^{-1}$) and the anion a barrier of 85 kJ mol$^{-1}$ (20 kcal mol$^{-1}$). In solution, solvation by a notional polar solvent lowers the numbers for the cation and anion to 115 and 70 kJ mol$^{-1}$ (27.5 and 17 kcal mol$^{-1}$), still large enough to retain configurational identity under normal conditions. 1,3-Disubstituted allyl systems therefore have three configurations, usually called W-shaped **2.65**, sickle-shaped **2.66**, and U-shaped **2.67**, which do not easily interconvert by rotation about the C—C bonds. Interconversion between the stereoisomeric allyl cations can take place by capture of a nucleophile, followed by rotation about the single bond, and regeneration of the cation by ionisation. Possibly because of the availability of these pathways, experimental measurements of the rotational barrier are lower for the cation than the theoretical value of 140 kJ mol$^{-1}$.

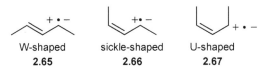

W-shaped      sickle-shaped      U-shaped
**2.65**            **2.66**                **2.67**

Allyl radicals also retain their configuration before being trapped by a reagent. Free energies of activation for a change from sickle-shaped to W-shaped of 66 kJ mol$^{-1}$ (16 kcal mol$^{-1}$) and 60 kJ mol$^{-1}$ (14 kcal mol$^{-1}$) have been measured for this process with substituents D and Me, respectively. The lower barrier in the radical, relative to that for the cation and anion, may be associated with the difficulty of localising charge on a carbon atom in the transition structure.

For the allyl anion itself, a good measurement is not really possible, because the free anion is not an accessible intermediate in solution. What we usually have is allyl-metal species, and interconversion between the corresponding geometries can take place by $\sigma$ coordination. If the coordination to the metal is $\eta^1$, it will weaken the $\pi$ bonding relative to the free anion, but if it is $\eta^3$ it will greatly strengthen it. Values of 45, 70, and 76 kJ mol$^{-1}$ (11, 17, and 18 kcal mol$^{-1}$) have been measured for allyl-lithium, potassium and caesium, respectively, with the last of these presumably a lower limit for the true barrier in a free allyl anion. One system free of this complication has been thoroughly studied: the azomethine ylids **2.68** and **2.69** are isoelectronic with an allyl anion, but do not have metal counterions. The free energy barrier to the conversion of the sickle-shaped isomer into the W-shaped isomer is 85 kJ mol$^{-1}$ (20.3 kcal mol$^{-1}$) and for the reverse reaction it is 84 kJ mol$^{-1}$ (20.1 kcal mol$^{-1}$), there being little difference in energy between them. Note that the ester groups greatly stabilise the anionic charge at C-1 and C-3, making rotation about the bond between C-1 and C-2 (or between C-2 and C-3) much easier than it would be in a free allyl anion.

MeO$_2$C

Ar —N CO$_2$Me
MeO$_2$C

120°

MeO$_2$C
Ar N
MeO$_2$C

Ar = p-MeOC$_6$H$_4$-

**2.68**                                              **2.69**

A number of other conjugated systems of three p orbitals show restricted rotation, although not to the same degree. Amides **2.70** typically have a barrier to rotation about the C—N bond of 80–90 kJ mol$^{-1}$ (19–21.5 kcal mol$^{-1}$), they have nearly trigonal nitrogen atoms, in contrast to amines, which have nearly tetrahedral nitrogen atoms, and the C—N bond is shortened because of the extra bonding provided by the $\pi$ overlap between the nitrogen lone pair and the $\pi$ bond. The comparatively rigid and planar conformation present in the amide system has profound consequences on the conformations of peptides and proteins.

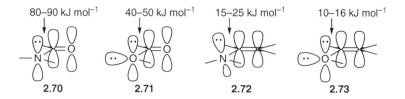

80–90 kJ mol$^{-1}$          40–50 kJ mol$^{-1}$          15–25 kJ mol$^{-1}$          10–16 kJ mol$^{-1}$

**2.70**                    **2.71**                    **2.72**                    **2.73**

The other systems, esters **2.71**, enamines **2.72**, and enol ethers **2.73**, similarly have restricted rotation about the bond drawn as a single bond, but the barrier is successively lower in each as the degree of $\pi$ bonding becomes less and less, and the degree of $\pi$ bonding localised at the double bond increases. This localisation also affects the lone pair, so that enamines, unlike amides, do not have a trigonal nitrogen atom, but a somewhat pyramidalised one, with the lone pair tilted slightly away from the vertical, relieving some of the eclipsing suffered by the alkyl substituents on the nitrogen atom.

Whereas the allyl anion, with a plane of symmetry through the central atom, has a node at that atom in $\psi_2$, amides, esters, enamines, enol ethers and enolate ions do not have a node precisely on the central atom. Taking planar *N*,*N*-dimethylvinyl-lamine and the enolate of acetaldehyde as examples, simple Hückel calculations give the $\pi$ orbitals in Fig. 2.15, which includes the allyl anion for comparison.

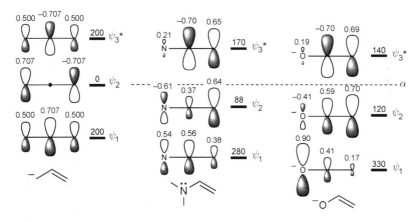

**Fig. 2.15**    $\pi$ Orbital energies and coefficients from simple Hückel calculations of the allyl anion, enamine and enolate ion (orbital energies in kJ mol$^{-1}$ relative to $\alpha$)

While the overlap between the atomic orbitals on the N or the O and the adjacent C are strongly bonding in $\psi_1$, they are antibonding in $\psi_2$. However, both $\psi_1$ and $\psi_2$ contribute to $\pi$ bonding between the two carbon atoms, and enamines and enolate ions have restricted rotation there. This is one reason why it is usually wise to draw enolate ions with the charge on oxygen rather than as carbonyl-stabilised carbanions—not only is more of the total charge on oxygen, but the degree of $\pi$ bonding is better illustrated this way.

One remaining detail to be explained is the relative energy of the two planar conformations available to amides and esters. Secondary amides adopt the *Z* con-formation **2.74a** rather than the *E* **2.74b**, and esters adopt the s-*trans* conformation **2.75a** rather than the s-*cis* **2.75b**. In both the esters and the amides, the conformations **2.74a** and **2.75a** benefit from the anti orientation of the carbon chains R$^1$ and R$^2$. In other words, the alkyl chain R$^1$ is effectively a larger substituent than the carbonyl oxygen, and the amide and ester alkyl groups R$^2$ prefer to be anti to R$^1$. However, this is not the whole story, because formate esters, with R$^1$ only a hydrogen atom, ought to be the other way round, and they are not. There is a stereoelectronic component as well, at least for the esters, which is identifiable as the generalised anomeric effect [see (Section 2.2.3.3) page 79]. In the s-*trans* conformation **2.75a**, a lone pair on the oxygen atom is favourably oriented anti to the C—O single bond of the carbonyl group, but in the s-*cis* conformation **2.75b** it is syn. This is responsible for the relatively high reactivity of lactones compared with open-chain esters, since lactones are forced to adopt the high-energy s-*cis* conformation.

$$\underset{\textbf{2.74a}}{\overset{R^1}{\underset{H-N}{\overset{}{\diagdown}}}\diagup}O \quad \rightleftarrows \quad \underset{\textbf{2.74b}}{\overset{R^1}{\underset{R^2-N}{\overset{}{\diagdown}}}\diagup}O \qquad \underset{\textbf{2.75a}}{\overset{R^1}{\underset{O}{\overset{}{\diagdown}}}\diagup}O \quad \rightleftarrows \quad \underset{\textbf{2.75b}}{\overset{R^1}{\underset{R^2-O}{\overset{}{\diagdown}}}\diagup}O$$

**2.3.1.3  Dienes.** In order to maintain the maximum level of $\pi$ bonding, butadiene, is planar, with the orbitals shown in Fig. 1.32. We estimated [see (Section 1.4.3) pages 30–31] that the conjugation between the two $\pi$ bonds lowered the energy by about 66 kJ mol$^{-1}$ (16 kcal mol$^{-1}$). We can see it in another way by noting that the $\pi$ bonding in $\psi_1$ between the p orbitals on C-2 and C-3 is between large lobes ($c_2 = c_3 = 0.600$), and the antibonding interaction in $\psi_2$ is between small lobes ($|c_1| = |c_2| = 0.371$). The planar conformations are called s-*trans* **2.76** and s-*cis* **2.77**, where the letter s denotes a conformation about a *single* bond. Experimentally, the activation energy for rotation about the bond between C-2 and C-3 is approximately 28 kJ mol$^{-1}$ (6.7 kcal mol$^{-1}$) going from s-*trans* to s-*cis*, and 16 kJ mol$^{-1}$ (3.8 kcal mol$^{-1}$) going from s-*cis* to s-*trans*, low enough for rotation to take place rapidly at room temperature, but different enough to ensure that most of the molecules will be in the s-*trans* conformation. Since the difference in energy between these two conformations is 12 kJ mol$^{-1}$ (2.9 kcal mol$^{-1}$), there is about 1% of the s-*cis* conformation at room temperature.

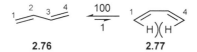

**2.76**            **2.77**

There are two reasons for the preference for the s-*trans* conformation. The more obvious is that the hydrogen atoms at C-1 and C-4 which are *cis* to the other double bond are sterically quite close in the s-*cis* conformation **2.77**, and repel each other. However, the difference in energy between *cis*- and *trans*-2-butene, which have similar, although not the same, differences in steric compression, is only about 4 kJ mol$^{-1}$ (1 kcal mol$^{-1}$). Another reason can be found in the $\pi$ system (exaggerated in Fig. 2.16), where the p orbitals on C-1 and C-4 are closer in space in the s-*cis* conformation than they are in the s-*trans*, and all the other orbital interactions, C-1 with C-2, C-1 with C-3, and their symmetry counterparts, are all equal in the two conformations. The lobes on C-1 and C-4 in $\psi_1$ are small and bonding, but this attractive overlap is more than offset by the antibonding interaction between the large lobes in $\psi_2$, making the overall interaction repulsive ($|\Delta E_2| > |\Delta E_1|$).

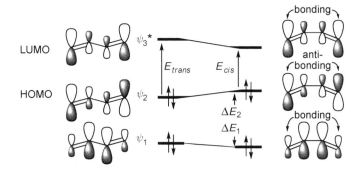

**Fig. 2.16**   Differences in π orbital energies for s-*trans* and s-*cis* butadiene

This perception also provides a simple explanation for an otherwise puzzling observation in UV spectroscopy. Dienes constrained to adopt an s-*cis* conformation by being endocyclic in a six-membered ring, absorb UV light at a longer wavelength than open-chain dienes with a comparable degree of substitution. Woodward's rules for UV absorption in dienes give a base value for s-*trans* dienes of 214 nm and for s-*cis* dienes of 253 nm. This absorption is a measure of the gap in energy between $\psi_2$ and $\psi_3^*$. In Fig. 2.16, we can see that $\psi_2$ is raised in energy in the s-*cis* conformation relative to the s-*trans*, and $\psi_3^*$ is lowered in energy, making the energy gap $E_{cis}$ less than $E_{trans}$.

***2.3.1.4   Lowering the Energy of the Transition Structure for Rotation.***   With longer conjugated systems the π stabilisation increases in the usual way, but each increment makes a smaller and smaller difference. In the transition structure for rotation, the full π stabilisation is divided into two, with each part having a shorter conjugated system. As a result, the barrier to rotation about the internal double bonds goes down as conjugated systems get longer. Carotenoids, for example, having eleven double bonds conjugated together, are notoriously susceptible to *cis-trans* isomerisation, and it seems likely that some of them are simply thermally induced rotations.

Moving on to the weaker π bonding in allyl systems, we deduced [see (Section 2.3.1.2) page 83] that the simple Hückel barrier to rotation is $0.828\beta$. By the same type of calculation we can estimate the barrier in the pentadienyl system: the full degree of π stabilisation (Fig. 1.35) is $2\beta + (2 \times 1.73\beta) = 5.46\beta$; the π stabilisation of the separate components for rotation between C-2 and C-3 is the sum of the energy of a π bond ($2\beta$) and of an allyl system ($2 \times 1.414\beta$), which comes to $4.82\beta$, and so the difference is now only $0.64\beta$. The experimental value in simple allyl systems is only a little above that which can be crossed at the normal temperatures of chemical reactions, and so we can expect that the longer conjugated systems with an odd number of atoms will rarely have stable configurations.

This effect is supplemented by terminal electronegative substituents, which increase the overall electron population at the extremities of the conjugated system, and reduce the effectiveness of overlap in the carbon chain in between. Thus the system of five conjugated p orbitals present in an alkene with an X-substituent at one end and a Z-substituent at the other (a 'push-pull' alkene), will have molecular orbitals related to the pentadienyl anion (Fig. 1.35). The nitroenamine **2.78**, which has one of the best donors and one of the best acceptors, has much weaker $\pi$ bonding between C-2 and C-3 than the drawing implies. Rotation about this bond is actually fast enough to make isolation of individual geometrical isomers impossible.

Another way of looking at the ease of rotation between C-2 and C-3 and the restriction between N-1 and C-2 is with the resonance structure **2.79**, which has the effect of expressing the reduction in double bond character. However, it is important to recognise the difference between the resonance structure and the transition structure for rotation **2.80**. The difference is that overlap of orbitals expressed as resonance cannot have any change in the position of the atoms, and it is correctly symbolised with the double-headed arrow. Rotation does have a change in the position of the atoms, and it is a 'reaction', symbolised with conventional reaction arrows. The cation-stabilising group at one end and the anion-stabilising group at the other stabilise the intermediate components, which are no longer conjugated in the transition structure. Such contributions to lowering the energy barrier will come from any stabilisation of the intermediate components in the transition structure—cation-stabilising, radical-stabilising or anion-stabilising, as appropriate—they will all lower the barrier. Increasing the stabilisation of the cationic centre in the transition structure, by having two donor substituents, as in the enediamine **2.81**, causes the two *N*-methyl groups to be coincident in the NMR spectrum even at $-63$ °C, because rotation about the formal C=C double bond is fast on the NMR timescale.

At the same time, rotation about the formally single bond between N-1 and C-2 in these compounds is more restricted than the drawing of a single bond implies, just as it was with amides. The two *N*-methyl groups in both enamines **2.63** and **2.82** have different chemical shifts and coalescence measurements show that the free energy of activation for rotation is 56 kJ mol$^{-1}$ (13 kcal mol$^{-1}$) for the former and 69 kJ mol$^{-1}$ (16.5 kcal mol$^{-1}$) for the latter. Decreasing the stabilisation of the anionic centre in the transition structure with a less powerful acceptor than a nitro group, as in the ester **2.83** reduces the barrier to rotation about the N—C bond to 58 kJ mol$^{-1}$ (14 kcal mol$^{-1}$).

56 kJ mol⁻¹ — structure **2.63** with CO₂Me groups

$$56 \text{ kJ mol}^{-1}$$

**2.63**  **2.82**  **2.83**  **2.84**

Extended conjugation through double bonds illustrates the principle of *vinylogy*, a word made up by combining vinyl and analogy. Vinylogous conjugated systems often have similar properties, both in the ground state and in reactivity, to the parent systems. The conjugated system of **2.83**, for example, is that of a vinylogous carbamate, in which the restricted rotation about the N—C bond is similar to but smaller than that of the corresponding carbamate **2.84**.

### 2.3.2 Preferred Conformations from Conjugation in the σ Framework

We have already seen [see (Section 2.2.3.3) page 79] how the generalised anomeric effect sometimes causes a chain of atoms to adopt a seemingly more hindered gauche conformation **2.47**, **2.48** and **2.49** rather than the more usual zigzag arrangement. Similar hyperconjugative interactions in neutral molecules between two σ bonds can lead them to adopt less obvious conformations in which the effect of conjugation overrides steric and electrostatic effects.

1,2-Difluoroethane might be expected to adopt the zigzag conformation **2.85**, both because the dipoles from the C—F bonds would be opposed, and because the two larger groups would be further apart. However, it does not—it adopts the conformation **2.86**, with an enthalpy advantage of 2.5–3.8 kJ mol⁻¹ (0.6–0.9 kcal mol⁻¹) as well as a small favourable entropy factor, since there are two gauche conformations and only one anti. The enthalpy advantage in this conformation stems from the antiperiplanar conjugation of the C—H bonds with the vicinal C—F bonds. Hyperconjugation will be energy-lowering with an interaction diagram like that in Fig. 2.9, but with the low-lying antibonding orbital σ*$_{CF}$, with the large coefficient on the carbon atom, taking the place of the empty p$_C$ orbital. A similar explanation accounts for the fact that *cis*-difluoroethene **2.88** is lower in energy than its *trans* isomer **2.87**, in contrast to most other alkenes.

**2.85**  **2.86**  **2.87**  **2.88**

With a more powerful donor than an H—C bond, cyclohexyl esters carrying a β silyl group demonstrate the preference for the donor and the acceptor bonds to be anti. The equilibrium proportion of the alcohol **2.89** (R = H) is in favour of the diequatorial isomer, but with esters (R = acyl) the equilibrium shifts to favour the diaxial conformation **2.90**. Furthermore, the equilibrium constant correlates with how good the carboxylate ion is as a leaving group as measured by the $pK_a$ of $RO^-$.

preferred for R = H

**2.89**

preferred for R = acyl

**2.90**

## 2.4　Other Noncovalent Interactions

We began in Chapter 1 by considering the strongest forces involved in bonding, the covalent bonds themselves, and worked our way down from strong $\sigma$ bonds to the weaker $\pi$ bonds. In this Chapter, we have looked at the weaker $\pi$ interactions of covalent bonds with each other and with p orbitals, and have come down to a level at which they provide only a delicate balance affecting the shapes that molecules adopt. There are other forces at work, both within a molecule and affecting how one molecule can interact with another, which also stem from the electron distribution. Weak though some of them are, these forces have profound consequences on the degree and sites of solvation, on intermolecular forces affecting the conformations of polymers, protein folding, crystal packing, and molecular recognition between an enzyme and its substrate, and between a receptor and its agonist.

### 2.4.1　The Hydrogen Bond

**2.4.1.1　X—H · · · X Bonds.**　Strong hydrogen bonds are found when a hydrogen atom bonded to one electronegative atom is close to another electronegative atom. It is found at its strongest and most simple in the $HF_2^-$ ion, which has been estimated to have a gas phase energy below that of the separate components of 167 kJ $mol^{-1}$ (39 kcal $mol^{-1}$), and at its most famous in the strong AT and GC pairing of bases in the helical structure of DNA.

The molecular orbitals in the $HF_2^-$ system in Fig. 2.17 resemble those of the allyl anion—a low-energy orbital with no nodes, and an orbital with a node at the central atom. The node at the hydrogen atom leaves it with no interactions with the two fluorine atoms, which are far enough apart to be essentially nonbonding. For this arrangement to be stabilised, $\psi_1$ and $\psi_2$ must together be lower in energy

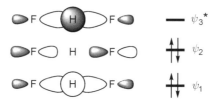

**Fig. 2.17**　The molecular orbitals of the symmetrical hydrogen bonds in $HF_2^-$

than the corresponding orbitals in the separate components HF and F⁻. Electronegative elements will lead to high electron populations on the atoms at the two ends of the three-atom system, incidentally making it difficult to locate hydrogen atoms by X-ray crystallography, and leading to the low field at which they come into resonance in $^1$H-NMR spectra. Electronegative elements have compact orbitals in $\psi_2$, making residual repulsion between them lower. This explains why strong hydrogen bonds are those involving electronegative elements, and why a linear array is best—any bending decreases the bonding in $\psi_1$ and increases the antibonding in $\psi_2$.

The same set of orbitals applies to the bridged B—H—B bonds in diborane **2.91** and the bridged C—H—C bond in the cyclodecyl cation **2.92**, except that in these systems there are only two electrons to feed into the orbitals, leaving $\psi_2$ empty. Since high electronegativity for the two atoms at the ends principally exerts its effect in keeping $\psi_2$ as nonbonding as possible, high electronegativity is no longer a requirement when $\psi_2$ is empty.

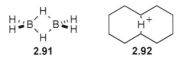

**2.91**                    **2.92**

An alternative perception for the nature of conventional hydrogen bonding, not in conflict with the molecular orbital picture, is that it stems from a Coulombic attraction between the negative charge of the lone pairs of the one electronegative element, and the partial positive charge left on the hydrogen atom by the polarisation of the bond towards the other electronegative element.

***2.4.1.2  C—H · · · X Bonds.***   When the hydrogen atom is attached to carbon, the degree of hydrogen bonding is much smaller than with conventional hydrogen bonds, usually much less than 17 kJ mol$^{-1}$ (4 kcal mol$^{-1}$). These kinds of hydrogen bonds manifest themselves in small shifts in spectroscopic properties, such as a lowering in the C—H stretching frequency in their infrared spectra, in small downfield shifts when $^1$H-NMR spectra are taken in oxygen-containing solvents, and in preferred conformations seen in solid-state structures.

***2.4.1.3  X—H · · · π Bonds.***   Similarly weak hydrogen bonds can also be formed from protons bound to electronegative elements coordinating to the p orbitals of C=C π bonds. The strength has been estimated to be anything up to 17 kJ mol$^{-1}$ (4 kcal mol$^{-1}$). The effect is seen in lower frequency O—H stretching in the infrared spectra of alcohols, correlating with how closely the hydrogen atom sits to a double bond or an aromatic ring, and most dramatically in the upfield shift of 1.5 ppm for the phenolic OH proton when the spectrum is taken in benzene, as a consequence of the proton sitting in the shielding region of the ring current **2.93**.

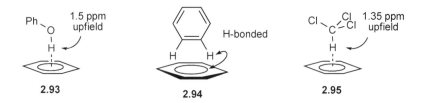

**2.93**          **2.94**          **2.95**

***2.4.1.4   C—H··· π Bonds.***   Even weaker hydrogen bonds, never more than 4 kJ mol$^{-1}$ (1 kcal mol$^{-1}$), can be detected between C—H bonds and C=C $\pi$ bonds. These interactions are seen in solid-state structures like that of benzene with its edge-to-face arrangement **2.94**, and in some noticeable upfield shifts in the $^{1}$H-NMR spectra on changing solvents from carbon tetrachloride to benzene. Larger shifts in the NMR spectra are seen with the more acidic C—H bonds, with chloroform showing an upfield shift of 1.35 ppm at infinite dilution, because of the formation of a bond to the centre of the $\pi$ system **2.95**.

### 2.4.2   Hypervalency

There are many molecules which clearly have more than the octet of electrons found in traditional Lewis structures. These include such molecules as $PF_5$, $SF_4$, $PhICl_2$, and $XeF_2$, such ions as $SiF_5^-$ and $PCl_6^-$, and intermediates like those involved in $S_N2$ reactions taking place at any of the elements below the first row in the periodic table. Such molecules have been called hypervalent, or to have expanded valence shells. Hypervalent molecules almost always have a high proportion of electronegative elements among their ligands.

   The standard method of explaining how such molecules can be stable is to invoke the interactions of filled p or hybrid orbitals on the ligands with an empty d orbital on the central element. Like any interaction of filled with unfilled orbitals, interactions with empty d orbitals are bound to be stabilising, but d orbitals are too high in energy relative to the p orbitals for their interaction to have any significant effect.

   It is better to see hypervalent bonding as the consequence of orbital interactions. Essentially, the central atom Y is involved in normal bonding to $n-2$ of its ligands. In addition, orbitals on each of the two remaining ligands X interact with an unused p orbital on the central element Y to create a set of three molecular orbitals (Fig. 2.18). To be overall bonding, $\psi_2$ must be largely nonbonding, which

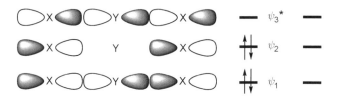

**Fig. 2.18**   The key molecular orbitals of hypervalent bonding

it will be if the ligands are kept well apart, and are electronegative elements, just as they are in hydrogen bonding.

Thus a pentacovalent phosphorus compound will have the three least electronegative ligands R in the plane of a trigonal bipyramid **2.96**, with bonds formed from the 3s and the $3p_x$ and $3p_y$ orbitals, and the two most electronegative ligands disposed linearly at the apices, with bonds formed from the two lower orbitals in Fig. 2.18. Furthermore, the electronegative ligands X will carry more negative charge, and will repel each other best if they are both apical. In consequence of both orbital and charge effects, the apical bonds are weaker than the basal, and these are the ones that make and break during reactions.

The same pattern is seen in the transition structures for nucleophilic substitution reactions. Apical entry of a nucleophile and apical departure will have the lowest-energy transtition structure **2.97** for the $S_N2$ reaction at carbon, and can give an intermediate with a lifetime **2.98** at silicon. The preference for electronegative elements to retain their apical positions ensures inversion of configuration when the nucleophile Nu and nucleofugal group X are both electronegative. When they are not both electronegative, apical attack **2.99** may be followed by a *pseudorotation* to give a different intermediate **2.100**, interchanging the positions of the ligands. Apical departure of X then explains the retention of configuration that is often seen in silicon chemistry when the leaving group is not conspicuously electronegative, and hence does not stabilise the arrangement in which it is apical.

### 2.4.3 Polar Interactions, and van der Waals and other Weak Interactions

***2.4.3.1 Coulombic Forces.*** If two molecules, or two parts of one molecule, have charges or dipoles, the sites of opposite charge attract each other, and sites with the same charge repel each other. These forces can be large and have a conspicuous influence on molecular properties. The same electrostatic forces also come into play in a number of weaker interactions. Furthermore, polar attractions from one polar molecule to another, or from one strongly hydrogen-bonding molecule to another, lead such molecules to aggregate, and to exclude nonpolar molecules. This is the basis for the well-known hydrophobic effect, in which nonpolar molecules stick together to avoid being in water.

***2.4.3.2 Dipole-Dipole Attraction.*** If only one of the molecules is charged or a dipole, it can still respond to weak polar forces in molecules not traditionally thought of as being polar. Two examples of weak dipolar interactions are seen in propanal **2.101** and propanol **2.102**, in which the lower-energy gauche conformations **2.101a** and **2.102a** have the methyl group close to the oxygen atom. The polarisation of the H—C bonds in the methyl groups leaves a weak positive charge on the outside and this is attracted to the partial negative charge on the oxygen atoms, lowering the energy of these conformations below those of the anti.

2.101a            2.101b            2.102a            2.102b

### 2.4.3.3  van der Waals Attraction.

The interaction of filled orbitals of one reagent with the filled orbitals of another is inherently energy-raising, and this is countered by a small attractive force from the interactions of filled orbitals with unfilled orbitals, which are inherently energy-lowering. Both forces fall off exponentially with distance, but they are not the only interactions between non-polar molecules. If we look at one electron in one of two nonpolar molecules, it will repel an electron in the other. At any given moment, if the first electron is on the side of the molecule facing the other molecule, it will cause its opposite number to spend more of its time on the far side of the second molecule. The electrons are said to be *correlated*. As the two electrons spend, on average, more of their time far away from each, the two molecules experience a small attractive dipolar electrostatic attraction. This only comes into play at short distances, with the energy falling off as the inverse sixth power of the distance apart. The resultant attractive force is known as a dispersion force, or more commonly as the van der Waals attraction. It is responsible, for example, for the weak force that helps to keep liquid hydrocarbons in the liquid state, and that helps them to aggregate in polar solvents. In liquid helium, at very low temperatures to be sure, the *only* attractions holding the atoms together are the van der Waals forces.

### 2.4.3.4  π-π Interactions and π Stacking.

Aromatic rings show an aptitude to aggregate that is not simply explained by van der Waals attraction. The phenomenon shows up in such important areas of molecular recognition as the stacked interactions between the aromatic rings in the DNA double helix, in intercalation by drugs and carcinogens into the DNA stack, in the aggregation of the chlorin rings in the chloroplast, in the tertiary structure of proteins, and in many host-guest supramolecules. An edge-to-face stacking arrangement is common, because of hydrogen bonding **2.94**, but a face-to-face stacking arrangement is not uncommon. However, a stack of nonpolar aromatic rings perfectly lined up directly on top of each other is straightforwardly repulsive (Fig. 2.19a)—one π system repels the other, and the van der Waals attractions are not powerful enough to stabilise this arrangement. Aromatic rings stacked above one another are stabilised when one π system is offset relative to the other (Fig. 2.19b), because there is an electrostatic attraction from the positively charged σ framework with the negatively charged π cloud, especially since the positive charge is largely on the peripheral and therefore exposed hydrogen atoms. π Stacking is more common

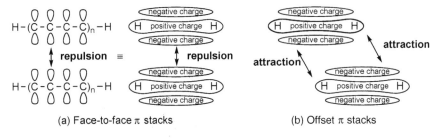

**Fig. 2.19**   $\pi$ Stacking

with the larger aromatic systems like porphyrins, probably because of the greater area for the van der Waals forces to work on, than they are with simple benzenes, where the arrangement **2.94** and the offset $\pi$ stack are close in energy.

## 2.5   Exercises

1. Given that two $\pi$ bonds conjugated together have a lower energy than two separate $\pi$ bonds (Fig. 1.32) and a C—H bond conjugated with a $\pi$ bond also lowers the energy (Fig. 2.11), explain why two C—H bonds conjugated to each other are not stabilised.

2. By considering the energies of the interacting p orbitals, explain why tri-methylborate **2.103** is much less Lewis acidic than boron halides **2.104**.

$$MeO-B{<}^{OMe}_{OMe} \qquad X-B{<}^{X}_{X}$$

**2.103**  **2.104**

3. Explain why silylamines are weaker bases than ammonia and why hexa-phenyldisiloxane $(Ph_3Si)_2O$ is linear.

4. Explain why it is easy to remove a proton from the methyl group attached to the boron atom in the trialkylborane **2.105**.

**2.105**

5. Considering the anomeric effect, and allowing for the effect of 1,3-diaxial repulsions, predict the lowest energy conformation for the dimethylacetal of formaldehyde $(MeO)_2CH_2$.

6. Reduction of butadiene with sodium in liquid ammonia gives more *cis*-2-butene than *trans*-2-butene, typically in a ratio of about 60:40. Identify all the intermediates. Consider the conformational equilibrium in the diene, its barrier to rotation about the central bond, the corresponding barrier to rotation in each of the intermediates and in the product, and hence identify at which stage the configuration was determined. Explain why the higher energy *cis*-isomer is the major product.

7. Explain why, when the substituent Y in an allylic ether is an alkyl group, the preferred conformation is **2.106**, but when it is a carbonyl or nitrile group, the preferred conformation is **2.107**.

**2.106**

preferred for Y = alkyl

**2.107**

preferred for Y = CO$_2$Et or C≡N

8. The photochemical promotion of a lone pair electron in a carbonyl group to $\pi^*_{C=O}$ is forbidden, yet molecular vibrations allow it to occur as a broad band of low intensity. In a saturated ketone $\lambda_{max}$ for the n→π* transition is typically at ~280 nm ($\varepsilon$ ~20). Explain why it is: (a) at shorter wavelength in an α-chloroketone **2.108**; (b) at longer wavelength in an α,β-unsaturated ketone **2.109**; and (c) at even longer wavelength in an acylsilane **2.110**.

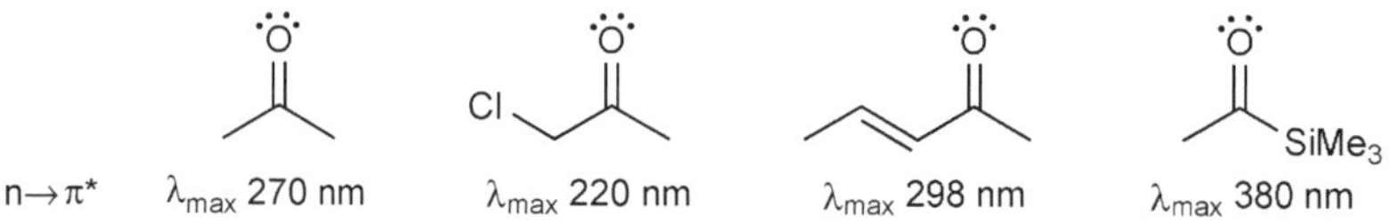

n→π*    $\lambda_{max}$ 270 nm      $\lambda_{max}$ 220 nm      $\lambda_{max}$ 298 nm      $\lambda_{max}$ 380 nm

**2.108**      **2.109**      **2.110**

9. Explain why the trimethylsilyl substituents in the allene **2.111** enable the phenyl and the methyl groups to exchange places rapidly, as shown by the coalescence at −90 °C of the signals from the trimethylsilyl groups in the $^1$H-NMR spectrum.

36 kJ mol$^{-1}$

**2.111a**      **2.111b**

# 3 Chemical Reactions—How Far and How Fast

## 3.1 Factors Affecting the Position of an Equilibrium

All the attractive and repulsive forces discussed in Chapter 2 affect not only the shape a molecule adopts, but also how favourably it can interact with another molecule, as we saw in the stacking of aromatic rings. However, for insight into the nature of the events leading to reaction, we need to look further, first to find whether two molecules colliding can possibly create one or more molecules lower in energy than the starting materials—the thermodynamics—and then whether there is an energetically accessible pathway between the starting materials and the products—the kinetics.

Thus we can understand easily enough that the reaction between bromine and ethylene giving dibromoethane is exothermic—it replaces one $\pi$ bond (C=C) with two $\sigma$ bonds (C—Br) at the expense of a weak $\sigma$ bond (Br—Br). However, it is not always obvious how strong the bonds will be when one molecule combines with another to form a single new molecule, or what happens to the energy if we exchange parts of one molecule with parts of another. A useful addition to understanding this sort of problem has been Pearson's concept of hard and soft acids and bases (HSAB).

## 3.2 The Principle of Hard and Soft Acids and Bases (HSAB)[11]

Lewis acids and bases (including $H^+$ and $OH^-$) can be classified as belonging, more or less, to one of two groups, one called hard and the other called soft. The striking observation is, and this is the basis of the classification, that hard acids form stronger bonds with hard bases, and soft acids form stronger bonds with soft bases. For example, a hard acid like the proton is a stronger acid than the silver cation, $Ag^+$, when a hard base like a hydroxide ion is used as the reference point;

*Molecular Orbitals and Organic Chemical Reactions: Student Edition*    Ian Fleming
© 2009 John Wiley & Sons, Ltd

but if a softer base like iodide ion had been used, we would have come to the opposite conclusion. This situation is summarised in the rule: *hard-likes-hard* and *soft-likes-soft*.

Pearson's classification and rank ordering of acids and bases did not, and at that stage was not intended, to explain hardness and softness. To explain it we look at the equilibrium between a Lewis acid and a Lewis base and the salt they form. The position of the equilibrium is affected both by charge and by orbital interactions. A hard acid bonds strongly to a hard base largely by electrostatic interactions. Hard acids and bases have the HOMO of the base and the LUMO of the acid far apart in energy. This leads to an *ionic* bond with little overlap [see (Section 1.7.2) page 46]. Also, the smaller the ion or molecule, the harder it is, because the charges can get closer in the ionic bond. A high positive charge on the acid and a high negative charge on the base will also contribute to the strength of the ionic bond. However, a soft acid bonds strongly to a soft base because the orbitals involved are close in energy. As we saw in Chapter 1 (see pages 46 and 47), we get the maximum overlap for *covalent* bonding when the interaction is between orbitals of similar energy. We can also see that in practice we shall not often have pure hardness and pure softness; rather, there is a continuum, and the C—Cl bond in Fig. 1.45 is a case where the bond strength comes from both types of interaction. In summary, a hard acid is small, has a high positive charge and a high-energy LUMO, and a hard base is small, has a high negative charge and a low-energy HOMO. Thus a proton is a hard acid and the silver cation is a soft acid; a fluoride ion is a hard base and an iodide ion is a soft base.

To go into this idea quantitatively, we need definitions of hardness and softness, and a rank order for acids and bases on a scale of hardness. This has been done in two ways: one based on molecular orbital theory, and the other on density functional theory.

We define two parameters: the absolute electronegativity, $\chi$, which is approximately the same as electronegativity as Mulliken originally defined it for atoms, namely the average of the ionisation potential $I$ and the electron affinity $A$ (Equation 3.1). The other is called the absolute hardness, $\eta$, which is identified with the difference in energy between the ionisation potential $I$ and the electron affinity $A$ (Equation 3.2).

$$\chi = \frac{(I + A)}{2} \qquad\qquad 3.1$$

$$\eta = (I - A) \qquad\qquad 3.2$$

Hardness in this definition is identical to the energy change for the reaction in Equation 3.3. The radical R· has zero hardness when the disproportionation has no change in energy. This also identifies the maximum degree of softness, which is defined as the reciprocal of hardness. This definition fits the earlier qualitative approach to explaining hardness and softness as being associated with polarisability—those species with a large difference between $I$ and $A$ were the ones that were not easily polarised.

$$R\cdot \quad + \quad R\cdot \quad \rightarrow \quad R^+ \quad + \quad R^- \qquad 3.3$$

The second approach to defining the absolute hardness $\eta$ has a companion parameter taken from density functional theory, called the electronic chemical potential $\mu$. The value of $\mu$ is essentially the same as the negative of $\chi$, as defined in Equation 3.1, and the value of $\eta$ is essentially the same as in the more approximate definition in Equation 3.2. Tables 3.1–3.5 record some useful values for radicals, molecules and ions based on this definition.

We shall return to the concepts of hardness and softness when we come to discuss kinetics, since it is there that we shall use them most tellingly.

**Table 3.1**  Absolute hardness $\eta$ (in eV) for some radicals[12]

| Radical | I | A | $\chi$ | $\eta$ |
|---|---|---|---|---|
| F | 17.42 | 3.40 | 10.41 | 14.0 |
| H | 13.59 | 0.74 | 7.17 | 12.9 |
| OH | 13.17 | 1.83 | 7.50 | 11.3 |
| $NH_2$ | 11.40 | 0.74 | 6.07 | 10.7 |
| $HO_2$ | 11.53 | 1.19 | 6.36 | 10.3 |
| CN | 14.02 | 3.82 | 8.92 | 10.2 |
| CHO | 9.90 | 0.17 | 5.04 | 9.7 |
| Me | 9.82 | 0.08 | 4.96 | 9.7 |
| Cl | 13.01 | 3.62 | 8.31 | 9.4 |
| NO | 9.25 | 0.02 | 4.63 | 9.2 |
| Ph | 8.95 | 0.10 | 5.20 | 8.9 |
| Et | 8.38 | −0.39 | 4.00 | 8.8 |
| $PH_2$ | 9.83 | 1.25 | 5.54 | 8.6 |
| Br | 11.84 | 3.36 | 7.60 | 8.5 |
| SH | 10.41 | 2.30 | 6.40 | 8.1 |
| $CH_2{=}CH$ | 8.95 | 0.74 | 4.85 | 8.2 |
| $CF_3$ | 9.25 | >1.10 | >5.18 | <8.1 |
| $Pr^i$ | 7.57 | −0.48 | 3.55 | 8.1 |
| $NO_2$ | >10.10 | 2.30 | >6.20 | >7.8 |
| MeCO | 8.05 | 0.30 | 4.18 | 7.8 |
| SeH | 9.80 | 2.20 | 6.00 | 7.6 |
| I | 10.45 | 3.06 | 6.76 | 7.4 |
| $Bu^t$ | 6.93 | −0.30 | 3.31 | 7.2 |
| $CCl_3$ | 8.78 | 1.90 | 5.35 | 6.9 |
| $PhCH_2$ | 7.63 | 0.88 | 4.26 | 6.8 |
| $SiH_3$ | 8.14 | 1.41 | 4.78 | 6.7 |
| PhO | 8.85 | 2.35 | 5.60 | 6.5 |
| PhS | 8.63 | 2.47 | 5.50 | 6.2 |
| $SiCl_3$ | 7.92 | 2.50 | 5.20 | 5.4 |
| Li | 5.39 | 0.62 | 3.00 | 4.8 |

**Table 3.2** Absolute hardness $\eta$ (in eV) for some molecules

| Molecule | I | A | $\chi$ | $\eta$ |
|---|---|---|---|---|
| HF | 16.0 | −6.0 | 5.0 | 22.0 |
| $CH_4$ | 12.7 | −7.8 | 2.5 | 20.5 |
| $BF_3$ | 15.8 | −3.5 | 6.2 | 19.3 |
| $H_2O$ | 12.6 | −6.4 | 3.1 | 19.0 |
| MeF | 12.5 | −6.2 | 3.2 | 18.7 |
| $N_2$ | 15.58 | −2.2 | 6.70 | 17.8 |
| $CO_2$ | 13.8 | −3.8 | 5.0 | 17.6 |
| $H_2$ | 15.4 | −2.0 | 6.7 | 17.4 |
| $NH_3$ | 10.7 | −5.6 | 2.6 | 16.3 |
| HCN | 13.6 | −2.3 | 5.7 | 15.9 |
| HCl | 12.7 | −3.3 | 4.7 | 16.0 |
| $Me_2O$ | 10.0 | −6.0 | 2.0 | 16.0 |
| CO | 14.0 | −1.8 | 6.1 | 15.8 |
| MeCN | 12.2 | −2.8 | 4.7 | 15.0 |
| MeCl | 11.2 | −3.7 | 3.8 | 14.9 |
| $MeNH_2$ | 9.0 | −5.3 | 1.9 | 14.3 |
| $HC{\equiv}CH$ | 11.4 | −2.6 | 4.4 | 14.0 |
| $HCO_2Me$ | 11.0 | −1.8 | 4.6 | 12.8 |
| $Me_3N$ | 7.8 | −4.8 | 1.5 | 12.6 |
| $CH_2{=}CH_2$ | 10.5 | −1.8 | 4.4 | 12.3 |
| $H_2S$ | 10.5 | −2.1 | 4.2 | 12.6 |
| $Me_2S$ | 8.7 | −3.3 | 2.7 | 12.0 |
| $PH_3$ | 10.0 | −1.9 | 4.1 | 11.9 |
| $O_2$ | 12.2 | 0.4 | 6.3 | 11.8 |
| $CH_2{=}O$ | 10.9 | −0.9 | 5.0 | 11.8 |
| $Me_3P$ | 8.6 | −3.1 | 2.8 | 11.7 |
| MeBr | 10.6 | ∼−1.0 | 4.8 | 11.6 |
| $Me_2NCHO$ | 9.1 | −2.4 | 3.4 | 11.5 |
| MeCHO | 10.2 | −1.2 | 4.5 | 11.4 |
| $BCl_3$ | 11.6 | 0.33 | 5.97 | 11.3 |
| $SO_2$ | 12.3 | 1.1 | 6.7 | 11.2 |
| $CCl_4$ | 11.5 | ∼0.3 | 5.9 | 11.2 |
| $Me_2CO$ | 9.7 | −1.5 | 4.1 | 11.2 |
| $CH_2{=}CHCN$ | 10.91 | −0.21 | 5.35 | 11.1 |
| $SO_3$ | 12.7 | 1.7 | 7.2 | 11.0 |
| $O_3$ | 12.8 | 2.1 | 7.5 | 10.7 |
| $MeNO_2$ | 11.13 | 0.45 | 5.79 | 10.7 |
| HI | 10.5 | 0.0 | 5.3 | 10.5 |
| Benzene | 9.3 | −1.2 | 4.1 | 10.5 |
| $HNO_3$ | 11.03 | 0.57 | 5.80 | 10.5 |
| Pyridine | 9.3 | −0.6 | 4.4 | 9.9 |
| Butadiene | 9.1 | −0.6 | 4.3 | 9.7 |
| $CS_2$ | 10.08 | 0.62 | 5.35 | 9.5 |
| $PCl_3$ | 10.2 | 0.8 | 5.5 | 9.4 |
| $:CH_2$ | 10.0 | 0.6 | 5.3 | 9.4 |

**Table 3.2**   (*continued*)

| Molecule | I | A | $\chi$ | $\eta$ |
|---|---|---|---|---|
| MeI | 9.5 | 0.2 | 4.9 | 9.3 |
| $Cl_2$ | 11.6 | 2.4 | 7.0 | 9.2 |
| $PhCH{=}CH_2$ | 8.47 | −0.25 | 4.11 | 8.7 |
| $PBr_3$ | 9.9 | 1.6 | 5.6 | 8.3 |
| $Br_2$ | 10.56 | 2.6 | 6.6 | 8.0 |
| $S_2$ | 9.36 | 1.66 | 5.51 | 7.7 |
| $I_2$ | 9.4 | 2.6 | 6.0 | 6.8 |

**Table 3.3**   Absolute hardness (in eV) for some bases

| Base | $I_{B+}$ | $A_{B+}$ | $\eta_B$ |
|---|---|---|---|
| $F^-$ | 17.42 | 3.40 | 14.0 |
| $H_2O$ | 26.6 | 12.6 | 14.0 |
| $NH_3$ | 24 | 10.2 | 13.8 |
| $H^-$ | 13.59 | 0.75 | 12.8 |
| CO | 26 | 14.0 | 12.0 |
| $OH^-$ | 13.0 | 1.83 | 11.2 |
| $NH_2^-$ | 11.3 | 0.74 | 10.6 |
| $CN^-$ | 14.2 | 3.6 | 10.6 |
| $H_2S$ | 21 | 10.5 | 10.5 |
| $PH_3$ | 20 | 10.0 | 10.0 |
| $N_3^-$ | 11.6 | 1.8 | 9.8 |
| $Cl^-$ | 13.01 | 3.62 | 9.4 |
| $NO_2^-$ | 12.9 | 3.99 | 8.9 |
| $ClO^-$ | 11.1 | 2.2 | 8.9 |
| $Br^-$ | 11.84 | 3.36 | 8.5 |
| $SH^-$ | 10.4 | 2.3 | 8.1 |
| $Me^-$ | 9.82 | 1.8 | 8.0 |
| $I^-$ | 10.45 | 3.06 | 7.4 |

**Table 3.4**   Absolute hardness (in eV) for some acids

| Acid | $I_A$ | $A_A$ | $\chi_A$ | $\eta_A$ |
|---|---|---|---|---|
| $H^+$ | $\infty$ | 13.59 | $\infty$ | $\infty$ |
| $Al^{3+}$ | 120 | 28.4 | 74.2 | 91.6 |
| $Li^+$ | 75.6 | 5.39 | 40.5 | 70.2 |
| $Mg^{2+}$ | 80.1 | 15.03 | 47.6 | 65.1 |
| $Na^+$ | 47.3 | 5.14 | 26.2 | 42.2 |

(*continued overleaf*)

**Table 3.4** (*continued*)

| Acid | $I_A$ | $A_A$ | $\chi_A$ | $\eta_A$ |
|---|---|---|---|---|
| $Ca^{2+}$ | 51.2 | 11.87 | 31.6 | 39.3 |
| $Fe^{3+}$ | 56.8 | 30.6 | 43.7 | 26.2 |
| $Zn^{2+}$ | 39.7 | 17.96 | 28.8 | 21.7 |
| $Cu^{2+}$ | 36.8 | 20.29 | 28.6 | 16.5 |
| $Hg^{2+}$ | 34.2 | 18.75 | 26.5 | 15.5 |
| $Ag^+$ | 21.5 | 7.57 | 14.6 | 13.9 |
| $CO_2$ | 13.8 | 0.0 | 6.9 | 13.8 |
| $Pd^{2+}$ | 32.9 | 19.42 | 26.2 | 13.5 |
| $Cu^+$ | 20.3 | 7.72 | 14.0 | 12.6 |
| $AlCl_3$ | 12.8 | $\sim1$ | 6.9 | 11.8 |
| $SO_2$ | 12.3 | 1.1 | 6.7 | 11.2 |
| $Br^+$ | 21.6 | 11.8 | 16.7 | 9.8 |
| $Cl_2$ | 11.4 | 2.4 | 6.9 | 9.0 |
| $I^+$ | 19.1 | 10.5 | 14.8 | 8.6 |
| $I_2$ | 9.3 | 2.6 | 6.0 | 6.7 |

**Table 3.5** Empirical hardness (in kJ mol$^{-1}$) for some cationic acids

| Acid | $D°_{AF}$ | $D°_{AI}$ | $\Delta_{FI}$ |
|---|---|---|---|
| $MeCO^+$ | 502 | 209 | 293 |
| $CHO^+$ | 510 | 217 | 293 |
| $H^+$ | 568 | 297 | 272 |
| $Ph^+$ | 518 | 268 | 251 |
| $Bu^{t+}$ | 451 | 209 | 242 |
| $CH_2{=}CH^+$ | 497 | 263 | 234 |
| $Li^+$ | 573 | 343 | 230 |
| $Na^+$ | 514 | 288 | 226 |
| $Pr^{i+}$ | 447 | 222 | 226 |
| $Et^+$ | 447 | 222 | 226 |
| $CH_2{=}CHCH_2^+$ | 410 | 184 | 226 |
| $Me^+$ | 456 | 234 | 222 |
| $c\text{-}C_3H_5^+$ | 464 | 247 | 217 |
| $CN^+$ | 468 | 305 | 163 |
| $NO^+$ | 234 | 84 | 150 |
| $Cs^+$ | 493 | 343 | 150 |
| $I^+$ | 280 | 150 | 130 |
| $Cu^+$ | 426 | 314 | 113 |
| $Ag^+$ | 351 | 251 | 100 |
| $HO^+$ | 217 | 234 | $-17$ |

## 3.3    Transition Structures

Imagine two molecules combining with each other in a simple, one-step, exothermic reaction leading to two possible products A and B (Fig. 3.1a). Chemists have long appreciated that the more exothermic reaction, that leading to the product B, is usually the faster—it has been called the rate-equilibrium relationship, and is related to the reactivity-selectivity principle. The explanation is easy enough—whatever features lead the product B to be lower in energy than the product A will have developed in the transition structure to some extent. Thermodynamics does affect kinetics—a source of endless confusion.

Because it is not always true nor is it enough, we need to know more about what else affects the energy of the transition structure. We bear in mind that influences from both sides of the reaction coordinate affect the transition structure. Perturbation theory gives us one way to learn something about the reactant side—we treat the interaction of the molecular orbitals of the two components as a perturbation on each other. The perturbation is similar to that which leads to bonding and antibonding orbitals in interaction diagrams like those in Chapter 1, where two separate orbitals interact to create molecular orbitals. However, as the perturbation increases, it ceases to be merely a perturbation, and the mathematical basis of the theory fails to accommodate so large a change. We do not, therefore, have direct access to a good picture of the transition structure in this way; nevertheless, we do get an estimate of the *slope* of an early part of the path along the reaction coordinate leading up to the transition structure (labelled Path A and Path B in Fig. 3.1a). Unless something unusual happens nearer the transition structure, the slopes will predict which of the two transition structures is the easier to get to—on the whole, the steeper path is likely to lead to the higher-energy transition structure.

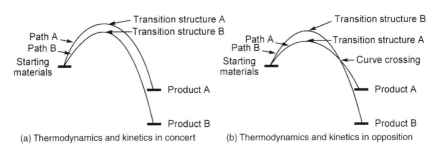

(a) Thermodynamics and kinetics in concert    (b) Thermodynamics and kinetics in opposition

**Fig. 3.1**    The energy along two possible reaction coordinates

The situation shown in Fig. 3.1a is the common one: the higher-energy transition structure leads to the higher-energy product, and the better orbital interaction matches it. However, there are some situations where there is a crossing of the curves (Fig. 3.1b). Some of the most interesting mechanistic problems arise when the more exothermic reaction is not the faster—in other words, when the thermodynamics and the kinetics do not go together. Hitherto organic chemists have concentrated on the influence the product has on the transition structure, but we now have, in perturbation theory, a useful tool for examining the reactant side of the reaction coordinate. If the orbital-interaction for path A is stronger than that for path B, as in Fig. 3.1b, it can help to explain the anomalous order of the two transition structures.

The Hammond postulate[13] says that the transition structure for an exothermic reaction (Fig. 3.2a) is closer in energy to the energy of the starting materials, and so it has more of the character of the starting materials, or, looking at the distances along the reaction coordinate, A < B. Equally, the transition structure for an endothermic reaction (Fig. 3.2b) is product-like, because B < A. We can therefore expect that the nature of the products will be particularly influential in affecting the rates of endothermic reactions (Fig. 3.2b), but that orbital interactions will be particularly influential in exothermic reactions (Fig. 3.2a).

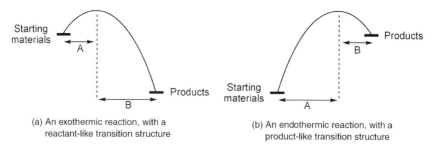

(a) An exothermic reaction, with a reactant-like transition structure

(b) An endothermic reaction, with a product-like transition structure

**Fig. 3.2**   The Hammond postulate

## 3.4   The Perturbation Theory of Reactivity

Now let us look at the perturbation which the reacting molecules exert upon each other's orbitals. Let the two reacting molecules have orbitals, filled and unfilled, as shown in Fig. 3.3. As the two molecules approach each other, the orbitals interact. Thus we can take, let us say, the highest occupied orbital (HOMO) of the molecule on the left and the highest occupied orbital of the molecule on the right and combine them in a bonding and an antibonding sense, just as we did when making a π bond from two isolated p orbitals.

The formation of the bonding orbital is, as usual, exothermic ($E_1$), but the formation of the antibonding orbital is endothermic ($E_2$), because there are two electrons which must go into it. The energy put into the antibonding combination is greater than that released from the bonding combination, and similarly for all combinations of fully occupied orbitals, the sum of which provide the repulsion experienced by one molecule when it is brought close to another. Combinations of

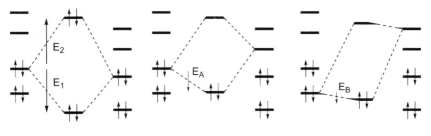

(a) The interaction of one filled
    orbital with another

(b) The interaction of the
    HOMO with the LUMO

(c) The interaction of a lower filled
    MO with a higher unfilled MO

**Fig. 3.3**    The interactions of the molecular orbitals of one molecule with the molecular orbitals of another

unfilled orbitals with other unfilled orbitals will have no effect on the energy of the system, because there are no electrons in them.

The interactions which do have an important energy-lowering effect are the combinations of filled orbitals with unfilled orbitals. Thus, in Fig. 3.3b and Fig. 3.3c, we have such combinations, and in each case we see the usual drop in energy in the bonding combination, and rise in energy in the antibonding combination. We can also see in Fig. 3.3b that it is the interaction of the HOMO of the left-hand molecule and the LUMO of the right-hand molecule that leads to the largest drop in energy ($2E_A > 2E_B$). The interaction of other occupied orbitals with other unoccupied orbitals, as in Fig. 3.3c, is less effective, because the closer the interacting orbitals are in energy, the greater is the splitting of the levels [see (Section 1.7.2) pages 46 and 47]. Now we can see why it is the HOMO/LUMO interaction which we look at, and why these orbitals, *the frontier orbitals*, are so important. The other occupied orbital/unoccupied orbital interactions contribute to the energy of the interaction and hence to lowering the energy of the transition structure, but the effect is usually less than that of the HOMO/LUMO interactions.

The HOMO/HOMO interactions (Fig. 3.3a) are large compared with the HOMO/LUMO interactions (Fig. 3.3b)—both $E_1$ and $E_2$ in Fig. 3.3a are larger than $E_A$ in Fig. 3.3b. This is because HOMO/HOMO interactions are between orbitals of comparable energy, whereas the HOMO of one molecule and the LUMO of another are usually well separated in energy. (In the mathematical form of perturbation theory, the former are first-order interactions, whereas the latter are second order.) Although the bonding ($E_1$) and antibonding ($E_2$) interactions cancel one another out to some extent, the net antibonding interaction between two molecules will be large—many such orbitals interact in this way, and their interactions are first order. The second-order interactions, like those of Fig. 3.3b and 3.3c, even though they are entirely bonding in character and reduce the activation energy, are relatively small. The HOMO/LUMO interaction is merely the largest of a lot of small interactions.

## 3.5   The Salem-Klopman Equation

Using perturbation theory, Klopman and Salem[14] derived an expression for the energy ($\Delta E$) gained and lost when the orbitals of one reactant overlap with those of another. Their equation has the following form:

$$\Delta E = - \underbrace{\sum_{ab}(q_a + q_b)\beta_{ab}S_{ab}}_{\text{first term}} + \underbrace{\sum_{k<l}\frac{Q_k Q_l}{\varepsilon R_{kl}}}_{\text{second term}} + \underbrace{\sum_{r}^{\text{occ.}}\sum_{s}^{\text{unocc.}} - \sum_{s}^{\text{occ.}}\sum_{r}^{\text{unocc.}} \frac{2(\Sigma_{ab}c_{ra}c_{sb}\beta_{ab})^2}{E_r - E_s}}_{\text{third term}} \qquad 3.4$$

where $q_a$ and $q_b$ are the electron populations (often loosely called electron densities) in the atomic orbitals a and b, $\beta$ and $S$ are the resonance and overlap integrals in Equations 1.3–1.5, $Q_k$ and $Q_l$ are the total charges on atoms k and l, $\varepsilon$ is the local dielectric constant, $R_{kl}$ is the distance between the atoms k and l, $c_{ra}$ is the coefficient of atomic orbital a in molecular orbital r, where r refers to the molecular orbitals on one molecule and s refers to those on the other, $E_r$ is the energy of molecular orbital r and $E_s$ is the energy of molecular orbital s.

The derivation of this equation involves approximations and assumptions. It is valid only when $S$ is small. The integral $S$ for a C—C bond being formed by p orbitals overlapping in a $\sigma$ sense reaches a maximum value of 0.27 at a distance of 1.74 Å and then rapidly falls off. Thus, any reasonable estimate of the distance apart of the atoms in the transition structure cannot fail to make $S$ small. The integral $\beta$ is roughly proportional to $S$, so the third term of Equation 3.4 is a second-order term. With $S$ always small, the higher-order terms are very small indeed, and we neglect them.

(i)   *The first term* is the first-order, closed-shell repulsion term, and comes from the interaction of the filled orbitals of the one molecule with the filled orbitals of the other (Fig. 3.3a). The interaction of a filled orbital with a filled orbital leads to a small antibonding effect, but there are *many* filled orbitals interacting strongly with *many* filled orbitals, and the total effect is the sum of many small ones. The overall effect of the first term of Equation 3.4 is, therefore, rather unpredictable, but it seems that adding up a lot of small items often averages out the total effect. Thus, if a molecule can be attacked at two possible sites, we can hope that the first term will be nearly the same for attack at each site. Similarly, if there are two possible orientations in a cycloaddition, the first term may not be very different in the two orientations. This appears not to be the case for the other two terms, and it is therefore with them that we shall mainly be concerned in explaining differential reactivity. In practice it is not obvious that we can rely on a benign first term, but we often do. We shall, therefore, be largely ignoring the first term from now on, because frontier orbital theory is mainly used to explain features of *differential* reactivity. We are on weak ground in doing so, and we should not forget it.

(ii) *The second term* is simply the Coulombic repulsion or attraction. This term, which contains the products of the total charges, $Q$, on each atom, is obviously important when ions or polar molecules are reacting together. Its contribution to the energy is inversely proportional both to the dielectric constant and to the distance apart of the two charges.

(iii) *The third term* represents the interaction of all the filled orbitals with all the unfilled of correct symmetry (Fig. 3.3b and 3.3c). It is the second-order perturbation term, and is only true if $E_r \neq E_s$. (When $E_r = E_s$, the interaction is better described in charge-transfer terms, and the perturbation is then first order of the form $\Sigma_{ab}2c_{ra}c_{sb}\beta_{ab}$.) Here we can see again that it is the HOMO and the LUMO which are most important—they are the orbitals with the smallest value of $E_r - E_s$, and hence they make a disproportionately large contribution to the third term of Equation 3.4.

---

In summary:

*As two molecules collide, three major forces operate.*

  (i) *The occupied orbitals of one repel the occupied orbitals of the other.*

 (ii) *Any positive charge on one attracts any negative charge on the other (and repels any positive).*

(iii) *The occupied orbitals (especially the HOMOs) of each interact with the unoccupied orbitals (especially the LUMOs) of the other, causing an attraction between the molecules.*

---

Let us use a reaction to apply these ideas: the allyl anion **3.1**, reacting with the allyl cation **3.2** shows all the features. The major contributions to bond-making will be the powerful charge-charge interaction (the second term of Equation 3.4) *and* the strong interaction from the HOMO of the anion and the LUMO of the cation ($E_1$ in Fig. 3.4). In contrast, the interaction of the HOMO of the cation and the LUMO of the anion is much less effective ($E_2$ in Fig. 3.4), because $E_r - E_s$ is relatively so large. This reaction provides a simple illustration of how the ideas behind Equation 3.4 work, and it also shows how it comes about that in general *the important frontier orbitals for a nucleophile reacting with an electrophile are* $HOMO_{(nucleophile)}/LUMO_{(electrophile)}$ and not the other way round.

**3.1**      **3.2**

Having identified the causes for the ease of a reaction like this, we must next use the ideas behind Equation 3.4 to identify the *sites* of reactivity in each of the reacting species. To find the contribution of the Coulombic forces, we need the total electron population on each atom. For the allyl anion, the excess $\pi$ charge (Fig. 1.29) is

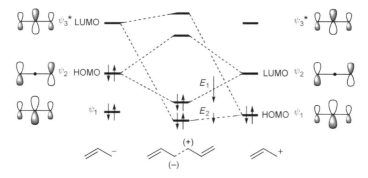

**Fig. 3.4** Orbital interactions in the reaction of the allyl anion with the allyl cation

concentrated on C-1 and C-3, and it is therefore here that positively charged electrophiles will attack. For the allyl cation, the $\pi$ electron *deficiency* is concentrated on C-1 and C-3, where the electron population is lowest, and it is therefore here that a negatively charged nucleophile will attack. Thus when the reaction takes place, the charge-charge attraction represented by the second term of Equation 3.4 will lead C-1 ($=$ C-3) of the allyl anion to react with C-1 ($=$ C-3) of the allyl cation, and C-2 will have little nucleophilic or electrophilic character.

When we add the contribution from the frontier orbitals, the picture is even more striking. The HOMO of the anion has coefficients at C-1 and C-3 of $\pm 0.707$, and similarly the LUMO of the cation has coefficients at C-1 and C-3 of $\pm 0.707$. In both frontier orbitals the coefficient on C-2 is zero. Thus the frontier orbital term is overwhelmingly in favour of reaction of C-1 (C-3) of the anion with C-1 (C-3) of the cation.

We have now seen how the attraction of charges and the interaction of frontier orbitals combine to make a reaction between two such species as the allyl anion and allyl cation both fast and highly regioselective. We should remind ourselves that this is not the whole story: another reason for both observations is that the reaction is very exothermic: the energy of a full $\sigma$ bond is released with cancellation of charge, which would not be the case if reaction took place at C-2 on either component. Thus we are in the situation of Fig. 3.1a—the Coulombic forces and the frontier orbital interaction on one side, and the stability of the product on the other, combine to lower the energy of the transition structure.

Indeed, you may well feel that there was little point in looking at the frontier orbitals in an imaginary reaction like this, even though it is hugely exothermic, when bonding between C-2 of the anion and C-2 of the cation would be absurd. The purpose of undertaking this exercise was twofold: in the first place, it shows that we get the same answer by considering the frontier orbitals as we do from a product development argument; and secondly, it shows how the allyl anion and allyl cation are nucleophilic and electrophilic, respectively, at both C-1 and C-3 *without* our having to draw the *canonical structures* used in valence bond theory.

The drawings chemists use for their structures will inevitably be crude representations—we shall always have to make some kind of localised drawing, whether it is of a benzene ring, or an enolate ion, or whatever. At the same time, we make mental reservations about how accurately such drawings represent the truth. If our mental reservations are made within the framework of molecular orbital theory, we shall have a better and more reliable picture of organic chemistry at our disposal.

## 3.6   Hard and Soft Nucleophiles and Electrophiles

The *rates* with which nucleophiles attack one electrophile are not necessarily a good guide to the rates with which the same nucleophiles will attack other electrophiles, just as there is no single measure of *thermodynamic* acidity or basicity. Following the principle of hard and soft acids and bases, we categorise nucleophiles (bases) and electrophiles (acids) as being hard or soft. Other things being equal, hard nucleophiles react faster with hard electrophiles, and soft nucleophiles with soft electrophiles.

We can now return to the Salem-Klopman equation [Equation 3.4, see (Section 3.5) page 106], where we see that the second term represents the attraction between the two molecules from a hard-hard interaction, and the third term represents the contribution to bonding from a soft-soft interaction. We saw from our consideration of the imaginary reaction of the allyl anion (the base or nucleophile) with the allyl cation (the acid or electrophile) that the important frontier orbital of a nucleophile is the relatively high-energy HOMO, and the important frontier orbital of an electrophile is the relatively low energy LUMO. Since hard reagents are characterised by having a large separation between the HOMO and the LUMO, hard nucleophiles (Tables 3.3–3.5) are generally those which are negatively charged and have relatively low-energy HOMOs—typically the anions centred on the electronegative elements. Likewise, hard electrophiles are generally those which are positively charged and have relatively high-energy LUMOs, typically the cations of the more electropositive elements. Thus their reactions with each other are fast because each makes a large contribution to the second term of Equation 3.4. However, soft nucleophiles have high-energy HOMOs and soft electrophiles have low-energy LUMOs, and their reactions with each other are fast because each makes a large contribution to the third term of Equation 3.4.

To take a simple example, the hydroxide ion is a hard nucleophile, because it has a charge, and because oxygen is a small, electronegative element. Accordingly, it reacts faster with a hard electrophile like a proton than with a soft electrophile like bromine. However, an alkene is a soft nucleophile, because it is uncharged and has a high-energy HOMO. Thus it reacts faster with an electrophile which has a low-energy LUMO, like bromine or the silver cation, than it does with a proton.

The rates of most reactions are affected by contributions from both of these terms of Equation 3.4, with one often being more important than the other. It is important to realise, for example, that a hard nucleophile may react faster with a soft electrophile than a soft nucleophile with the same soft electrophile. Thus the hydroxide ion almost certainly reacts faster with the silver ion than ethylene does. This is because the hydroxide ion is, for several reasons, more generally reactive than ethylene. Hardness and softness are most useful when they are used to differentiate finer grades of reactivity.

In summary:

*Hard nucleophiles have a low-energy HOMO and usually have a negative charge.*
*Soft nucleophiles have a high-energy HOMO but do not necessarily have a negative charge.*
*Hard electrophiles have a high-energy LUMO and usually have a positive charge.*
*Soft electrophiles have a low-energy LUMO but do not necessarily have a positive charge.*

(i)   *A hard-hard reaction is fast because of a large Coulombic attraction.*
(ii)  *A soft-soft reaction is fast because of a large interaction between the HOMO of the nucleophile and the LUMO of the electrophile.*
(iii) *The larger the coefficient in the appropriate frontier orbital (of the atomic orbital at the reaction centre), the softer the reagent.*

## 3.7   Other Factors Affecting Chemical Reactivity

Of the many factors controlling chemical reactivity some are obviously involved in the derivation of Equation 3.4, but some are not. *Strain* in the $\sigma$ framework, whether gained or lost, is not included, except insofar as it affects the energies of those orbitals which are involved. Factors which affect the *entropy* of activation are not included, nor are *solvent effects.*

We cannot, then, expect this approach to understanding chemical reactivity to explain everything. Most attempts to check the validity of frontier orbital theory computationally indicate that the sum of all the interactions of the filled with the unfilled orbitals swamp the contribution from the frontier orbitals alone. Even though the frontier orbitals make a weighted contribution to the third term of the Salem-Klopman equation, they do not account quantitatively for the many features of chemical reactions for which they seem to provide such an uncannily compelling explanation. Organic chemists, with a theory that they can handle easily, have fallen on frontier orbital theory with relief, and comfort themselves with the suspicion that something deep in the patterns of molecular orbitals must be reflected in the frontier orbitals in some disproportionate way.

# 4 Ionic Reactions—Reactivity

The task of understanding chemical reactions is never over—how many times have you read that something is not yet fully understood? Nothing is ever fully understood. However, the general outline of understanding has been in place for some time, guided by the principles that pairs of electrons, illustrated with curly arrows, flow from the less electronegative elements towards the more electronegative, that $\pi$ bonds are more reactive than $\sigma$ bonds, and that conjugation is energy-lowering, especially if it is aromatic.

One problem with this simple picture is that it seems inherently unlikely that pairs of electrons should act in concert—a pair of electrons in a single orbital spend as much time as far away from each other as possible, so why should a pair of electrons act together to move from one bond into another? It seems more reasonable for them to move one at a time. The transfer of one electron from one molecule to another is well known—it is the basis for one-electron oxidation and one-electron reduction, with many examples in electrochemistry, in sodium-in-ammonia reductions, and in inorganic redox reactions—but is it a common pathway in ionic organic chemistry, or something that only happens in favourable circumstances?

## 4.1 Single Electron Transfer (SET) in Ionic Reactions

Let us look at the key steps in three of the most fundamental reaction types in ionic organic chemistry—the $S_N2$ reaction **4.1** → **4.2**, nucleophilic attack on an X=Y double bond **4.3** → **4.4**, where X and Y may be any combination of C, N, O or S, and electrophilic attack on a C=C double bond **4.5** → **4.6**. A high proportion of ionic organic chemistry is covered by these three reactions or their reverse. The standard formulation for the first has the electron pairs moving in concert from the nucleophile into the Nu—C bond and the simultaneous movement of the electrons in the C—X bond onto the leaving group. For nucleophilic attack on a $\pi$ bond, a lone pair provides the electrons for a new $\sigma$ bond Nu—X, and a pair of electrons from the $\pi$ bond move onto the atom Y which is the more electronegative, or the one better able to stabilise a negative

*Molecular Orbitals and Organic Chemical Reactions: Student Edition*   Ian Fleming
© 2009 John Wiley & Sons, Ltd

charge. The subsequent fate of the anion **4.4** will depend upon the rest of the structure. For electrophilic attack by a π bond **4.5**, the pair of electrons moves to make a σ bond to the electrophile, and leaves behind a carbocation **4.6**. The fate of this cation also depends upon the rest of the structure—the two most common fates being the loss of a proton or the capture of a nucleophile. For now, we are concerned with the first steps, the attack of the nucleophile or electrophile, because, more often than not, these are rate-determining.

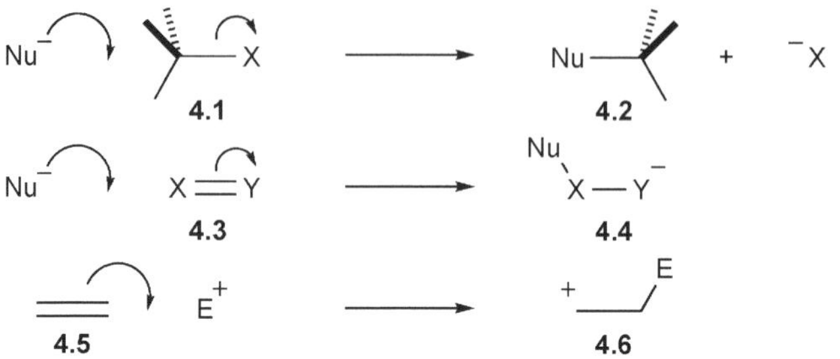

In the most simple alternative mechanisms involving single electron transfer (SET), the two reagents in each case transfer one electron from the partner with the lower ionisation potential (higher-energy HOMO) to the one with the higher electron affinity (lower-energy LUMO). This may or may not be part of a loose association called a charge-transfer complex **4.7**, **4.10** and **4.12**, but in all three types of reaction, a pair of radicals **4.8**, **4.11** and **4.13** is created. These radicals may or may not be charged, depending upon the charge carried by each of the original reagents. In the $S_N2$ case, the radical anion **4.8** can lose the halide ion $X^-$ to give the radical **4.9**. In all three cases, the radical pairs **4.9**, **4.11** and **4.13** can be expected to combine together rapidly. Radical-radical couplings are rare, because radicals, usually created in low concentration, do not live long enough to meet another radical, but in this case they are created as a pair within a solvent cage, effectively in high concentration. With the concentration problem overcome, radical-radical couplings are inherently fast, because they are so exothermic. The rate-determining step is therefore likely to be the transfer of the electron, and the product **4.2**, and the intermediates **4.4** and **4.6**, are the same as those in the conventional mechanism.

There is good evidence that some nucleophilic substitution reactions do involve SET, but the best established use a  slightly different mechanism. These are the

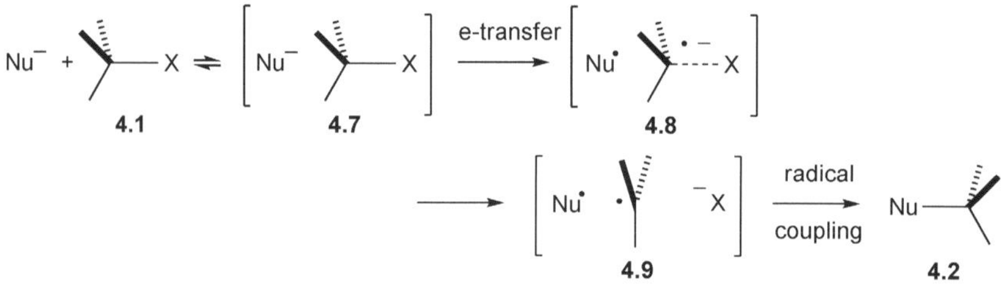

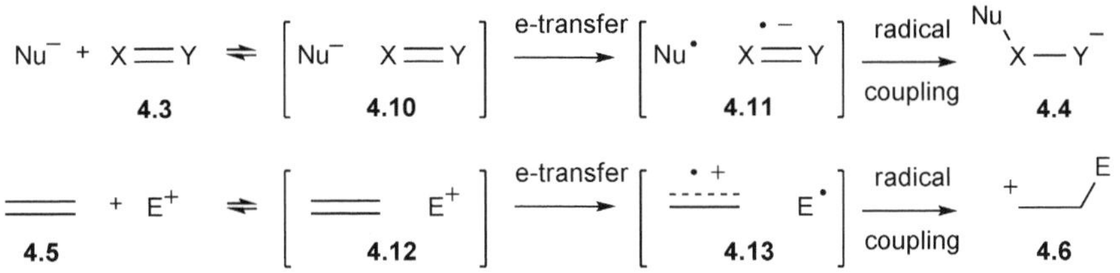

$S_{RN}1$ reactions, with the subscript RN standing for radical nucleophilic. They begin with an initiation step in which an electron is transferred to the electrophile R-X to give a radical anion R-X$^{\bullet-}$, which breaks apart to give the radical R$^\bullet$ and the anion X$^-$. The radical R$^\bullet$ combines with the nucleophile Nu$^-$ to give a radical anion R-Nu$^{\bullet-}$, and this transfers its electron to the electrophile R-X, giving the radical anion R-X$^{\bullet-}$ and the product R-Nu, making the sequence self sustaining. There is a further subtlety, in which the departure of the nucleofugal group X$^-$ is concerted with the electron transfer, making the radical anion RX$^{\bullet-}$ a transition structure rather than an intermediate.

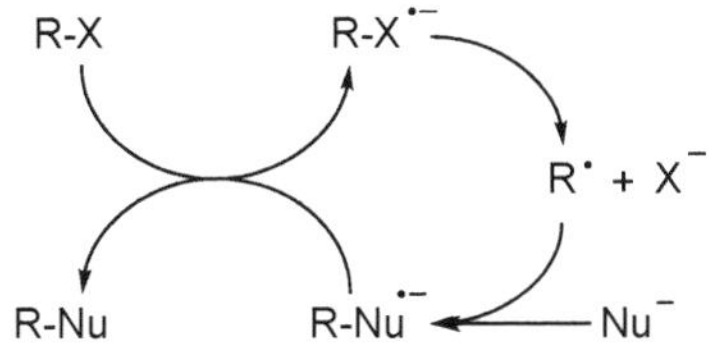

The $S_{RN}1$ reaction

Examples are the reaction of the nitronate anion **4.14** with *p*-nitrobenzyl chloride **4.15**, and the reaction of the pinacolone enolate **4.16** with bromo-benzene **4.17**. The former might have been a straightforward $S_N2$ reaction, but actually takes the $S_{RN}1$ pathway because the nitro groups make the electron transfer exceptionally easy. The latter cannot take place by a conventional $S_N2$ reaction, because aryl (and vinyl) halides are not susceptible to direct displacement, and the $S_{RN}1$ pathway overcomes this difficulty.

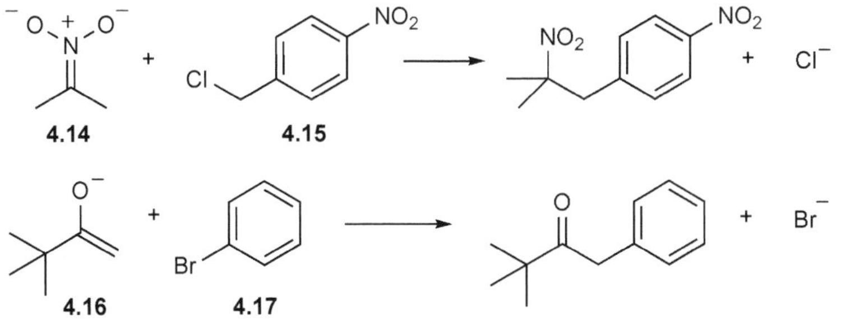

Given that SET pathways are involved in some nucleophile-electrophile combinations, ought we to consider that they are the most likely pathway in all of them? This is a seductive proposition, because a unified mechanism for a wide range of reactions has an instant appeal. Most of the well established SET pathways have been found with reagents conspicuously carrying radical-stabilising substituents or for substrates incapable of following the traditional mechanism. In consequence, it might be that SET pathways are followed only in those substrates well adapted to SET. This may be right, but it is not a good argument. If the radicals inside the cage are well stabilised, they are more likely to live long enough to escape, and to give us opportunities to detect them, but if the radicals in the cage are not well stabilised, their coupling is likely to be so fast that they can evade the radical probes with which we try to detect them.

We shall not be able to come to a firm conclusion here. Nevertheless, we do need to address the problem of how to use molecular orbital theory, and the principle of hard and soft acids and bases, to explain reactivity and selectivity. The traditional two-electron mechanism and the SET mechanisms will need different wording, and maybe a different explanation. Fortunately, this is not a great problem. The aspects of molecular orbital theory that we shall invoke work in much the same way in both families of mechanism. Thus, a nucleophile will be more nucleophilic if its available pair of electrons is in a high energy orbital, whether those electrons are used directly to make a bond or one of them is transferred on its own. Electrophilicity, likewise, will be greater if the reagent has a low-energy LUMO, whether this is encouraging direct attack by a pair of electrons or the acquisition of a single electron. For this parallel to work, the electron transfer must be slower than the coupling of the radicals, which it usually is. Regioselectivity in product formation, however, may need different explanations, since the new bond is formed in the conventional mechanism in the first step, but in the SET mechanism it is formed in the radical coupling step or, in $S_{RN}1$ reactions, the step in which a radical attacks the nucleophile. Mechanistic discussion in the rest of this book will be phrased using the conventional mechanism, which has yet to be displaced from most discourse about most reactions, but the possibility that the pathway may be by SET is worth keeping in mind.

## 4.2   Nucleophilicity

### 4.2.1   Heteroatom Nucleophiles

Just as there is no single measure of acidity and basicity, there is no single measure of nucleophilicity and electrophilicity—the rank order of nucleophiles changes when the reference electrophile changes. A hard nucleophile like a fluoride ion reacts fast with a silyl ether in an $S_N2$ reaction at the silicon atom, which is relatively hard, but a soft nucleophile like triethylamine does not. In contrast, triethylamine reacts with methyl iodide in an $S_N2$ reaction at a carbon atom, but fluoride ion does not. These examples, which are all equilibria, are governed by

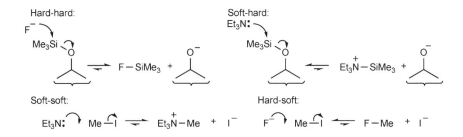

thermodynamics, but there are similar examples illustrating divergent patterns of nucleophilicity in the rates of reactions. A scale of nucleophilicity, therefore, requires at least two parameters, and these were first provided empirically by Edwards with Equation 4.1.

$$\log \frac{k_{Nu}}{k_0} = \alpha E_N + \beta H \qquad 4.1$$

where $k_{Nu}$ is the rate constant (or equilibrium constant) for a reaction, $k_0$ is the rate constant (or equilibrium constant) for water as the nucleophile, and $E_N$ and $H$ are the two scales of nucleophilicity. The $E_N$ scale is measured by the rate of reaction with methyl bromide, and the $H$ scale is $(pK_a + 1.74)$, where 1.74 is a correction for the $pK_a$ of $H_3O^+$. Thus the two scales reflect both softness $(E_N)$ and hardness $(H)$, although these terms were not in use when Edwards formulated his equation. The extent to which the hard or soft character of the electrophile contributes will affect the relative sizes of $\alpha$ and $\beta$, which will be different for each reaction, a high value of $\alpha/\beta$ implying a soft electrophile.

Another way of looking at the same problem uses the Salem-Klopman equation (Equation 3.4). Using only the HOMO of a nucleophile and the LUMO of an electrophile, Klopman simplified Equation 3.4 to Equation 4.2:

$$\Delta E = \frac{Q_{Nu}Q_{El}}{\varepsilon R} + \frac{2(c_{Nu}c_{El}\beta)^2}{E_{HOMO(Nu)} - E_{LUMO(El)}} \qquad 4.2$$

Starting with this simplification, Klopman worked out the contribution of the frontier orbital terms to the nucleophilicities and electrophilicities of a range of inorganic bases and acids. From the known ionisation potentials and electron affinities, and correcting for the effect of solvation, he calculated values $(E^{\ddagger}$, Table 4.1, listed with hardness at the top and softness at the bottom) for the effective energy of the HOMO of the nucleophiles and the LUMO of the electrophiles. The results agree well with Pearson's original, empirically derived order of softness, and, if allowance is made for the absence of solvation, with his more theoretically derived order in Tables 3.3 and 3.4. The higher the value of $E^{\ddagger}$ for the HOMO of a nucleophile, the softer it is, and the higher the

**Table 4.1**  Calculated softness character for inorganic nucleophiles and electrophiles (hard at the top and soft at the bottom)[14]

| Nucleophile | Effective HOMO $E^{\ddagger}$ (eV) | Electrophile | Effective LUMO $E^{\ddagger}$ (eV) |
|---|---|---|---|
| $F^-$ | $-12.18$ | $Al^{3+}$ | 6.01 |
| $H_2O$ | $-10.73$ | $La^{3+}$ | 4.51 |
| $HO^-$ | $-10.45$ | $Ti^{4+}$ | 4.35 |
| $Cl^-$ | $-9.94$ | $Mg^{2+}$ | 2.42 |
| $Br^-$ | $-9.22$ | $Ca^{2+}$ | 2.33 |
| $CN^-$ | $-8.78$ | $Fe^{3+}$ | 2.22 |
| $HS^-$ | $-8.59$ | $Cr^{3+}$ | 2.06 |
| $I^-$ | $-8.31$ | $Ba^{2+}$ | 1.89 |
| $H^-$ | $-7.37$ | $Cr^{2+}$ | 0.91 |
| | | $Fe^{2+}$ | 0.69 |
| | | $Li^+$ | 0.49 |
| | | $H^+$ | 0.42 |
| | | $Ni^{2+}$ | 0.29 |
| | | $Na^+$ | 0 |
| | | $Cu^{2+}$ | $-0.25$ |
| | | $Tl^+$ | $-1.88$ |
| | | $Cu^+$ | $-2.30$ |
| | | $Ag^+$ | $-2.82$ |
| | | $Tl^{3+}$ | $-3.37$ |
| | | $Hg^{2+}$ | $-4.64$ |

**Table 4.2**  Nucleophilicity of inorganic ions towards electrophiles as a function of $E_{HOMO} - E_{LUMO}$

| Nucleophile | $\Delta E$ calculated for | | | | Found | | |
|---|---|---|---|---|---|---|---|
| | $E_{HOMO}$ | $E_{LUMO}$ $-7$ eV | $E_{LUMO}$ $-5$ eV | $E_{LUMO}$ $+1$ eV | $k \times 10^4$ Equation 4.3 | Edwards' $E_N$ Equation 4.4 | $pK_a$ Equation 4.5 |
| $F^-$ | $-12.18$ | 1.06 | 0.82 | 0.54 | 0 | 1.0 | 3.2 |
| $HO^-$ | $-10.45$ | 1.49 | 1.01 | 0.58 | 0 | 1.65 | 15.7 |
| $Cl^-$ | $-9.94$ | 1.54 | 0.97 | 0.52 | 0.001 | 1.24 | – |
| $Br^-$ | $-9.22$ | 1.75 | 0.98 | 0.48 | 0.23 | 1.51 | $-4.3$ |
| $CN^-$ | $-8.78$ | 2.30 | 1.17 | 0.56 | 10 | 2.79 | 9.1 |
| $HS^-$ | $-8.59$ | 2.64 | 1.25 | 0.55 | too fast | too fast | 7.1 |
| $I^-$ | $-8.31$ | 2.52 | 1.07 | 0.45 | 6900 | 2.06 | $-7.3$ |

value of $E^{\ddagger}$ for the LUMO of an electrophile, the harder it is. Table 4.1 is therefore a useful practical list.

Klopman used Equation 4.2 to estimate the relative nucleophilicities of a range of anionic nucleophiles: $I^-$, $Br^-$, $Cl^-$, $F^-$, $HS^-$, $CN^-$ and $HO^-$ towards different electrophiles. He assumed unit charges and unit values for the coefficients, $c$. For the $E_{HOMO(Nu)}$, he used the values of $E^{\ddagger}$ in Table 4.1. Since it is roughly proportional to the overlap integral $S$, the value of $\beta$ changes from one element to another, but it can be calculated. This left only the energy of the LUMO of the electrophile, $E_{LUMO(El)}$, as an unknown on the right-hand side of the equation. Klopman therefore calculated the $\Delta E$ values in Table 4.2 for a series of imaginary electrophiles with different values of the LUMO energy.

(i) Setting $E_{LUMO(El)}$ at $-7$ eV made the $E_{HOMO} - E_{LUMO}$ term small, and hence the frontier orbital term, the second term of Equation 4.2, made a large contribution to $\Delta E$. The order of the values $\Delta E$ was $HS^- > I^- > CN^- > Br^- > Cl^- > HO^- > F^-$, which is the order of nucleophilicities observed for the attack of these ions on peroxide oxygen (Equation 4.3).

$$\text{Nu}^- \quad \text{HO} - \text{OH} \longrightarrow \text{Nu} - \text{OH} + {}^-\text{OH} \qquad 4.3$$

(ii) Setting $E_{LUMO(El)}$ at $-5$ eV, the order of nucleophilicities is slightly changed, because the frontier orbital term makes a smaller contribution to $\Delta E$. The order of $\Delta E$ values is now $HS^- > CN^- > I^- > HO^- > Br^- > Cl^- > F^-$, which parallels the Edwards $E_N$ values for the nucleophilicities of these ions towards saturated carbon (Equation 4.4).

$$\text{Nu}^- \quad \longrightarrow \quad \text{Nu} - \quad + \quad {}^-\text{X} \qquad 4.4$$

(iii) Finally, setting $E_{LUMO(El)}$ high at $+1$ eV, the frontier orbital term is made unimportant, and the order of $\Delta E$ values is governed almost entirely by the Coulombic term of Equation 4.2: $HO^- > CN^- > HS^- > F^- > Cl^- > Br^- > I^-$. This is the order of the $pK_a$s of these ions, in other words, of the extent to which the equilibrium lies to the right in Equation 4.5.

$$\text{Nu}^- \quad \text{H} - \overset{+}{\text{OH}}_2 \rightleftharpoons \text{Nu} - \text{H} + \text{H}_2\text{O} \qquad 4.5$$

Simply by adjusting the relative importance of the two terms of Equation 4.2, Klopman duplicated the otherwise puzzling changes of nucleophilicity as the electrophile changes. The proton is a hard electrophile because it is charged and small. Hence, a nucleophile can get close to it in the transition structure, and $R$ in

Equation 4.2 is made small. In contrast, the oxygen-oxygen bond has no charge, and, being both weak and between electronegative atoms, it has a relatively low-lying $\sigma^*$ LUMO, making it a soft electrophile. Similarly, with nucleophiles such as $F^-$, $Cl^-$, $Br^-$, and $I^-$, the energy of the HOMO will rise as we go down the periodic table, and with nucleophiles like $Cl^-$, $HS^-$, and $R_2P^-$, it will rise as we move to the left in the periodic table.

### 4.2.2 Solvent Effects

Solvents have long been implicated in the divergent scales of nucleophilicity and electrophilicity, because small, charged nucleophiles and electrophiles are often highly solvated. Gas-phase $S_N2$ reactions, where there is no solvent, are quite different from their solution counterparts, but gas-phase reactions do show a pattern in which matching the nucleophile to the leaving group can lead to higher rates, reminiscent of hard-hard and soft-soft pairings, and so it seems that solvent effects are not the whole story. The orbitals of a solvent interacting with those of the nucleophile and the electrophile are responsible for some of the energy of solvation, and are amenable to treatment by perturbation theory. If the orbitals are close enough in energy for a first-order treatment to be appropriate, reaction would occur; so solvation is a second-order interaction. The second term of Equation 4.2 is therefore an approximation, and the major orbital interactions are between the HOMO of the solvent and the LUMO of an electrophile, and between the LUMO of the solvent and the HOMO of a nucleophile. Using this idea, and using ionisation potentials and electron affinities as measures of the energies of the HOMO and the LUMO of a range of solvents, Dougherty has been able to explain some otherwise puzzling changes of solvating power.

### 4.2.3 Alkene Nucleophiles

Alkenes, with a $\pi$ orbital as the HOMO and no charge, are inherently soft nucleophiles, and their nucleophilicity ought to bear some relationship to the energy of their HOMOs. The relative rates of attack by different alkenes on several electrophiles have been measured, but most systematically, Mayr has measured the nucleophilicity of a wide range of alkenes **4.18** being attacked by a family of diarylmethyl cations **4.19**.[15]

The order of nucleophilicity, graphically illustrated in Fig. 4.1, matches well with expectation—the better donor substituents are the more effective at increasing the rate of reaction. However, since this is an endothermic reaction, the transition

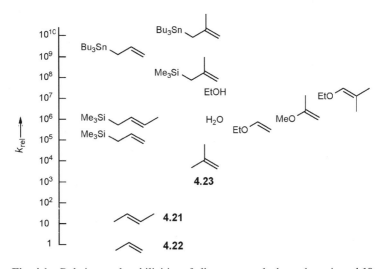

**Fig. 4.1** Relative nucleophilicities of alkenes towards the carbocations **4.19**

structures are product-like (Fig. 3.2b). In consequence, the rates correlate well with the stabilities of the carbocations produced **4.20**, but not well with the ionisation potential (and hence the HOMO energy) of the alkene. Donor substituents $R^1$ and $R^2$, of course, stabilise the cation **4.20**, and can be expected to lower the energy of the transition structure. Nevertheless, the orbitals of the starting material do play a part, as we can see for those alkenes having substituents $R^3$ on the carbon atom under attack. The substituent $R^3$ does not contribute much to the stabilisation of the cation **4.20**, but it does raise the energy of the HOMO of the alkene **4.18**. On the whole, a substituent on the carbon atom under attack ($\beta$) increases the rate of attack, with 2-butene **4.21** about one order of magnitude more reactive than propene **4.22**. However, the difference in rates for 2-butene and propene is not as large as having the extra methyl group on the $\alpha$ carbon, as in isobutylene **4.23.**

In summary:

*C- and X-Substituents raise the energy of the HOMO and increase nucleophilicity.*

*Z-Substituents lower the energy of the HOMO and decrease nucleophilicity.*

### 4.2.4 The α-Effect

The solvated proton is a hard electrophile, little affected by frontier orbital interactions. For this reason, the $pK_a$ of the conjugate acid of a nucleophile is a good measure of the rate at which that nucleophile will attack other hard electrophiles. We shall see that carbonyl groups are fairly hard, but somewhat responsive to frontier orbital effects. Thus a thioxide ion, $RS^-$, is more nucleophilic towards a carbonyl group than one would expect from its $pK_a$: a Brønsted plot of the log of the rate constant for nucleophilic attack on a carbonyl group against the $pK_a$ of the nucleophile is a good straight line only when the nucleophilic atom is the same. In other words, there is a series of straight lines, one for oxygen nucleophiles, one for sulfur nucleophiles, and yet another for nitrogen nucleophiles.

Some nucleophiles stand out because they do not fit on Brønsted plots: they are more nucleophilic than one would expect from their $pK_a$ values. These nucleophiles, $HO_2^-$, $ClO^-$, $HONH_2$, $N_2H_4$, and $R_2S_2$, have a nucleophilic site which is flanked by a heteroatom bearing a lone pair of electrons. The orbital containing the electrons on the nucleophilic atom overlaps with the orbital of the lone pair, raising the energy of the HOMO relative to its position in the unsubstituted nucleophile. Consequently, the denominator of the third term of Equation 3.4 is reduced, and the importance of this term is increased. The result is an increase in nucleophilicity, called the $\alpha$-effect, which can be quite large (Table 4.3). The order of the effect appears to be right: the LUMO of the triple bond of the nitrile will be lower than that of the double bond of the carbonyl group, which will be lower than that of the $\sigma$ bond of the bromide. Hence the frontier orbital term is most enhanced in the case of the nitrile, and least enhanced for the bromide and arenesulfonates.

**Table 4.3**   Relative nucleophilicity of hydroperoxide ion and hydroxide ion

| Electrophile | $k_{HOO^-}/k_{HO^-}$ | Electrophile | $k_{HOO^-}/k_{MeO^-}$ |
|---|---|---|---|
| PhC≡N | $10^5$ | $ArSO_2OMe$ | 6–11 |
| $p\text{-}O_2NC_6H_4CO_2Me$ | $10^3$ | | $k_{HOO^-}/k_{HO^-}$ |
| $PhCH_2Br$ | 50 | $H_3O^+$ | $10^{-4}$ |

However, the energy of the HOMO in a range of $\alpha$-nucleophiles does not actually correlate well enough with their nucleophilicity for this explanation of the effect to be satisfactorily settled. The problem raised by the small $\alpha$-effect in reactions with alkyl halides, where the LUMO energy is not conspicuously low, and yet they are soft electrophiles, raises the question of what happens when both charge and frontier orbital terms are small, and what happens when they are both large? Prediction is not simple in this situation, but Hudson has suggested the two rules in the box, which are usually but not invariably observed.

> (i) When both charge and orbital terms are *small*: nucleophiles and electrophiles will be *soft* (that is, orbital control is more important).
> (ii) When both charge and orbital terms are *large*: nucleophiles and electrophiles will be *hard* (that is, charge control is more important).

Superoxide anions, $RO_2^{\bullet-}$ are exceptionally powerful nucleophiles benefiting from an $\alpha$-effect and possibly from the availability of an unpaired electron in an SET mechanism.

## 4.3 Ambident Nucleophiles

### 4.3.1 Thiocyanate Ion, Cyanide Ion and Nitrite Ion (and the Nitronium Cation)

Thiocyanate ions, cyanide ions and nitrite ions can each react with an electrophile $R^+$, depending upon its nature and the conditions, to give either of two products: a thiocyanate **4.24** or an isothiocyanate **4.25** from the thiocyanate ion, an isonitrile **4.26** or a nitrile **4.27** from cyanide ion, and an alkyl nitrite **4.28** or a nitroalkane **4.29** from nitrite ion. Each is nucleophilic at more than one site, and nucleophiles like these are called *ambident*.

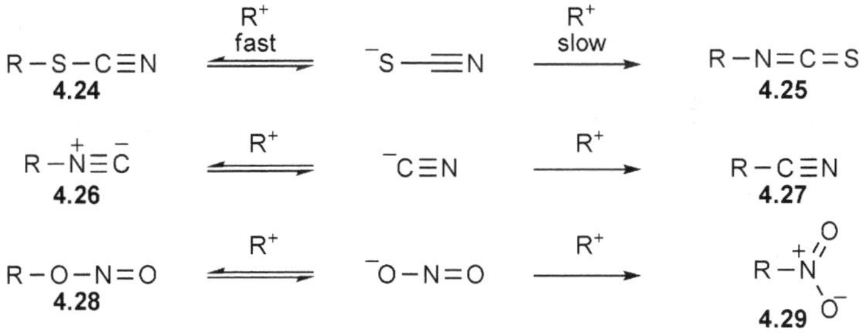

The principle of hard and soft acids and bases ought to apply. The thiocyanate ion will be softer on the sulfur atom and harder on the nitrogen atom, the cyanide ion will be softer on the carbon atom and harder on the nitrogen atom, and the nitrite ion will be harder on the oxygen atom and softer on the nitrogen atom. We might expect that harder electrophiles will give the isothiocyanates **4.25**, the isonitriles **4.26** and the nitrites **4.28**. However, other factors are at work, and this pattern is unreliable. Earlier attempts to use these expectations to explain the patterns of reactivity in this area have been overtaken by more recent work by Mayr.[16]

The thiocyanate **4.24** is the kinetically preferred product in alkylations by alkyl halides undergoing $S_N2$ reactions and with carbocationic electrophiles in $S_N1$

reactions, in spite of the fact that a carbocation, being charged, is a hard electrophile and the sulfur is the soft end of the nucleophile. This is partly explained by the relatively small change in bond lengths and in electronic reorganisation required in going from the thiocyanate ion to the thiocyanate product. Thiocyanates **4.24** more or less rapidly isomerise to isothiocyanates **4.25**, which are the thermodynamically preferred products, the tertiary alkyl thiocyanates rearranging by an $S_N1$ pathway, and primary alkyl thiocyanates by an $S_N2$ pathway.

Cyanide ions react with the soft alkyl halides in $S_N2$ reactions and with the hard carbocations in $S_N1$ reactions to give, almost always, the nitrile **4.27**, which is thermodynamically preferred. Isonitrile products are formed along with the nitrile products when the cation is so reactive that the rate has reached the diffusion-controlled limit, and the reversible reaction that would equilibrate the products is too slow. One consequence when reactions are as fast as this is that there is a barrierless combination of ions, and selectivity is not then controlled by the kinetic factors associated with the principle of hard and soft acids and bases.

Silver cyanide sometimes leads to isonitrile products when potassium cyanide gives nitriles. A telling example is the formation of the isonitrile **4.32** in the reaction with silver cyanide, whereas potassium cyanide gives the nitrile **4.31**. Since both reactions take place with overall retention of configuration by way of an episulfonium ion **4.30**, it cannot be that the silver ion induces an $S_N1$ reaction. The relatively soft silver is attached to the carbon of the cyanide ion, leaving the nitrogen end free to be nucleophilic, whereas potassium is not so attached. Similarly, trimethylsilyl cyanide sometimes gives isonitrile products with carbocations stabilised only by alkyl groups, perhaps because the silyl group is attached to the carbon atom at the time of reaction, and the isonitriles rearrange too slowly.

Nitrite ions generally give nitroalkanes **4.29** as the major products, which are also thermodynamically more stable than alkyl nitrites **4.28**. Again, with carbocationic electrophiles, it is only when the reaction rate has reached the diffusion-controlled limit that alkyl nitrites can be detected or even be the major products. Nitrite ions also give mixtures of nitroalkanes and nitrites in $S_N2$ reactions, even though the alkyl electrophiles are relatively soft. A hint that the principle of hard and soft acids and bases might have some role in explaining ambient reactivity with nitrite ion comes from the $S_N2$ reactions with methyl iodide (relatively soft), with the trimethyloxonium ion (harder) and with methyl triflate (hardest) giving mixtures of methyl nitrite and nitromethane. There is an increase in the ratio of methyl nitrite to nitro-methane with the harder electrophiles, rising from 30:70, through 50:50, to 59:41, respectively. This is a small effect, and little weight can be placed on any explanation; nevertheless, we can see that the pattern is in line with the different contributions made to Equation 4.2. The contribution from the first term is greater for the harder electrophiles because of the greater charge on the carbon atom in methyl triflate than in the less polarised methyl iodide.

For the contribution from the second term of Equation 4.2, it is the HOMO of the nitrite ion that we need. The two lower $\pi$ orbitals of the $NO_2$ system in Fig. 4.2 resemble $\psi_2$ in the allyl system, with large coefficients on the oxygens and a node on the nitrogen, but the orbital above them, the HOMO in the nitrite anion, resembles $\psi_3*$ in the allyl system, with a large coefficient on the nitrogen. The bent shape of the HOMO is another example of Jahn-Teller distortion (see pages 32, 36 and 42). The nitrogen will be the nucleophilic site when the second term of Equation 4.2 is the more important contributor to $\Delta E$. An alternative, or supplementary explanation is that the solvent gathers around the site of higher charge, the oxygen atoms in the case of nitrite, and that this hinders the reaction there, and the nitrogen, with little charge, must be less crowded by solvent molecules.

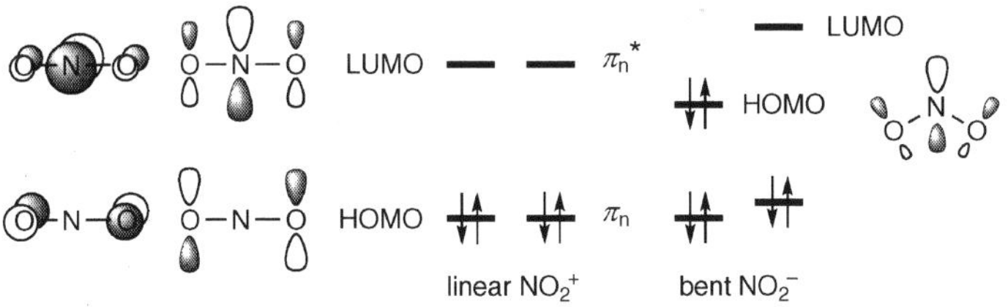

**Fig. 4.2**   Frontier orbitals of the nitronium ion and the nitrite ion

In nitration, where we have an ambient cation, the important frontier orbital will be the LUMO of $NO_2^+$, and this is a similar orbital to the HOMO of $NO_2^-$, except that the cation is linear. The nitronium ion, $NO_2^+$, always reacts on nitrogen, both with soft nucleophiles like benzene and with hard ones like water.

### 4.3.2   Enolate Ions

The most important ambident nucleophile is the enolate ion **4.33**. We have seen the $\pi$ orbitals of an enolate ion calculated in Fig. 2.15. The size of the lobes can be taken as roughly representing the size of the $c$-values (or $c^2$-values) of the atomic orbitals which make up the molecular orbitals. The system is closely related to that of the allyl anion in Fig. 1.28, but the effect of the oxygen atom is to polarise the electron distribution, with the lowest-energy orbital $\psi_1$ strongly polarised towards oxygen, and the next orbital up in energy $\psi_2$ polarised away from oxygen. The effective $\pi$ charge on each atom in the ion is proportional to the sum of the squares of the $c$-values for the filled orbitals, namely $\psi_1$ and $\psi_2$. Using the values from Fig. 2.15, the total charge on the oxygen atom is $0.9^2 + (-0.41)^2 = 0.98$ and on the carbon atom it is $0.17^2 + 0.7^2 = 0.52$. However, the $c$-values in the HOMO are the other way round, $-0.41$ on oxygen and $0.70$ on carbon.

**4.33**

With charged electrophiles, then, the site of attack will be oxygen, as is indeed the case, kinetically, with protons and carbocations. With electrophiles having little charge and relatively low-lying LUMOs, the reaction will take place at carbon. In other words, hard electrophiles react at oxygen and soft electrophiles at carbon. Solvent, of course, also hinders reaction at the sites of highest solvation, which will generally be the atoms carrying the highest total charge.

We can also explain why the nature of the leaving group on an alkyl halide (or tosylate, for example) affects the proportion of $C$- to $O$-alkylation in such enolates as the sodium salt **4.34** derived from ethyl acetoacetate: the harder the leaving group (i.e. the more acidic the conjugate acid of the leaving group), the lower is the proportion of $C$-alkylation (Table 4.4). The softer the leaving group, the lower will be the energy of the LUMO. In addition, the harder the leaving group, the more polarised is its bond to carbon, and hence the more charge there will be on carbon in the transition structure. This is the same phenomenon as the effect electronegative ligands have on the hardness of

**Table 4.4**   The proportion of $C$- to $O$-alkylation as a function of the leaving group X in EtX

| Nucleofugal group, $X^-$ | $I^-$ | $Br^-$ | $TsO^-$ | $EtSO_4^-$ | $CF_3SO_3^-$ |
|---|---|---|---|---|---|
| $k_C/k_O$ | >100 | 60 | 6.6 | 4.8 | 3.7 |

a Lewis acid. As a result, the Coulombic term of Equation 4.2 will grow in importance with the hardness of the leaving group, making *O*-alkylation easier, and the frontier orbital term will grow with its softness, making *C*-alkylation easier. We must not forget that solvent effects may be the dominant influence in the regioselectivity. There is evidence that both the alkylation and the acylation of enolate ions in the gas phase take place, more often than not, on oxygen, the site that solvents protect.

### 4.3.3 Allyl Anions

The allyl anion itself presents no problems: C-1 and C-3 are overwhelmingly the nucleophilic sites both for hard and soft electrophiles. A substituent on C-1 makes the allyl anion **4.35** an ambident nucleophile, since attack can now take place at C-1, the $\alpha$ position, to give the alkene **4.36**, or at C-3, the $\gamma$ position, to give what is usually the thermodynamically favoured product **4.37**. The substituent can be C, Z or X, the geometry can be W-shaped or sickle-shaped, and the regioselectivity for each can be the same for hard and soft electrophiles, or different for each. There are further complications—allyl anions are not usually free anions, but can have a metal covalently bonded either to C-1 or C-3, and the nature of the metal and the position it is attached to, rather than any inherent selectivity in the free anion, may determine the regioselectivity. Steric interactions between a large substituent and a large electrophile can change selectivity from $\alpha$ attack to $\gamma$ attack. Allyl anions, especially those with Z-substituents, are well enough stabilised for the attack on some electrophiles to be reversible, especially with aldehydes and ketones, the halogens, and sulfenyl halides. As a consequence, some results are thermodynamic and not kinetic, but it is not always easy to tell which. What is clear is that aldehydes and ketones often show different regioselectivity from other electrophiles, either because of reversibility, or, more likely, because the metal coordinates to the carbonyl group delivering the electrophile to the allylic position relative to the metal. Finally, solvents, and leaving groups affect the ratio, and the presence of other substituents at C-1, C-2 and C-3 add even more variables. The outcome, not surprisingly, is that firm rules about regioselectivity are not yet in place. The following brief discussion highlights only some of the more significant cases.

### 4.3.3.1  X-Substituted Allyl Anions.

Allyl anions with alkyl substituents almost always react with carbonyl electrophiles at the more substituted $\alpha$ position, as in the reaction of the prenyl Grignard reagent with aldehydes to give the product **4.39**, presumably because the metal attaches itself preferentially to the less-substituted end of the allyl system and then delivers the electrophile in a six-membered transition structure **4.38**. In contrast, alkylation of a similar anion with an alkyl halide gives mainly the product **4.40** of $\gamma$ attack, which is normal for an X-substituted allyl anion when a cyclic transition structure is not involved.

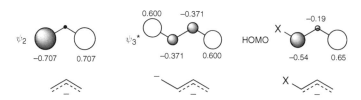

The coefficients in the HOMO can be estimated using the simple arguments developed earlier [see (Section 2.1.2.3) page 64], as in Fig. 4.3. The HOMO of an X-substituted allyl anion will have some of the character of $\psi_2$ of the allyl anion, which is symmetrical, but it will also have some of the character of a carbanion conjugated to an allyl anion, in other words $\psi_3^*$ of butadiene, which has the larger coefficient on C-4, corresponding to the $\gamma$ position.

**Fig. 4.3**  Estimating the coefficients of the HOMO of an X-substituted allyl anion

### 4.3.3.2  C-Substituted Allyl Anions—Pentadienyl Anions.

Allyl anions with C-substituents also suffer both $\alpha$ attack and $\gamma$ attack, as illustrated by the reactions of the open-chain C-substituted anions **4.41**, and **4.42**. Simple predictions based on the $\pi$ orbitals suggest that the C-substituted system should be equally reactive at the $\alpha$ and $\gamma$ carbons, since mixing any amount of $\psi_3$ of the pentadienyl anion into the HOMO of the allyl anion will leave the coefficients at the $\alpha$ and $\gamma$ carbons equal. Even more puzzling is the case of the cyclohexadienyl anion **4.43**, the last intermediate in Birch reduction. It is attacked almost exclusively at the central carbon atom, C-3, which corresponds to $\alpha$ attack on a C-substituted allyl anion. This gives the energetically less favourable product **4.44** with the two double bonds out of conjugation, and in spite of a 2:1 statistical preference for attack at C-1. This selectivity seems to be largely independent of the nature of the electrophile or of the metal. It is a highly exothermic reaction—just the kind which should show the influence from the interaction of the orbitals of the starting materials, rather than the influence from the relative energies of the two possible products.

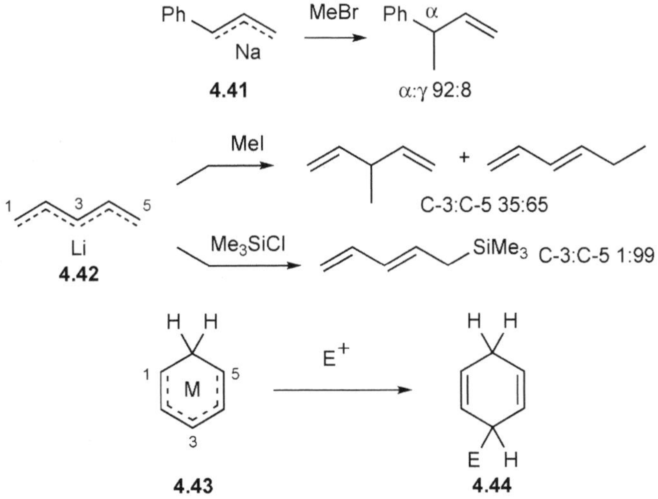

The sums of the squares of the coefficients of the filled orbitals $\psi_1$, $\psi_2$ and $\psi_3$ for the pentadienylsystem [Fig. 1.35, see (Section 1.4.3) page 31] are equal on C-1, C-3 and C-5. The frontier orbital coefficients for the HOMO $\psi_3$ are also equal on C-1, C-3 and C-5. However, the simple Hückel calculation which gives these values neglects the fact that C-3 is flanked by two trigonal carbon atoms, but that C-1 and C-5 have one trigonal carbon adjacent to them and one tetrahedral. Total $\pi$-electron populations and HOMO coefficients have been calculated allowing for the overlap of the C—H bonds at C-6 with the $\pi$ system (i.e. hyperconjugation), and these give the values shown in Fig. 4.4a and b, respectively. The presence of a larger coefficient on C-3 is supported by ESR measurements on the radical corresponding to this anion, which show a larger coupling to the hydrogen on C-3 than to those on C-1 and C-5 (Fig. 4.4c). Thus both the experiment and the calculation imply that there is a larger coefficient and a higher total charge at C-3 than at C-1.

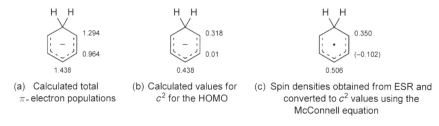

(a) Calculated total
π-electron populations

(b) Calculated values for
$c^2$ for the HOMO

(c) Spin densities obtained from ESR and
converted to $c^2$ values using the
McConnell equation

**Fig. 4.4**   Electron distribution in the cyclohexadienyl system

### 4.3.3.3   Z-Substituted Allyl Anions—Dienolate Ions.   Z-Substituted allyl anions **4.45a** are typified by dienolate ions **4.45b**, which is how they are best drawn. They almost always react faster at the α carbon than at the γ carbon, both with soft and hard electrophiles.

Using a model like that in Chapter 2 [see (Section 2.1.2.3) page 64], the extra conjugation for a Z-substituted allyl anion will mix $\psi_3$ of the pentadienyl anion with $\psi_2$ of butadiene (Fig. 4.5), which suggests that the γ carbon at the terminus ought to be the more nucleophilic site. Clearly this is not borne out in practice, and the measurements of the total electron population do not agree with this diagnosis—the $^{13}$C-NMR chemical shifts of the dilithium dienolate show that the charge on C-2, the α carbon, is actually twice that at C-4. Evidently Hückel theory is inadequate in this case, and it is the total charge distribution that explains the α-selectivity.

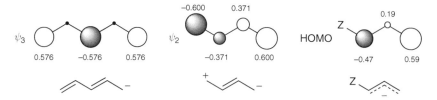

**Fig. 4.5**   Crude estimate of the coefficients of the π orbitals of a 1-Z-substituted allyl anion

Changing the conjugated system from a dienolate anion to a dienol ether changes the regioselectivity—silyl enol ethers **4.46** react mainly at the γ position. The change from the oxyanion substituent to the less powerful donor, the silyl ether, effectively changes the orbitals from being closer to those of a pentadienyl anion to being closer to those for a diene, and γ attack becomes more favourable. Thus the reactivity at C-4 shown by silyl enol ethers is, in a sense, the expected behaviour—it is the electron distribution and hence the reactivity at C-2 in the dienolate anions that is harder to explain.

### 4.3.4 Aromatic Electrophilic Substitution

Aromatic rings, except for highly symmetrical systems like benzene itself, are ambident nucleophiles. In electrophilic aromatic substitution, the rate-determining step is usually the attack of the electrophile on the $\pi$ system, to create the Wheland intermediate with a tetrahedral carbon atom and a cyclohexadienyl cation (or other conjugated cation from nonbenzenoid rings). As with the closely similar reaction between an alkene and a carbocation [see (Section 4.2.3) page 118], the first step is endothermic, and we can expect that the argument based on the product side of the reaction coordinate will be strong and satisfying, and so it is.

The concept of 'localisation energy' is used to account for the rates, and sites, of electrophilic substitution. It is a calculated value of the endothermicity in a reaction and is therefore part of the argument based on product development control. The plot of localisation energy against rate constant is a good straight line. Similarly, the preferred site of attack can usually be accounted for quite satisfactorily by looking at each of the possible cyclohexadienyl cation intermediates, and assessing which is the best stabilised—the better stabilised the intermediate, the faster the reaction leading to it. We shall see some of this in more detail later [see (Section 4.3.4.2) page 131]. Thus we have no need of further explanation. Nevertheless, arguments based on the starting material side of the reaction coordinate are worth looking at.

#### 4.3.4.1  Frontier Orbitals of Unsubstituted Aromatic Molecules.  The problem with using frontier orbital theory with aromatic compounds is that there are two high-energy molecular orbitals close in energy, and the HOMO itself is not of such overriding importance. In benzene itself, the HOMO is a degenerate pair $\psi_2$ and $\psi_3$ (Fig. 1.36), which, taken together with equal weight, make the frontier electron population on each atom equal. Fukui explicitly examined this problem, defining the *frontier electron population*, to include orbitals equal in energy to or just below the HOMO itself. He estimated the effective $\pi$-electron population ($f$) in aromatic rings using Equation 4.6:

$$f = 2\frac{c_3^2 + c_2^2 e^{-D\Delta\lambda}}{1 - e^{-D\Delta\lambda}} \qquad 4.6$$

where $c_3$ and $c_2$ are the coefficients in $\psi_3$ and $\psi_2$, respectively, $\Delta\lambda$ is the difference in energy between $\psi_3$ and $\psi_2$, and D is a constant (3 is used in fact) representing some kind of measure of the contribution of $\psi_2$ to the overall effect. As it ought, this expression gives the higher-energy orbital $\psi_3$ slightly greater weight.

The preferred site of electrophilic attack in the nitration of the aromatic molecules is indicated in Fig. 4.6 by an arrow. The numbers on the structures drawn in Fig. 4.6 are, for the first seven compounds, the coefficients of the HOMO of the molecule, or, for the remaining five, the frontier electron population ($f$) calculated using an equation like Equation 4.6 modified to deal with the presence of heteroatoms. Except for the anomalous pyrrocoline, the arrow comes from the largest (or larger) number. Thus, in all these cases, we have the situation (Fig. 3.1a) where the lower-energy product and the lower-energy approach to the transition structure are connected smoothly by what is evidently the lower-energy pathway. We can feel confident, in a situation like this, that we have a fairly good qualitative picture of the influences which bear on the transition structure.

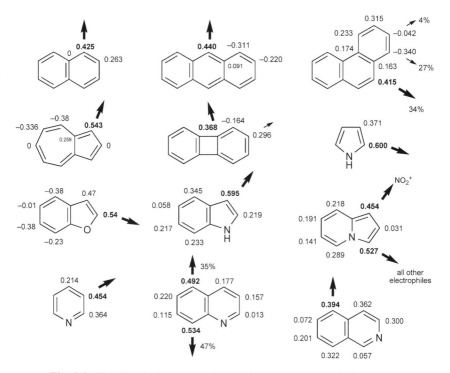

**Fig. 4.6**   Frontier electron populations and the sites (arrowed) of nitration

Fukui also suggested a parameter $s_r$ called the *superdelocalisability*, defined in Equation 4.7:

$$s_r = \sum_j^{occ.} \frac{c_j^2}{E_j}$$

4.7

in which $c_j$ is the coefficient at the atom r in the filled orbital j, and $E_j$ is the energy of that orbital. This expression bears an obvious relation to the third term of Equation 3.4. A plot of the rate constant for nitration at particular sites in a range of aromatic hydrocarbons against $s_r$ gave a good correlation over several powers of ten in rate constant.

### 4.3.4.2  Molecular Orbitals of the Intermediates in Electrophilic Attack on Monosubstituted Benzenes.

The standard explanation for the regioselectivity in aromatic electrophilic substitution of monosubstituted benzene rings, appropriately enough for an endothermic rate-determining step, assesses the relative stability of the possible cyclohexadienyl cations, and assumes that the transition structures leading to them will be in the same order. With X-substituted benzenes, substitution takes place in the *ortho* and *para* positions. As a simple model for the orbitals of X-substituted benzenes, we use the benzyl anion **4.47**. The intermediates produced by *ortho* and *para* attack **4.48** and **4.49** are lower in energy than the intermediate produced by *meta* attack **4.50**, because we can easily draw the intermediates with the charges cancelled, whereas it is not possible to cancel the charges with the *meta* intermediate **4.50**. However, we ought to be clear that this valence bond description is a superficial argument (which fortunately works).

Localised drawings (and curly arrows) work as well as they do because they *illustrate the electron distribution in the frontier orbital*, and for reaction kinetics it is the frontier orbital that is most revealing. However, in the present case, we are using a thermodynamic argument, for which we need to know the energy of *all* of the filled orbitals, and not just one of them. We assume that the σ framework is little affected by the substituent, and assess the π energies of the three possible intermediates **4.48–4.50**. They are shown for each of the three intermediates in Fig. 4.7. Note that the larger the β value the greater the π stabilisation and the lower the π energy. Although the calculations do not give good absolute values for the energies, they get the relative energies right, which is all we need be concerned with.

We can see in Fig. 4.7 that the main reason why the total π stabilisation of **4.50** is less ($2 \times 3.08\beta$) than that of **4.48** ($2 \times 3.50\beta$) and **4.49** ($2 \times 3.45\beta$) is that the highest filled orbital, $\psi_3$ in **4.50**, is not lowered in energy, whereas $\psi_3$ is lowered in the intermediates **4.48** and **4.49**. There are no π-bonding interactions between any

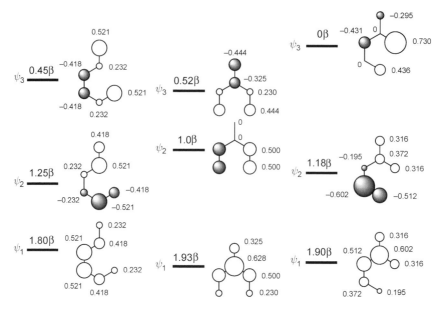

**Fig. 4.7** Coefficients and energies of the $\pi$ molecular orbitals of the intermediates in the electrophilic substitution of the benzyl anion at the *ortho*, *para* and *meta* positions

of the adjacent atoms in $\psi_3$—it is a nonbonding molecular orbital. This is, of course, the same point that the drawings were making, but we should be sure that the two lower-energy orbitals in **4.50** do not compensate for the high energy of $\psi_3$. In fact, they do, but not by much: we can see that $\psi_1$ of **4.50** is actually lower in energy (larger $|\beta|$) than $\psi_1$ of **4.48**, and $\psi_2$ of **4.50** is lower than $\psi_2$ of **4.49**. We now have a more thorough version of the original explanation for *ortho/para* substitution in X-substituted benzenes.

In a similar way, we can use the benzyl cation as a model for a benzene ring having a Z-substituent. Again we would have three possible intermediates, the $\pi$ systems of which are the same as the ones we have just been looking at, except that this time only $\psi_1$ and $\psi_2$ are filled in each case. We have already observed that the sum of the $|\beta|$ values for $\psi_1$ and $\psi_2$ is greater for the intermediate **4.50** which is the result of attack in the *meta* position. This, then, is the product-development argument for *meta* substitution in Z-substituted benzenes.

*4.3.4.3 Pyrrole, Furan and Thiophen.* Electrophilic substitution in these three heterocyclic rings takes place faster at the 2-position than at the 3-position. The standard explanation for attack at C-2 is based on the relative energies of the intermediates **4.51** and **4.52**. They both have conjugated systems of four p orbitals, with the heteroatom at the end in **4.51** and inside in **4.52**. It is not obvious why the former should be lower in energy than the

latter, although calculations agree that it is. Without the benefit of a calculation, one is reduced to saying that the conjugated system **4.51** is linearly
conjugated, with the lone pair on the nitrogen atom overlapping with the $\pi$
system of an allyl cation **4.51b**, whereas the conjugated system **4.52** has the
lone pair on the nitrogen atom overlapping with the $\pi$ bond and an isolated
cation **4.52b**. This essentially valence-bond description hardly amounts to a
satisfying explanation, although it does resemble the argument that a linearly
conjugated system (like **4.51**) is lower in energy than a cross-conjugated
system (like **4.52**).

The frontier orbital argument, weak as it is, is clear in predicting attack at C-2. The
$\pi$ molecular orbitals of pyrrole are representative of all three—they have been
shown earlier [see (Section 1.7.5) page 52], where $\psi_3$ is the HOMO **4.53** (with a
node running through the heteroatom, it is the same as $\psi_2$ of butadiene). An
estimate of the total $\pi$ charge **4.54** can be calculated from the sums of the squares
of the coefficients.

### 4.3.4.4 Pyridine N-oxide.

A special case of aromatic electrophilic substitu-
tion is provided by the ambient reactivity of pyridine $N$-oxide **4.55**. Klopman
used Equation 3.4 to calculate the relative reactivity ($\Delta E$ values) for electrophilic
attack at the 2-, 3- and 4-positions as it is influenced by the energy of the LUMO of
the electrophile. He obtained a graph (Fig. 4.8) which shows that each position in
turn can be the most nucleophilic. At high values of $E_r - E_s$ (hard electrophiles),
attack should take place at C-3; at lower values of $E_r - E_s$, it should take place at
C-4; and, with the softest electrophiles, it should take place at C-2. Attack at each
of these sites is known: the hardest electrophile $SO_3$ does attack the 3-position, the
next hardest ($NO_2^+$) the 4-position, and the softest ($HgOAc^+$) the 2-position.

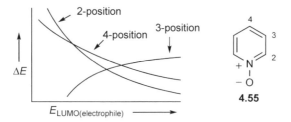

**Fig. 4.8**   Electrophilic substitution of pyridine $N$-oxide

However, this reaction is really more complicated. The sulfonation almost certainly takes place on the $O$-protonated oxide, and this must affect the relative reactivity of the 2-, 3- and 4-positions. The value of the exercise is in the way in which it shows how a single nucleophile, can, in principle, be attacked at different sites, depending upon the energy of the LUMO of the electrophile.

## 4.4   Electrophilicity

### 4.4.1   Trigonal Electrophiles

Trigonal carbon electrophiles react with nucleophiles with the formation of a tetrahedral product or intermediate, and this step, often rate-determining, may be followed by the expulsion of a nucleofugal group, if there is one. Carbonyl groups, the most important trigonal electrophiles, are electrophilic in the order: acid chlorides > aldehydes > ketones > esters > amides > carboxylate ions, an order which is easily explained. Setting aside the acid chloride for the moment, the $\pi$ energy of the conjugated system in the starting materials is lowered by the substituent X, with the oxyanion ($X = O^-$) most effective, a methyl group the least effective, and a hydrogen atom making no contribution at all. This allows us to rank the $\pi$ energy of the starting materials in order on the left in Fig. 4.9. In the

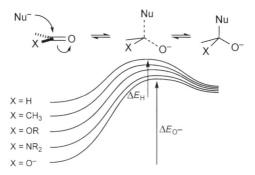

**Fig. 4.9**   Relative energies of starting materials and transition structures for nucleophilic attack on carbonyl groups

tetrahedral intermediate, this overlap is removed or rather it is reduced to an anomeric effect which, being in the $\sigma$ system, is less. This means that the energies of the intermediates, although probably in the same order, are closer together. The activation energy for disrupting the $\pi$ stabilisation is therefore least for the aldehyde $\Delta E_H$ and greatest for the carboxylate ion $\Delta E_{O^-}$.

It is also possible to account for the order of electrophilicity using only the energy of $\pi^*_{CO}$, which is the LUMO for each of the carbonyl compounds. Using the arguments developed in Chapter 2, C-substituents lower the energy of the LUMO a little, X-substituents raise the energy of the LUMO, and Z-substituents substantially lower the energy of the LUMO.

Thus an $\alpha$-diketone is unmistakably more electrophilic than an ordinary ketone—the energy of the starting material is raised by the conjugation of the two carbonyl groups [see (Section 2.1.6) page 69], and the energy of the LUMO is lowered, since the extra carbonyl group is a Z-substituent. A simple conjugated ketone, however, has the LUMO energy lowered by conjugation, but so is the energy of the starting material lowered, and the two effects work in opposition. In practice ketones like acetophenone are usually less reactive towards nucleophiles than simple ketones like acetone. Acid chlorides are more electrophilic than aldehydes, even though there is stabilising conjugation between the lone pair on the chlorine atom and the carbonyl $\pi^*$ orbital. There are three contributions to this anomaly: (i) the $\pi$ stabilisation is small, because chlorine is both one row lower in the periodic table and more electronegative, and consequently the energy match between the $p_{Cl}$ orbital and the $\pi^*_{CO}$ is poor; (ii) the $\pi$ stabilisation is offset by strong inductive electron withdrawal along the C—Cl bond, raising the electrophilicity of the carbonyl group; and (iii) the anomeric stabilisation [see (Section 2.2.3.3) page 78] in the tetrahedral intermediate is greater—conjugation of the oxyanion with the C—Cl $\sigma^*$ orbital pulls down the energy of the transition structure. Although these effects are also present with esters, they do not override the conjugation of the lone pair on the oxygen stabilising the starting material, but there is a more delicate balance with acid anhydrides, which are similar in electrophilicity to aldehydes.

In one famous case, stabilisation in the intermediate plays a larger role in determining electrophilicity than the differences in the energy of the starting materials. In contrast to the order of electrophilicity in alkyl halides in $S_N1$, $S_N2$, E1 and E2 reactions ($I^- > Br^- > Cl^- >> F^-$), aromatic halides are electrophilic in the opposite order. 2,4-Dinitrofluorobenzene (**4.56**, X = F) reacts 600 times faster with methoxide ion than 2,4-dinitrochlorobenzene, and 3100 times faster than 2,4-dinitroiodobenzene. The explanation is that the anomeric effect between the methoxy oxygen and the C—Halogen bond in the intermediate **4.58** combines with stabilisation from negative hyperconjugation between the C—Halogen $\sigma^*$ orbital and the cyclohexadienyl anion $\pi$ system. Both interactions lower the energy more for the C—F bond than for the other halogens, whereas the overlap of the lone pair on the halogens with the $\pi$ system of the starting material **4.56** is relatively small. Thus the energy picture is that seen in Fig. 4.10. Since the reaction is endothermic, the structure of the intermediate **4.58** significantly affects the energy of the transition structure **4.57**, and the activation energy for the

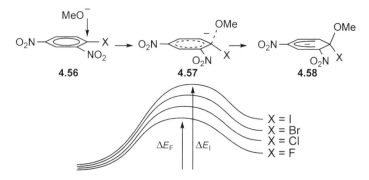

**Fig. 4.10** Relative energies of transition structures and tetrahedral intermediates for nucleophilic attack on aryl halides

fluoride $\Delta E_F$ is less than that for the iodide $\Delta E_I$. The normal order is of course restored when the second step becomes rate-determining, as it does with highly activated systems.

### 4.4.2 Tetrahedral Electrophiles

Some of the same features affect the electrophilicity of tetrahedral electrophiles undergoing $S_N2$ reactions. Thus some donor substituents on the carbon being attacked reduce the electrophilicity of alkyl halides, where the order of reactivity is methyl > ethyl > isopropyl > *tert*-butyl. The transition structure of an $S_N2$ reaction is inherently electron rich, and donor substituents slow it down. Equally, steric hindrance to attack, the traditional explanation for this reactivity order, is greater for the more substituted alkyl halides.

In contrast, Z- and C-substituents conjugated to the site of attack increase the rates of $S_N2$ reactions. The Z-substituent may partly operate by lowering the LUMO energy of the starting material, but most probably both the C- and Z-substituents are primarily operating to lower the energy of the transition structures **4.59** and **4.60**, respectively. The forming and breaking bonds have four electrons in total and any delocalisation of these in allylic overlap lowers the energy of the transition structure. This electronic explanation is dramatically supported by the difference in the rate of nucleophilic attack on the benzyl sulfonium salts **4.61** and **4.62**, which react at relative rates of 8000:1. Whereas the open-chain system **4.61** can easily allow the forming and breaking bonds to overlap with the $\pi$ system of the benzene ring, the cyclic sulfonium salt **4.62** cannot.

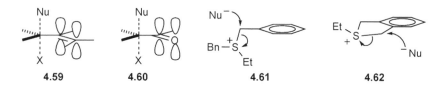

X-Substituents easily accelerate $S_N1$ reactions by stabilising the carbocation, but they can also sometimes accelerate $S_N2$ reactions, even though the transition structure seems to have an excess of electrons at the site of substitution. The transition structures can adjust to take energetic advantage from more or less stretching of the C—X bond, with the extremes being the perfect $S_N2$ and perfect $S_N1$ reactions. Any feature that energetically helps the stretching of the C—X bond, such as an X-substituent on the carbon undergoing attack, will lower the energy of that transition structure. Thus we have the unusual feature that both X- and Z-substituents can accelerate reactions that are formally $S_N2$.

### 4.4.3  Hard and Soft Electrophiles

Electrophiles can be hard or soft—the hardest electrophiles are small, charged and have a relatively high-energy LUMO, and soft electrophiles are large, have little charge and have a conspicuously low-energy LUMO. The proton, because of its size and charge is very hard. Tables of acidity ($pK_a$ values) give a rank order of thermodynamic electrophilicity of protons attached to various ligands, but the only extensive tables related to electrophilicity for other electrophiles are the lists of hardness in Tables 3.3–3.5, where we see that large metal cations like $Hg^{2+}$ are relatively soft, in spite of their charge. $\pi$-Bonded species like C=C double bonds conjugated to Z-substituents are inherently soft, with low-energy LUMOs, as are other uncharged reagents like methyl iodide, sulfenyl halides and iodine. It is not therefore possible to compare in any absolute sense the electrophilicity of a soft electrophile like iodine and a hard one like an acid chloride.

There is a special problem in comparing the hardness and softness of alkyl halides and carbonyl compounds, the two most common carbon electrophiles. The LUMO in an $S_N2$ reaction is a $\sigma^*$ C—X antibonding orbital, with a LUMO higher in energy than the LUMO of a C=O $\pi$ bond, yet alkyl halides are classed as softer than carbonyl electrophiles. One big difference is that $S_N2$ reactions are exothermic (otherwise they would not take place) and the transition structure is early along the reaction coordinate, resembling the starting materials more than the product, whereas addition to a carbonyl group is endothermic, with a transition structure more like that of the tetrahedral intermediate. Arguments about the importance or otherwise of orbital interactions, and of hardness and softness, apply more effectively to $S_N2$ reactions than to carbonyl additions.

## 4.5  Ambident Electrophiles

The attack of a nucleophile on a conjugated system is susceptible to the same kind of analysis that we gave to the attack of an electrophile on a conjugated system. In most cases, all the molecular orbital factors, both those affecting the product stability and those in the starting materials, point in the same direction. We use the

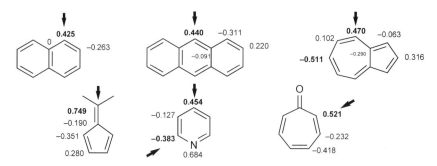

**Fig. 4.11**   LUMOs of some carbon electrophiles, and the sites of nucleophilic attack upon them

LUMO of the conjugated system (and the HOMO of the nucleophile, of course) as the important frontier orbitals, as in Fig. 4.11, which shows electrophilic reactivity at the site where the arrow points for a range of carbon electrophiles. In each case, there is a high coefficient of the LUMO at the site of attack; each of them also has a high total electron deficiency at this site; and, with the possible exception of pyridine, the tetrahedral intermediate obtained from such attack is lower in energy than attack at the alternative sites.

### 4.5.1   Aromatic Electrophiles

***4.5.1.1   The Pyridinium Cation.***   The pyridinium cation **4.63** is even more readily attacked by nucleophiles at C-2 and C-4 than pyridine. Looking at the product side of the reaction coordinate, the linearly conjugated intermediate **4.64**, which also benefits from an anomeric effect if the nucleophile is an electronegative heteroatom, is lower in energy than the cross-conjugated intermediate **4.66**, and will therefore be the product of thermodynamic control. Since this step is neither strongly endothermic nor always reversible, the orbitals and the charge distribution in the starting material may be important when the reaction is kinetically controlled. The *total* π electron deficiency **4.63** at C-2 of +0.241 and at C-4 of +0.165 indicate that charge control (in other words with hard nucleophiles) will lead to reaction at C-2. This is the case with such relatively hard nucleophiles as the hydroxide ion, amide ion, borohydride ion and Grignard reagents.

However, if we look at the LUMO, we find that it has the form **4.65**, namely that of $\psi_4^*$ of benzene, but polarised by the nitrogen atom. This polarisation has reduced the coefficient at C-3, and the coefficient at C-4 is larger than that at C-2, as can be seen from the simple Hückel calculation for pyridine itself in Fig. 4.11, which gives LUMO coefficients of 0.454 and −0.383, respectively, and an energy of 0.56β (compare benzene with 1β for this orbital). Thus, soft nucleophiles should attack at C-4, where the frontier orbital term is largest. Again this is the case: cyanide ion, dithionite, enolate ions, and hydride delivered from the carbon atom of the Hantzsch ester **4.67** react faster at this site.

### 4.5.1.2 ortho- *and* para-*Halogenonitrobenzenes.*

*ortho-* and *para*-Halogenonitrobenzenes are readily attacked by nucleophiles, as we saw in Fig. 4.10. The first step is usually rate-determining. Product development control should therefore have *ortho* attack faster than *para* attack, because the intermediate **4.68** with the linear conjugated system will be lower in energy, other things being equal, than the intermediate **4.69** with the cross-conjugated system.

**4.68**

**4.69**

The Coulombic term will also lead to faster reaction at the *ortho* than at the *para* position. The frontier orbital term, however, should favour attack at the *para* position. Thus the ESR spectrum of the benzyl radical, which has the odd electron in an orbital which ought to be a model for the LUMO of a Z-substituted benzene, shows that there is a larger coefficient in the *para* position than in the *ortho*.

There is some evidence which supports this analysis. With a *charged* activating group, as in the diazonium cations **4.70** and **4.71**, attack at the *ortho* position is faster than attack at the *para* position, because of the large Coulombic contribution. With the *uncharged* activating groups in the compounds **4.72** and **4.73**, the order is the other way round, and when the nucleophile is uncharged, the preference for *para* attack is even greater.

### 4.5.2 Aliphatic Electrophiles

#### 4.5.2.1 α,β-Unsaturated Carbonyl Compounds.

Most nucleophiles attack α,β-unsaturated ketones faster at the carbon atom of the carbonyl group (e.g. **4.75** → **4.74**) than at the β position. Attack at the β carbon (e.g. **4.75** → **4.76**) is commonly the result of a slower, *but thermodynamically more favourable*, reaction. For this mode of reaction, the first step must be reversible. Conjugate attack is therefore most straightforward when the nucleophile is a well-stabilised anion, making the first step easily reversible, as it is with cyanide ion.

Taking acrolein as an example of an α,β-unsaturated carbonyl compound, a better calculation than that used in Fig. 2.2 gives the frontier orbital coefficients and energies in Fig. 4.12, which shows that the coefficent in the LUMO is larger at the β position than at the carbonyl group. We can therefore expect that, if any nucleophile is going to attack directly at the β-carbon atom, it will be a soft nucleophile, responsive to the frontier orbital term. This simple analysis leaves out of consideration the fact that many additions to α,β-unsaturated carbonyl compounds need or take advantage of coordination to the oxygen atom by a metal cation or a proton, or even just a hydrogen bond. This is especially true for hydride or carbon nucleophiles. The orbitals of the reactive species are therefore more like those of protonated acrolein, for which the LUMO has the larger coefficient on the carbonyl carbon, not the β position (Fig. 4.12). Thus even soft nucleophiles can be expected to attack directly at the carbonyl group when Lewis or protic acid catalysis is involved. The effect of the Lewis acid on regioselectivity is seen with lithium aluminium hydride reacting with cyclohexenone—in ether, the ratio of direct to conjugate attack is 98:2, but if the lithium ion is sequestered by a cryptand, the selectivity changes to 23:77.

Moving to α,β-unsaturated esters, hydroxide ion and alkoxide ion (hard nucleophiles) react with ethyl acrylate **4.77** by direct attack at the carbonyl group to give ester hydrolysis and exchange, respectively, whereas the β-dicarbonyl enolate ion **4.78** (a soft nucleophile) undergoes a Michael reaction. There is no certainty in this latter reaction that the attack of the enolate anion on the carbonyl group is not a more rapid (and reversible) process.

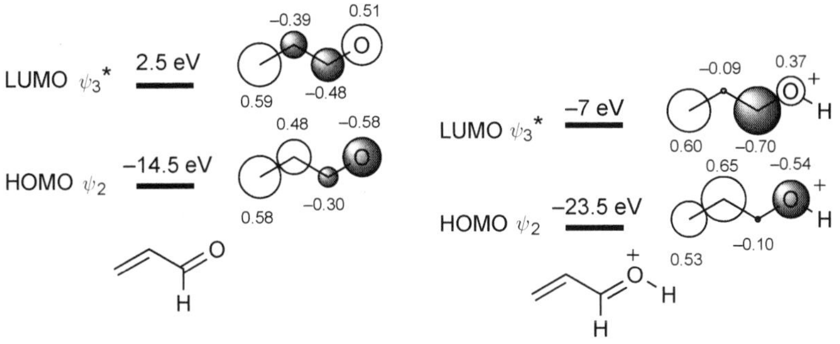

**Fig. 4.12** Frontier orbital energies and coefficients for acrolein and protonated acrolein

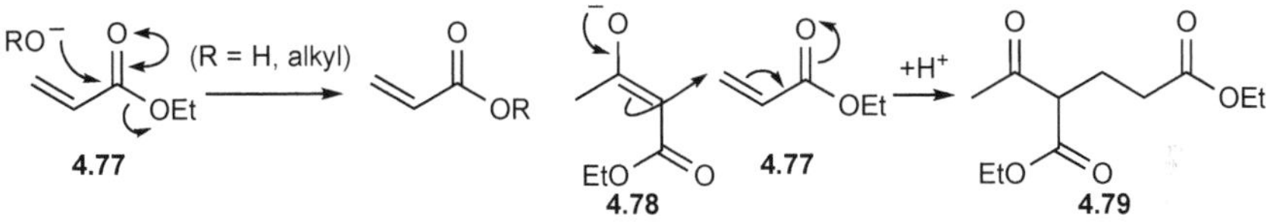

One case, however, is clear—ammonia and amines react with ordinary esters to give amides, and it is known that the attack at the carbonyl group is rate-determining and effectively irreversible above pH 7. Ammonia (neutral and therefore a relatively soft nucleophile) reacts in methanol with methyl acrylate **4.80** kinetically at the $\beta$ position to give the primary amine **4.81**, and reaction continues in the same sense to give successively the secondary amine **4.82** and tertiary amine **4.83**.

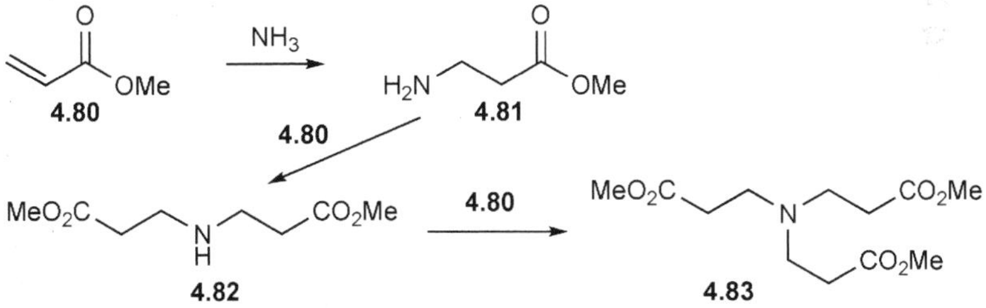

The more a carbonyl group is like that of protonated acrolein (Fig. 4.12), the more likely it is that all nucleophiles will attack directly at the carbonyl carbon atom. In agreement with this perception, and in contrast to its behaviour with methyl acrylate, ammonia reacts with acryloyl chloride at the carbonyl carbon atom to give acrylamide.

*4.5.2.2 Allyl Halides.* In bimolecular substitutions on allyl halides, direct displacement of the halide ion ($S_N2$) almost always occurs, and conjugate attack ($S_N2'$) is rare. It is perhaps significant that the few examples of conjugate reaction which have been observed are with strikingly soft nucleophiles such as phenylthi-oxide ion, cyanide ion, azide ion, and secondary amines, all in nonpolar solvents, and preferably when the allyl halide possesses some steric hindrance at the $\alpha$ position.

When the allylic system carries X-substituents, and the solvent is polar, the reaction may take a unimolecular path, and the reactions are then $S_N1$ and $S_N1'$. The regioselectivity will be determined by thermodynamic fac-tors if the only available nucleophile is a good nucleofugal group, with a product like prenyl chloride **4.85** having the more substituted double bond usually favoured. A corollary is that the thermodynamically less stable isomer **4.84** is the more reactive (by a factor of about 3 in ethanol at 25 °C in this case). However, if the reaction is not under thermodynamic control, the regioselectivity will be determined by the coefficients and charges at the $\alpha$ and $\gamma$ carbon atoms of the allyl cation. We can treat the X-substituted allyl cation **4.86** as resembling a protonated $\alpha,\beta$-unsaturated carbonyl compound. The donor substituent in protonated acrolein (on the right in Fig. 4.12) is a hydroxyl group, and it leads to a higher coefficient on the carbonyl carbon (equivalent to the $\alpha$ position in the allyl system). An allyl cation having a donor less effective than a hydroxyl ought there-fore to have a larger coefficient in the LUMO at the carbon atom adjacent to it than at the other end. The methyl groups in the prenyl cation **4.86** are not powerful donors, and yet, when either of the chlorides **4.84** or **4.85** is solvolysed in water, under conditions that do not equilibrate the products, the major product (85:15) is the tertiary alcohol **4.87**, showing that capture at the more sterically hindered site in the cation **4.86** is indeed faster than attack at the primary position giving the alcohol **4.88**.

### 4.5.2.3 Unsymmetrical Anhydrides.

*4.5.2.3 Unsymmetrical Anhydrides.* Nucleophilic attack on the carbonyl groups of unsymmetrically substituted maleic anhydrides **4.89** show some curious selectivities, with lithium aluminium hydride completely selective for attack at $C_\alpha$ when R is a methoxy group, and still quite high (88:12) when R is a methyl group. Both methoxy and methyl are X-substituents conjugated to the $\beta$ carbonyl, and ought to reduce the coefficient in the LUMO of the atomic orbital at this site. This leaves the $\alpha$ site the more electrophilic by default, in spite of the greater steric hindrance at this site. Calculations support this argument with larger coefficients at $C_\alpha$ in the LUMO for methoxy and methyl, and with the difference between them reduced for the less effective X-substituent.

LUMO coefficients

|       | $c_\alpha$ | $c_\beta$ |
|-------|------|-------|
| R=MeO | 0.39 | −0.34 |
| R=Me  | 0.36 | −0.35 |
| R=Ph  | 0.22 | −0.25 |

**4.89**

The story is even more remarkable for unsymmetrical succinic anhydrides **4.90**, where hydride reduction takes place with surprising selectivity at the obviously more hindered carbonyl group, giving the lactones **4.91** and **4.92** in a ratio of 95:5. The electronic difference between the two carbonyl groups is that $C_\alpha$ has a hyperconjugative interaction with a $CMe_2$ group and $C_\beta$ with a $CH_2$ group. Which has the larger hyperconjugative effect is an ongoing debate—the hydrogen atom is more electropositive than carbon, and so the bond is more polarised, and better able to stabilise electron deficiency. However, the methyl group has a greater stock of electrons that can participate in delocalisation. A calculation gave $C_\alpha$ the larger coefficient in the LUMO, implying that C—H is the better at hyperconjugation, and agreeing with the experimental result. With Grignard reagents, when steric effects are more important, attack takes place unexceptionably at $C_\beta$, and when the two substituents are chlorine atoms, effectively Z-substituents by negative hyperconjugation, attack is completely selective for $C_\alpha$.

LUMO coefficients

| $c_\alpha$ | $c_\beta$ |
|-------|--------|
| 0.599 | −0.591 |

**4.90**    LiAlH₄ or NaBH₄    **4.91**   95:5   **4.92**

*4.5.2.4  Arynes.*  The two p orbitals of benzyne are bent apart, making their interaction considerably less than the interaction of the two p orbitals forming the π bond of ethylene. The HOMO is therefore raised and the LUMO lowered in energy relative to the frontier orbitals of an alkene or a linear alkyne. Whereas nucleophilic attack on a benzyne is made favourable by the low energy of the LUMO, electrophilic attack, which ought also to be favourable, because of the raised energy of the HOMO, is not normally observed. The explanation can be found on the product side of the reaction coordinate. The product of nucleophilic attack is a phenyl anion, and the product of electrophilic attack a phenyl cation. The former is well known—trigonal anions, having a more exposed nucleus, are stabilised relative to tetrahedral anions, and they are under no particular strain from being bent. This is one of the reasons why acetylenes in general react more readily than alkenes with nucleophiles. However, trigonal cations like the phenyl cation are high in energy, and the bending in them raises their energy even more. The high energy of a diagonal or trigonal cation is probably the main reason why acetylenes in general are less reactive than alkenes towards electrophiles.

To be ambident, a benzyne must be unsymmetrical, and the regioselectivity will be determined by the electronic and steric effects of the substituent. A major factor is the relative stability of the regioisomeric products, with the benzyne **4.93** giving the lithium intermediate **4.94**, and the benzyne **4.95** giving the lithium intermediate **4.96**. These two substituents are both excellent at stabilising the neighbouring C—Li bond, the former by coordination, and the latter by conjugation between the

C—F bond and the C—Li bond. On the starting material side of the reaction coordinate, which ought to be important, since it is an exothermic reaction, the C—F bond is a Z-substituent on the benzyne triple bond. Following the device we used earlier (see pages 60–65) to deduce the pattern of coefficients in substituted alkenes, we can argue that the C—F bond has some of the character of a cation on carbon **4.97**, in which the empty p orbital will be conjugated to the in-plane bent $\pi$ bond. The LUMO will resemble that of an allyl cation, and will therefore have the larger coefficient on C-3 **4.98**.

With an alkyl group as the substituent, the product anions are not substantially different, nor are the coefficients on C-2 and C-3, and they suffer attack by amide ion equally at C-2 and C-3. With an oxyanion or amide anion substituent, attack at C-2 becomes quite substantial, perhaps avoiding the formation of a C—Li bond adjacent to the anion.

　　Pyridynes are inherently unsymmetrical. Nucleophiles readily attack a 2,3-pyridyne **4.100** entirely at C-2. A calculation shows that the coefficient in the LUMO of 2,3-pyridyne **4.101** is larger at C-2 than at C-3. Also, the total charge distribution **4.102** is such that C-2 bears a partial positive charge. We can rationalise the polarisation of the LUMO by comparing the lobes of the p orbitals in the plane of the ring **4.101** with the $\pi$ system of the allyl anion. The large coefficient in the LUMO of the allyl anion is on the central atom, just as it is here. The net result is that nucleophiles attack at C-2, because both Coulombic and frontier orbital forces favour attack at this site. They also react at the 2-position because the anion formed, a 3-pyridyl anion, stabilised by negative hyperconjugation with the C—N bond, is more stable than the alternative 2-pyridyl anion.

total charge distribution
**4.102**

In contrast, nucleophiles attack 3,4-pyridyne at C-4 only slightly faster than at C-3. 3,4-Pyridyne is less polarised, because the p orbital on nitrogen is not conjugated with the p orbitals on C-3 and C-4.

In spite of their high total energy, arynes in general are selective towards different nucleophiles; thus benzyne selectively captures the anion of acetonitrile in the presence of an excess of the dimethylamide ion used to generate both it and the benzyne. Nucleophilicity towards benzyne, determined by competition experiments, is in the order $RLi \sim RS^- > R_2N^- \sim RO^-$ and $I^- > Br^- > Cl^-$, which is an order of softness. The low energy of the LUMO of benzyne, coupled with its being uncharged, makes it a soft electrophile.

#### 4.5.2.5 *Substitution versus Elimination.*   Alkyl halides react with nucleophiles by undergoing substitution or elimination, which are in competition with each other. The usual pattern is for the more substituted alkyl halides to undergo elimination more easily than substitution, and for the less substituted to undergo substitution more easily than elimination. A major factor in determining this pattern is the greater level of steric hindrance at the carbon atom of the more substituted alkyl halides, while at the same time the hydrogen atoms remain inherently unhindered on the periphery (and there are usually more of them). Other factors favouring elimination are the relief of steric compression as tetrahedral carbons become trigonal, and the lower energy of the more substituted alkenes.

A more subtle factor affecting the ratio of substitution to elimination is the nature of the leaving group, and this is amenable to a treatment based on the molecular orbitals involved. The LUMO is the important frontier orbital for both $S_N2$ and E2 reactions. We have already seen that this is largely localised as $\sigma^*$ for the C—Cl bond in methyl chloride (Fig. 1.46a). In a substrate for elimination like ethyl chloride, the LUMO is not localised on $\sigma^*$ for the C—X bond, where X is the electronegative group. We can try to deduce what the LUMO will look like from the interaction of the orbitals of a methyl group and the orbitals of a CH$_2$X group. The orbitals of the methyl fragment, constructed from their component $C_{2s}$, $C_{2p}$ and $H_{1s}$ orbitals, mixed in appropriate proportions, make up the set in Fig. 4.13a.

We need to consider the antibonding orbitals, of which $\sigma^*_3$ and $\pi^*_z$ have appropriate symmetry to mix with the relevant orbitals of an XCH$_2$ group. The

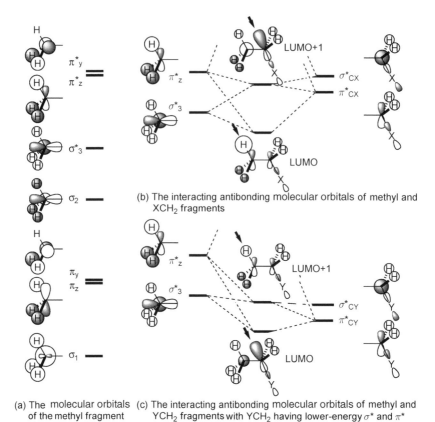

(a) The molecular orbitals of the methyl fragment

(b) The interacting antibonding molecular orbitals of methyl and XCH$_2$ fragments

(c) The interacting antibonding molecular orbitals of methyl and YCH$_2$ fragments with YCH$_2$ having lower-energy $\sigma^*$ and $\pi^*$ orbitals than XCH$_2$

**Fig. 4.13**   The LUMO of EtX and EtY where Y is more electronegative than X

$\sigma^*_3$ orbital is largely a mix of the 2s and 2p$_x$ orbitals, and the $\pi^*_z$ orbital is purely a 2p$_z$ orbital, both being mixed with the 1s orbitals on hydrogen. When these are to interact with the orbitals of an XCH$_2$ group, they mix with each other to some extent, because the symmetry has been broken. The $\sigma^*_3$ orbital acquires some 2p$_z$ character and the $\pi^*_z$ orbital acquires some 2s character. Because they both have a 2p$_z$ component, these two orbitals can mix with the $\pi^*$ and $\sigma^*$ orbitals of the XCH$_2$ group to create two orbitals labelled LUMO and LUMO+1 in Fig. 4.13b, together with higher-energy orbitals that we need not consider. In this case, the LUMO is closer in energy to the $\sigma^*_3$ orbital, and so has more H—C antibonding character than C—X antibonding character. The two lower gauche hydrogens have opposite signs in $\sigma^*_3$ and $\pi^*_z$ and nearly cancel, but the hydrogen atom anti-periplanar to the C—X bond has the same sign and is amplified. In addition, $\pi$ bonding is already present, and elimination is therefore favoured by attack where the bold arrow approaches. The LUMO+1 orbital, however, is closer in energy to the $\pi^*_{CX}$ orbital, and it has much more C—X antibonding character. Because it has also mixed with the $\sigma^*_{CX}$ orbital, which has a large 2s component, the upper lobe has been extended, and the lower reduced, making attack behind the C—X bond, where the bold arrow points, favourable. This argument suggests that, in the gas phase, and other things being equal, elimination is favoured in this substrate, because the LUMO is the lower energy of these two orbitals.

Now let us take a different substrate with a better leaving group Y, for which the energies of the $\pi^*_{CY}$ and $\sigma^*_{CY}$ orbitals are lower. A different picture emerges, in which the LUMO and the LUMO+1 orbitals more or less change places. In Fig. 4.13c, we see that the LUMO is now closer in energy to the $\pi^*_{CY}$ orbital and has therefore more C—Y antibonding character, with the large lobe the site of attack for substitution. The LUMO+1 orbital is now the one closer in energy to the $\sigma^*_3$ and $\pi^*_z$ orbitals, and it has more of the character suitable for elimination. Thus with lower energy C—Y antibonding orbitals, substitution should be favoured, since the orbital pattern in the LUMO favours it. This picture allows us to see how the nature of the leaving group can affect whether substitution or elimination will be favoured. In practice, the more electronegative the leaving group the higher the $S_N2$:E2 ratio (ROTs > RCl > RBr > RI > $RN^+Me_3$), in agreement with the analysis in Fig. 4.13, since the more electronegative the atom Y, the lower the energy of its antibonding orbitals.

Superimposed on this pattern is the effect of changing the nucleophile, which is called a base if it is removing a proton in an elimination reaction. Hindered bases will inherently attack the more exposed hydrogen atoms, encouraging elimination. The hyperconjugation between the anti-periplanar C—H and C—Cl bonds that is manifest in the LUMO of ethyl chloride also removes charge from the hydrogen atom, which, because it is so small, will have a relatively concentrated partial positive charge. Hard nucleophiles, therefore, are more likely to induce an E2 reaction than an $S_N2$ substitution, and soft nucleophiles to attack at carbon. This is the usual observation: the harder the nucleophile/base, the more elimination there is relative to substitution.

## 4.6   Carbenes

Carbenes are ambiphilic, having simultaneously both nucleophilic and electrophilic properties. The HOMO is largely a filled p orbital (labelled $\sigma_z$ in Fig. 1.14, but $\sigma_x$ in the drawings below) involved in some of the C—H bonding, but relatively high in energy, because of its closeness in energy to an isolated p orbital. The LUMO is an unfilled purely p orbital (labelled $2p_y$ in Fig. 1.14, but $p_z$ in the drawings below), which is therefore nonbonding. Thus the HOMO is high in energy, and the LUMO is low in energy, and, not surprisingly, carbenes are very reactive.

Substituents have a profound effect on the reactivity of carbenes. Donor substituents lower the energy if they are conjugated to the empty p orbital **4.103**, and electron-withdrawing substituents lower the energy if they are conjugated to the filled $p_x$ orbital **4.104**. Since these interactions leave the other frontier orbital more or less unchanged (it is orthogonal), the former still has a high-energy HOMO, and the latter still has a low-energy LUMO. They become, therefore, relatively nucleophilic and electrophilic, respectively.

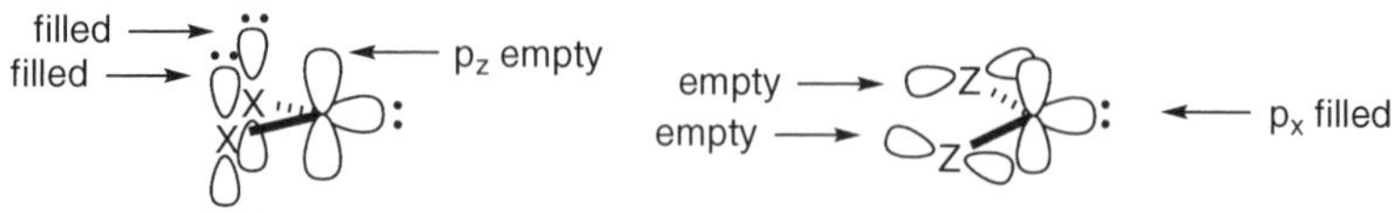

### 4.6.1 Nucleophilic Carbenes

In practice, donor substituents make it possible actually to isolate a range of carbenes **4.105**. With somewhat less stabilisation, the carbene **4.106**, although it is only found as a reactive intermediate, is exceptionally easy to form. It is the key intermediate in all the metabolic steps catalysed by thiamine coenzymes, and its reactions are characterised by its nucleophilicity. Similarly, dimethoxycarbene **4.107** reacts as a nucleophile with electrophiles like dimethyl maleate to give the intermediate **4.108**, and hence the cyclopropane **4.109**, but it does not insert into unactivated alkenes.

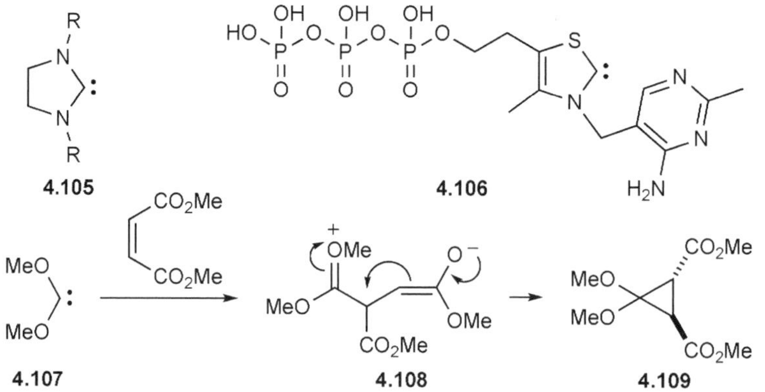

The well known cycloaddition of a carbene to an alkene, to be discussed again in the next chapter, can be viewed as the simultaneous interaction of the HOMO of the alkene with the LUMO of the carbene, and of the LUMO of the alkene and the HOMO of the carbene. Nucleophilic carbenes with a high-energy HOMO, interact strongly with molecules having a low-energy LUMO (Fig. 4.14a), which is why they react well with electrophilic alkenes. In the case of the very nucleophilic dimethoxycarbene, bond formation is so entirely dominated by the HOMO$_{(carbene)}$-LUMO$_{(alkene)}$ interaction that it gives the

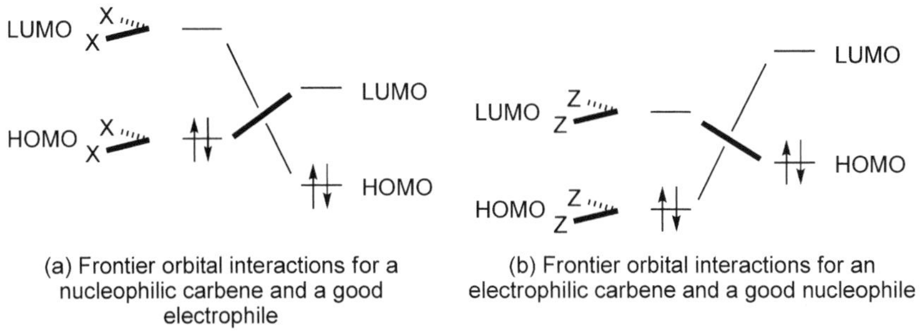

(a) Frontier orbital interactions for a
nucleophilic carbene and a good
electrophile

(b) Frontier orbital interactions for an
electrophilic carbene and a good nucleophile

**Fig. 4.14** Frontier orbital interactions for carbenes with electrophilic and nucleophilic reagents

zwitterionic intermediate **4.108** rather than undergoing a concerted cycloaddition.

### 4.6.2  Electrophilic Carbenes

Electrophilic carbenes, however, like the bis(methoxycarbonyl)carbene **4.110**, have a low-energy LUMO, and react with molecules like alkenes that have a high-energy HOMO (Fig. 4.14b), and cyclo-add stereospecifically. They also insert into C–H bonds, with high selectivity for tertiary C–H bonds, which argues for a substantial degree of cationic charge on the carbon in the transition structure, characteristic of electrophilic attack on the hydrogen atom. Dihalocarbenes are electrophilic in character, inserting easily and stereospecifically syn into C=C bonds, but not usually reacting with C–H bonds.

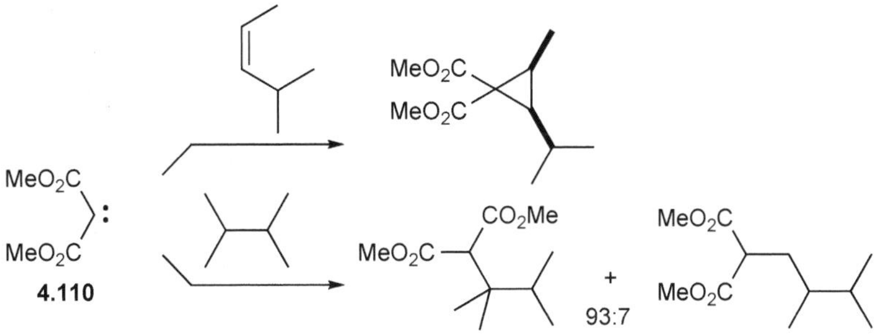

### 4.6.3  Aromatic Carbenes

The carbenes cyclopropenylidene **4.111** and cycloheptatrienylidene **4.112** have the empty p orbital conjugated with one and three $\pi$ bonds, respectively, making them aromatic like the cyclopropenyl **1.10** and tropylium cations **1.9**. The filled $p_x$

orbital is unchanged as a source of nucleophilicity, and these carbenes are notably nucleophilic, reacting with electrophilic alkenes like fumarate, but not with simple alkenes.

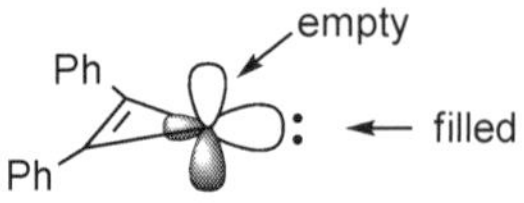

**4.111** 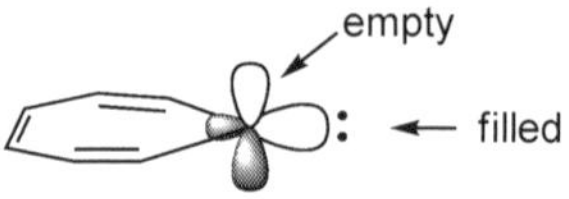     **4.112**

Cyclopentadienylidene is not so straightforward. It might change the normal configuration for a carbene **4.113** to that shown as **4.114** in order to allow the filled $p_z$ orbital to be conjugated with the two $\pi$ bonds like the cyclopentadienyl anion **1.8**, and the unfilled $p_x$ orbital would have to take up the orthogonal role. However, this is not without an energetic penalty, since it keeps the areas of high electron population close together on the left-hand side of **4.114**. This carbene is not notably electrophilic in its reactions with alkenes, but it is somewhat electrophilic, reacting with dimethyl sulfide, for example, to give the ylid **4.115**.

**4.113**                    **4.114** 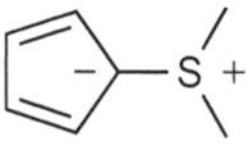                    **4.115**

The superficially similar carbene **4.116**, another carbene stable enough to be isolated, has the best of all worlds. With six electrons for the $\pi$ system coming from the double bond and the two nitrogen lone pairs, it has an aromatic sextet without having to fill the $p_z$ orbital. The $p_z$ orbital is empty and the $p_x$ orbital is filled, making this a nucleophilic carbene, which reacts with the electrophile carbon disulfide to give the zwitterion **4.117**.

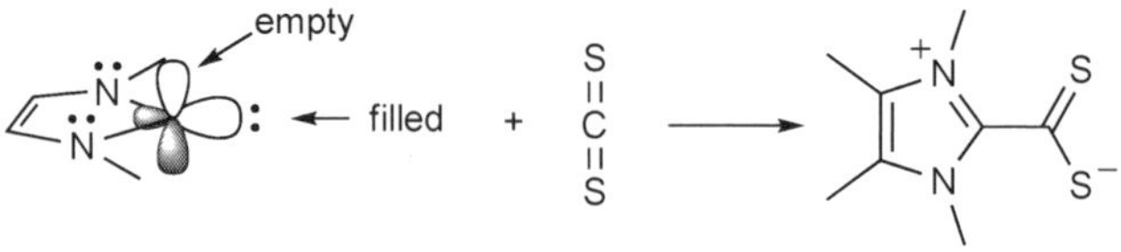

**4.116**                                        **4.117**

The account given so far seems to leave no room for anomalies, and yet they abound. Some nucleophilic carbenes do not react with some of the common electrophilic probes, and some electrophilic carbenes do not react with some of the nucleophilic probes. Furthermore, there is frequently only a poor correlation between the frontier orbital energies and the patterns of reactivity. The usual qualifications have to be invoked—that the frontier orbital theory is not a complete account of all the forces at work. One of the more obvious of the other forces is steric hindrance, of course, and another is that some carbenes are unselective, because they are so reactive that they are diffusion controlled.

## 4.7 Exercises

1. Account for the following pattern:

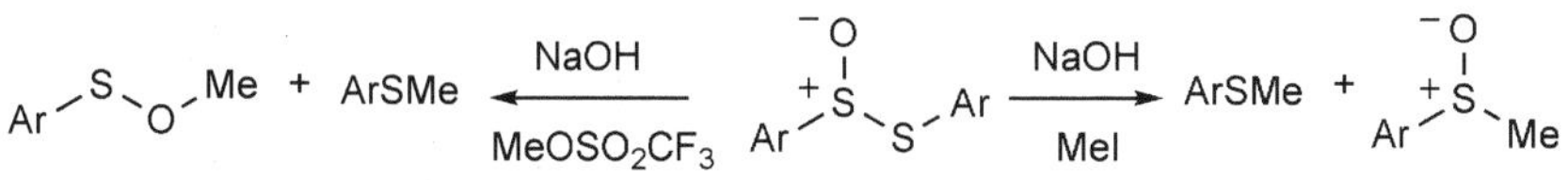

2. Explain why the barrier to rotation is higher in an amide **4.118** than it is in a hydroxamic acid **4.119**.

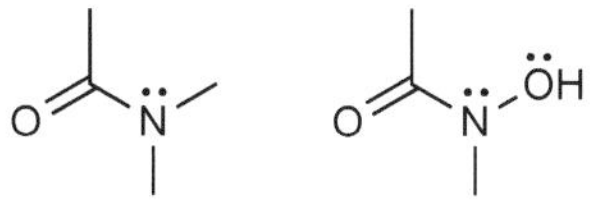

**4.118**            **4.119**

3. Why are oximes and hydrazones less electrophilic than other imines?

4. Explain why nitrobenzene, although it reacts predominantly in the *meta* position, gives more *ortho*-dinitrobenzene than *para*.

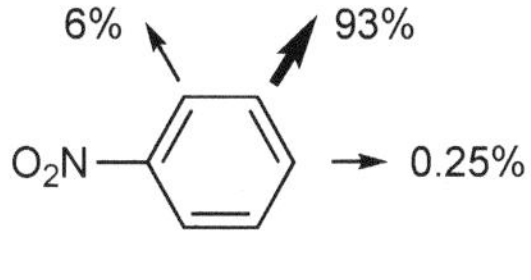

5. The dipole of pyrrole points in the direction **4.120**. The explanation usually offered is that the orbitals are polarised in the sense illustrated by the canonical structures **4.121** and **4.122**. This cannot be the reason, because this is only one of the π orbitals, the HOMO. We can see on the drawing **4.54** that the *total* π charge, derived from the coefficients on Fig. 1.54, is, as ought to have been expected, higher on the nitrogen atom and lower on the carbon atoms. The same must be true in the σ framework. Why then is the dipole pointing in the other direction?

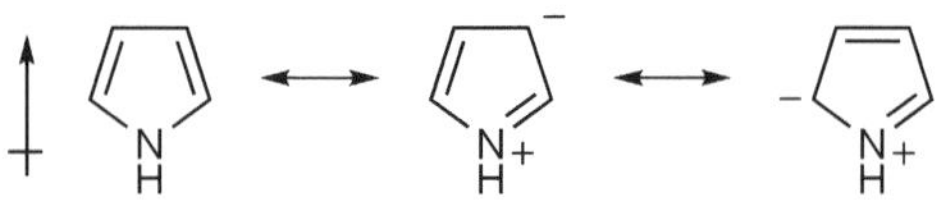

**4.120**          **4.121**          **4.122**

6. Explain the regiochemistry of this $S_N1''$ reaction:

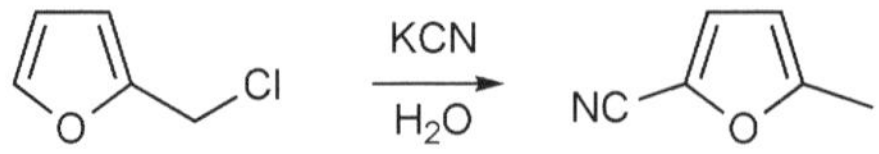

7. Explain why the carbene **4.123**, anomalously for an X-substituted carbene, inserts into an alkene:

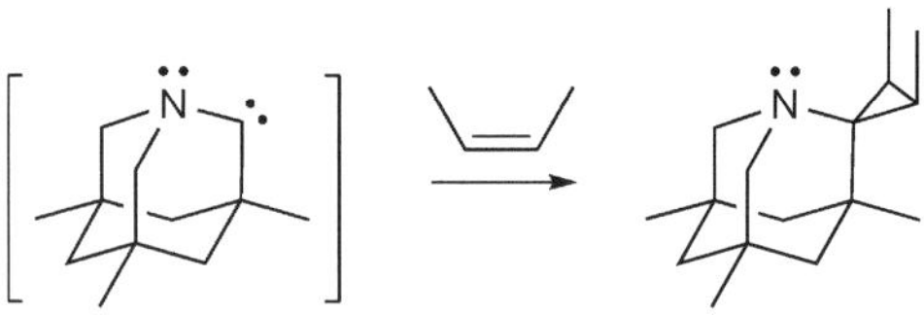

**4.123**

# 5 Ionic Reactions— Stereochemistry

To achieve control of stereochemistry, understanding is vital, and understanding requires a feeling for all the factors that influence the stereochemistry of organic reactions. We begin with two adjectives, stereoselective and stereospecific, which, with their derived adverbs, are much used and misused. They are defined following Zimmerman below, and used carefully in this book, because the distinction between them is useful.

The more encompassing term *stereoselective* simply means that more of one stereoisomer is produced than of one or more others. Thus the reduction of camphor **5.1** takes place mainly with attack of the hydride reagent on the less-hindered face, avoiding the C-8 methyl group, to give more isoborneol **5.2** than borneol **5.3**. The degree of stereoselectivity is expressed as the diastereoisomer ratio of isoborneol to borneol. It is helpful to normalise the numbers as percentages (90:10 in this case), without implying that the yield is 100%.

The less simple term *stereospecific* is used for those reactions where the configuration of the starting material and the configuration of the product are related in a mechanistically constrained way. Thus the diastereoisomeric bromides **5.4** and **5.6** give different alkenes **5.5** and **5.7** by anti elimination. Since each of these reactions produces more of one isomer than the other, they are also stereoselective. The characteristic feature of a stereospecific reaction is that one stereoisomer of the starting material gives one stereoisomer of the product, and a different stereoisomer of the starting material gives a different stereoisomer of the product.

**5.4**          **5.5**          **5.6**          **5.7**

This particular reaction was studied when analytical methods were not available to measure the probably small degree to which each isomer gave some of the other alkene, either by a different mechanism or by incomplete stereospecificity in the E2 reaction itself. No matter how much stereochemical leakage there is, as long as the diastereoisomer ratio is greater than 50:50, the reaction is still stereospecific. It is not helpful to use the word stereospecific to mean 100% stereoselective, as many people thoughtlessly do—a useful distinction is lost, and understanding suffers.

Unfortunately, there is a grey area. There are reactions that are, in their fundamental nature, the same as those we call stereospecific, but for which it is not possible to have two stereoisomers either of the starting material or of the product. Thus the addition of bromine to an isolated double bond is stereospecifically anti, but the corresponding addition to an acetylene cannot be *proved* to be stereospecifically anti by the usual criterion, because there is no possibility of having two stereoisomers of an acetylene. The same problem arises for reactions taking place in the opposite direction—in elimination reactions producing acetylenes, one vinyl bromide may react faster than the other, but they both produce the same acetylene.

This chapter is divided into two sections, largely separating stereospecific reactions from the merely stereoselective. The first deals with the ionic stereospecific reactions, and the explanations based on molecular orbital theory for the sense of the stereospecificity. The second deals with stereoselective reactions, in which a new stereocentre is created selectively under the influence of one or more existing stereochemical features, which is also sometimes a question of how the orbitals interact. The stereospecificity that is such a striking feature of pericyclic reactions is covered in the next chapter.

## 5.1   The Stereochemistry of the Fundamental Organic Reactions

### 5.1.1   Substitution at a Saturated Carbon

**5.1.1.1   The $S_N2$ Reaction.**   It is well known that bimolecular nucleophilic substitution (the $S_N2$ reaction) takes place with inversion of configuration. This is a stereospecific reaction because one enantiomer of the starting material gives largely one enantiomer of the product. A number of factors contribute to this well nigh invariable result, but the best explanation is simple: the transition structure for inversion will be a trigonal bipyramid **5.8**, with the electronegative elements in the apical positions, keeping the negative charges as far apart as possible.

**5.8**

We can also explain inversion of configuration in the $S_N2$ reaction by looking at the frontier orbitals, but it is a much weaker explanation. The appropriate frontier orbitals will be the HOMO of the nucleophile and the LUMO of the electrophile. Taking the orbitals of methyl chloride in Figs 1.45 and 1.47, we can see the LUMO is the $\sigma^*_{CX}$ orbital. The overlap is bonding when the nucleophile approaches the electrophile from the rear, but would be both bonding and anti-bonding of the nucleophile were to approach from the front.

In the absence of solvent, the gas-phase $S_N2$ reaction is different, but it still takes place with inversion of stereochemistry. There is a double well in the energy surface: the nucleophile and the alkyl halide combine exothermically with no energy barrier to give an ion-molecule complex. In a sense the naked nucleophile is solvated by the only 'solvent' available, the alkyl halide. The $S_N2$ reaction then takes place with a low barrier, and the product ion-molecule complex dissociates endothermically to give the products.

***5.1.1.2 The $S_E2$ Reaction.*** In electrophilic substitution, the substrate is usually an organometallic reagent, for which we can use the orbitals of methyl-lithium (Figs. 1.49 and 1.50) as the simplest version. The frontier orbitals for the $S_E2$ reaction will be the HOMO of the nucleophile (the $\sigma_{CLi}$ orbital strongly associated with C—M bonding) and the LUMO of the electrophile. In this case, the frontier orbital interaction (Fig. 5.1) can be bonding for attack on *either* side of the carbon atom.

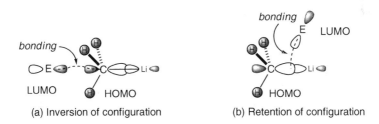

(a) Inversion of configuration      (b) Retention of configuration

**Fig. 5.1**  Frontier orbitals for the $S_E2$ reaction

In agreement, electrophilic substitution at a saturated carbon atom sometimes takes place with retention of configuration **5.10** → **5.9**, when it is called $S_E2$ret, and sometimes, but more rarely, with inversion of configuration **5.10** → **5.11**, when it is called $S_E2$inv. Retention of configuration is the more usual pattern for electrophilic attack on a C—M bond. This may simply be because electrophiles are attracted to the site of highest electron population, but explanations for changes from retention to inversion in going from one electrophile to another, from one metal to another, and from one substrate to another are far from clear. It is not uncommon to find even one system in which the two pathways are delicately balanced.

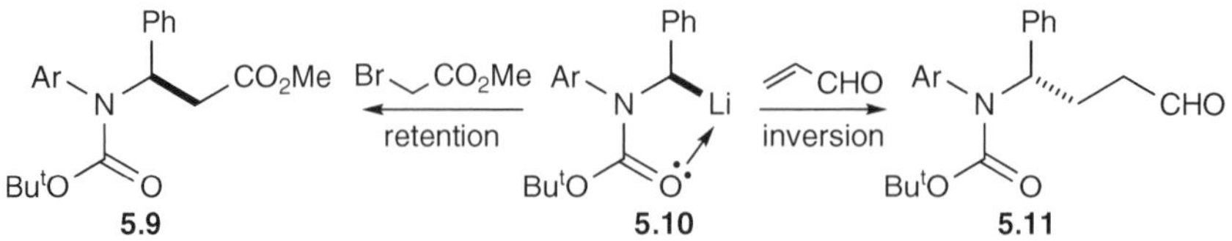

It has only recently become possible for synthetic chemists to use the stereochemistry that reactions like this possess, as seen with the reagent **5.10** created using butyllithium and (–)-sparteine. The explanation offered in this case is that reactive electrophiles, those not requiring Lewis acid catalysis, are apt to react with inversion of configuration, while those that need to coordinate to the metal to experience some Lewis acid catalysis, are apt to react with retention of configuration, because the electrophile is necessarily being held on the same side as the metal.

### 5.1.2   Elimination Reactions

**5.1.2.1   The E2 Reaction.**   $\beta$-Elimination, which is usually but not always stereospecifically anti, is the frequent accompaniment to substitution, as we saw earlier [see (Section 4.5.2.5) pages 145–147]. We have also already had [see (Section 2.2.3.4) page 81] some discussion about why anti arrangements are preferred in the anomeric effect, where we saw that it is not solely because it allows all the groups to be staggered and not eclipsed.

As with the anomeric effect, this is not the whole story for elimination reactions either, because there are systems where this factor is not present, and yet there is still a preference for anti elimination. Thus the anti elimination of the vinyl chloride **5.12** giving diphenylacetylene is over 200 times faster than the syn elimination of the vinyl chloride **5.13**, and this in spite of the almost certainly higher energy of the latter, which has the two large substituents, the phenyl groups, *cis*.

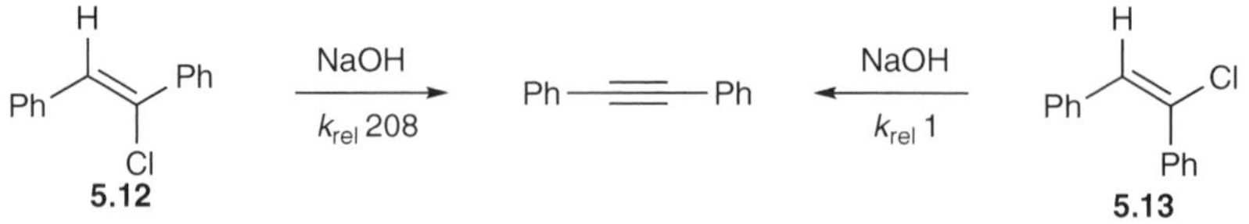

In one sense, the stereochemistry at the carbon carrying the nucleofugal group X in the anti-periplanar process **5.14** can be seen as an inversion of configuration, since the electrons supplied by the C—M bond, where M stands for any electrofugal group, flow into the $\pi$ bond of the product **5.15** from the side opposite to the

C—X bond. This is the simplest perception available to the organic chemist to account for why E2 reactions take place with anti-periplanar geometry.

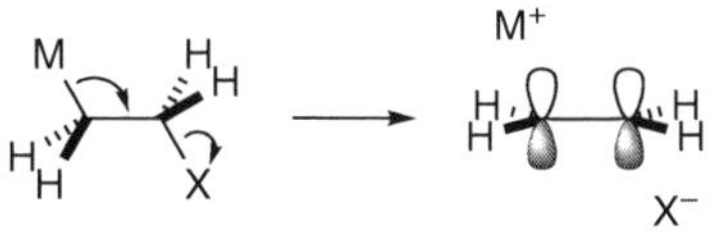

**5.14**  **5.15**

In the gas phase, there is again a well with a reactant complex before the transition structure for elimination. In the reactant complex for ethyl fluoride, the base, modelled by a fluoride ion, is bonded to the hydrogen atom that is about to leave, stretching that H—C bond, and allowing the C—F bond to stretch too. From here it is easy to see how the molecular orbitals flow into those of the product, supporting the picture of the event at the C—X bond as resembling an $S_N2$ pathway. The transition structure, in the absence of solvation, has the electrofugal hydrogen atom coordinated to both carbons, but both bonds are long and the C—F bond even longer. The corresponding transition structure for syn elimination is higher in energy, and the transition structure resembles that for carbanion formation ahead of elimination, in other words an E1cb mechanism.

***5.1.2.2 The E2′ Reaction.*** The stereochemistry of the E2′ process is even less well understood. It is exemplified by the decarboxylative eliminations of the vinylogous $\beta$-hydroxy acids **5.16** and **5.18**, which are both largely, although not exclusively, syn. The corresponding E2 reaction with $\beta$-hydroxy acids is highly anti selective, in the usual way for $\beta$ eliminations.

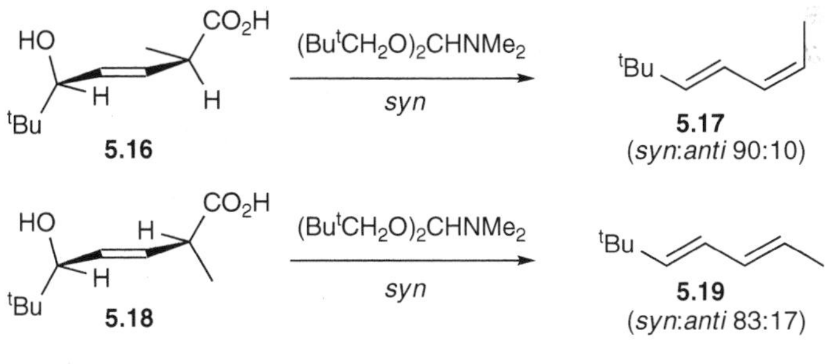

The tau bond model appears to provide a quick and easy explanation. An anti interaction between each of the breaking bonds and the lower tau bond leads overall to a syn selective reaction for each diastereoisomer.

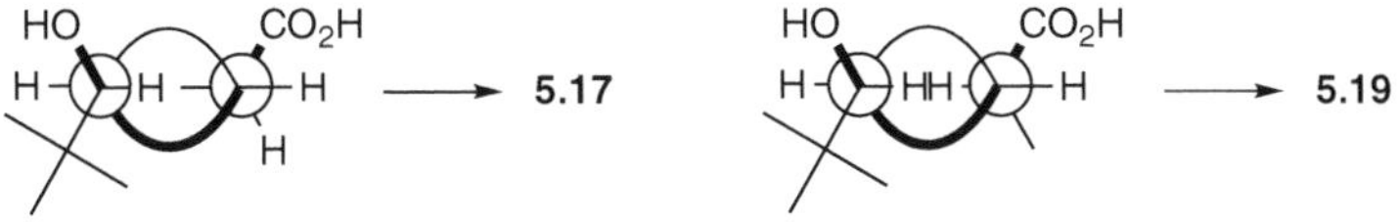

The change from anti for an E2 reaction to syn for an E2′ is a satisfying pattern—adding two electrons to the transition structure changes the stereochemistry. The same pattern is found for aromaticity, where each added pair of electrons changes the system from aromatic to antiaromatic, and back again, and we shall see the same change for pericyclic reactions (Chapter 6). There is a natural supposition that each added pair of electrons ought to cause stereochemistry to alternate, but it is not reliable here. Adding one more double bond for the E2″ reaction does not cause it to change back to being selectively anti. The tau bond model would support this expectation—successive anti overlap through the tau bonds down the chain **5.20** and **5.23** suggests that decarboxylative elimination should be anti. In practice, the base-induced elimination of the ethers **5.21** and **5.24** is largely syn, with the major products being the trienes **5.22** and **5.25**, respectively.

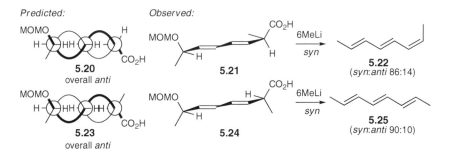

The tau bond model is an intriguing, but evidently defective approach to understanding the stereochemistry of elimination reactions. The problem therefore remains—there is no simple and satisfying way to explain the stereochemistry beyond the simple $\beta$ elimination.

### 5.1.3   Nucleophilic and Electrophilic Attack on a $\pi$ Bond

#### 5.1.3.1   Nucleophilic Attack on a $\pi$ Bond—The Bürgi-Dunitz Angle.
Nucleophilic attack on the $\pi$ bond of a carbonyl group is widely recognised to take place from above (or below) the plane of the double bond, but not directly down the axis of the $p_z$ orbital **5.26**. Bürgi and Dunitz deduced, from an examination of a large number of X-ray crystal structures, that the angle $\theta$ in the transition structure **5.27** was obtuse, typically close to 107° and not 90°. The angle $\theta$ is called the Bürgi-Dunitz angle.

It is a common misunderstanding to think that the Bürgi-Dunitz angle implies that the angles $\phi$ are acute. They can be sometimes, but they are not usually—the angles $\phi$ are also obtuse in the transition structure, but to a somewhat smaller extent. That both $\theta$ and $\phi$ will be obtuse is hardly surprising—as the reaction proceeds, the carbon atom of the carbonyl group is changing from trigonal to tetrahedral, and the transition structure is almost certain to have a geometry at this atom somewhere in between. Only at long distances, with little bonding developed, will $\phi$ be acute. This is borne out by the X-ray structures, which show that $\phi$ is less than 90° only when the nucleophile is more than 2.5 Å from the carbon atom. The essence of Bürgi and Dunitz's perception is that $\theta$ is a slightly larger angle than $\phi$.

There are several reasons. On the product side of the reaction coordinate, the tetrahedral intermediate will have a large repulsion between the charge developing on the oxygen atom and any charge on the nucleophile. On the starting material side (Fig. 5.2), the repulsive interaction of the filled orbitals with the filled orbitals **5.28**, the first term of the Salem-Klopman equation, will push the nucleophile away from the oxygen atom, because the HOMO of the carbonyl group has the larger coefficient there. The Coulombic forces alone, the second term of the equation, will lead the nucleophile to approach along the line of the C—O bond **5.29**. For the third term, the interaction of the HOMO of the nucleophile and the LUMO of the carbonyl group has a repulsion from the oxygen atom, because of the orbital of opposite sign on it **5.30**. All three factors make $\theta$ an obtuse angle.

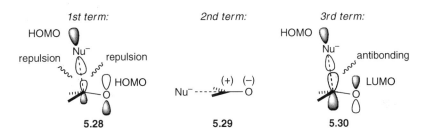

**Fig. 5.2**   The Salem-Klopman equation applied to the Bürgi-Dunitz angle

Superimposed on the Bürgi-Dunitz angle is an angle defined by $\psi$ in the view of an unsymmetrical carbonyl group seen from above **5.31**. This angle is called the Flippin-Lodge angle, and it is expected to be positive when the group $R^1$ is larger than the group $R^2$. A calculation, for example, makes it 7° for hydride attack on pivalaldehyde ($R^1 = Bu^t$, $R^2 = H$). It becomes more significant when one of the substituents R is an electronegative group, since this will carry negative charge repelling the nucleophile. Thus, esters and amides **5.32** have a positive angle $\psi$ if the steric repulsion from the R group is not too forbidding, close to 40° for an ester and 50° for an amide. The same angles, the Bürgi-Dunitz and the Flippin-Lodge, will have their counterparts for nucleophilic attack on a C=C bond, but the former at least ought to be muted, because all three factors **5.28–5.30** will be reduced when there is a carbon atom in place of the oxygen atom.

$$\text{O} \quad\quad\quad \text{O}$$

R$^1$ ↑ R$^2$          R$_2$N or RO ↑ R
  ψ                              ψ
  **5.31**                    **5.32**

### 5.1.3.2 Electrophilic Attack on a C=C Double Bond by Nonbridging Electrophiles.

Electrophilic attack by a proton or a cationic carbon on a C=C double bond may give an open cation **5.34** or a bridged cation **5.36**. We have met the problem of hyperconjugating and bridged cations before [see (Section 2.2.1.2) page 72], and the same problem arises here.

E$^+$                    E$^+$              E$^+$
  ↘                        ↑                  ↑
  ·····  →  E  ·····     ·····  →  E  ·····
         +                              +

**5.33**     **5.34**     **5.35**     **5.36**

For cations that are not bridged, the carbon atom being attacked is changing from trigonal to tetrahedral, and the angle analogous to the Bürgi-Dunitz angle can be expected to be obtuse in the transition structure. On the starting material side of the reaction coordinate (Fig. 5.3), the first term of the Salem-Klopman equation would push the electrophile away from the centre of the double bond **5.37** and discourage attack there, or anywhere else, but the other two terms would encourage attack from inside **5.38** (where the concentration of charge in the $\pi$ cloud is represented as a minus sign) and **5.39**. It seems likely that, while the early approach may be from inside **5.35**, the electrophile may have moved outside to give an obtuse angle by the time the transition structure has been reached. Thus the angle of approach in an electrophilic attack, acute or obtuse, will depend upon how early the transition structure is.

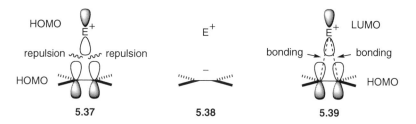

HOMO — E$^+$                    E$^+$                    E$^+$ — LUMO
repulsion ~~ ~~ repulsion                         bonding → ← bonding
HOMO ·····  ·····          ·····  −  ·····          ·····  ·····  HOMO

**5.37**                    **5.38**                    **5.39**

**Fig. 5.3**  The Salem-Klopman equation applied to electrophilic attack on a C=C bond

The stereochemistry of the second step of an addition initiated by a non-bridging electrophile like a proton will be controlled by which surface of the intermediate cation **5.34** is more easily attacked by the nucleophile. The addition of hydrogen chloride to an alkene is not stereospecifically anti, because the chloride does not necessarily attack the cation either specifically anti or syn to the proton, in contrast to addition initiated by bridging electrophiles like bromine, or metallic electrophiles like the mercuric ion, described below. The stereochemistry will depend on the substituents in the cation **5.34**, and how they influence the relative accessibility of the two surfaces when the nucleophile attacks.

### 5.1.3.3 Nucleophilic and Electrophilic Attack by One π Bond on Another.

A combination of nucleophilic and electrophilic attack on double bonds is the core of the aldol reaction, where both the nucleophile and the electrophile are π bonds. The ideas we have seen in the previous two sections can be combined to understand the transition structure **5.40** calculated for this reaction in the gas phase. This transition structure has obtuse approach angles both for the electrophilic and for the nucleophilic double bonds, the two reagents have all their substituents staggered, when viewed down the developing bond **5.40b**, and the two oxygen atoms are as far apart as possible, presumably repelling each other because of the partial negative charges they both carry. However, there are alternative conformations such as **5.41**, which maintain the obtuse angles and the staggered groups **5.41b**, and are not much higher in energy. The transition structure **5.40** is described as anti-periplanar and the transition structure **5.41** is described as synclinal.

5.40a      5.40b      5.41a      5.41b

Those aldol reactions in which a metal atom, like lithium or boron, is coordinated to both oxygens are certainly synclinal, since the metal coordinates to both oxygens and the transition structure is cyclic and usually chair-shaped—as first proposed by Zimmerman and Traxler. There are, however, many related reactions, when a C=C and C=O group or two C=C groups combine, in which this problem is less settled, either by theory or experiment. Examples are the reactions between enamines and Michael acceptors, and the Lewis acid-catalysed reactions between allylsilanes or allylstannanes and aldehydes, in none of which is there a cyclic component holding the reagents in a synclinal geometry. There is experimental evidence for synclinal and anti-periplanar preferences for various examples of these reactions, and we must conclude that there is only a small energy difference between them.

### 5.1.3.4 Electrophilic Attack on a C=C Double Bond by Bridging Electrophiles.

Heteroatom electrophiles, like peracids, sulfenyl halides and the halogens nearly always give bridged products in the first step. The difference between these electrophiles and the proton or carbon electrophiles discussed above is that the electrophilic atoms all carry a lone pair, so that the bridging bonds **5.43** have a total of four electrons. (The bridging in the structure **5.36** only had two electrons to share between the two bonds.) The factors from the Salem-Klopman equation illustrated as Fig. 5.3 now lead the electrophile straight onto the $\pi$ bond **5.42**, since they match the product-like character, instead of opposing it. Depending upon what E is, the intermediate **5.43** may be stable, or it may open to give the product of anti addition **5.44**, as a consequence of the preference for inversion of configuration in the $S_N2$-like ring-opening step.

The more available the lone pair, the more firmly is bridging followed. Epoxidation gives directly a bridged product, with no intermediate; the addition of sulfenyl halides is the electrophilic addition most strictly following the anti rule; whereas halogenation, with relatively tightly bound lone pairs, shows significant loss of stereospecificity, corresponding to some degree of preliminary opening of the bridged intermediate, especially when the alkene has a good cation-stabilising substituent like a phenyl group.

Hydroboration, oxymercuration, oxypalladation and other additions to alkenes in which the electrophilic heteroatom is electropositive, are less straightforward. They begin by the coordination of the metal to the alkene, but whether that coordination is best represented as a bridged structure is not clear. Whether it is represented as bridged or involved in hyperconjugation **5.46**, the transition structure for the next step has the nucleophile attacking with high levels of stereocontrol—syn in hydrometallation **5.45** and carbometallation, but anti in oxymetallation **5.46**. The hydro- and carbometallations are syn, because the nucleophile is delivered from the metal. The oxymetallations are anti, either because the nucleophile attacks a bridged intermediate, or because it attacks anti-periplanar to the donor substituent in the low-energy conformation **5.46**, in which the empty p orbital is stabilised by hyperconjugation with the M—C bond. This kind of addition is the reverse of a $\beta$-elimination, and responds to the same stereochemical constraints in favour of the anti-periplanar pathway. Just because a reaction is stereospecifically anti does not prove that it takes place by way of a bridged intermediate.

*5.1.3.5 Baldwin's Rules.* The direction of attack on $\sigma$ and $\pi$ bonds affects the ease with which rings can form. Baldwin pointed out that when a nucleophile is tethered to an electrophile, it matters whether the bond being attacked is part of the ring or outside it. He noted that essentially all the reactions in which the bond was outside the ring were straightforward. In contrast, when the bond was within the ring, there were some cases where ring formation appeared to be difficult, even when the ring being formed was not strained. Thus conjugate additions of the type **5.47** are easy and high yielding, but the superficially similar conjugate addition **5.48** does not take place; instead, the oxyanion attacks directly at the carbonyl group **5.49**.

| **5.47** | **5.48** | **5.49** |

Baldwin identified the problem as occurring most dramatically when a five-membered ring was being formed by attack on a double bond within the ring, as in **5.48**. He labelled this a 5-*endo-trig* process, with the 5 referring to the ring size, the endo referring to the double bond being within the ring, and the *trig* referring to the trigonal carbon under attack. Thus the easy reactions **5.47** and **5.49** are 5-*exo-trig*, with which there is evidently no difficulty.

The explanation for this difference comes when we look at the ease with which the nucleophilic atom in each case can reach the appropriate position in space for attack on the double bond. In both cases, the nucleophile must approach from above and behind the $\pi$ bond with approach angles resembling those in the transition structure **5.27**. For the 5-*exo-trig* reactions, the carbon under attack, C-1, will be on its way to becoming tetrahedral, and the chain of atoms attached to it, culminating in the oxyanion, can easily fold to put the oxyanion in a nearly ideal position **5.50**. For the 5-*endo-trig* process, the chain of atoms C-1, C-2 and C-3 must all be in the same plane **5.51**. The oxyanion is then only two atoms away from C-3 and it cannot reach to the position it needs to in order to attack at C-1. The chain is simply too short when it is trying to form a five-membered ring. Baldwin suggested that the problem is much less serious with a chain of six atoms, which is evidently just long enough to reach with a little distortion.

| **5.50** | **5.51** |

Similar arguments apply to reactions in which the double bond is the nucleophile. Thus 5-*exo-trig* enolate alkylations **5.52** are high yielding, but the similar

5-*endo-trig* enolate alkylation **5.53** does not take place, and *O*-alkylation **5.54** takes place instead. In ring-forming reactions the electrophile is not often going to be the kind that induces bridging, and an obtuse approach angle leading towards a tetrahedral intermediate **5.34** is likely. The geometric constraints for electrophilic attack will make the 5-*exo-trig* process **5.55** easy and the 5-*endo-trig* process **5.56** difficult, just as they did for nucleophilic attack. The *O*-alkylation **5.54** does not meet the problem because there are lone pairs on the oxygen atom which can easily line up behind the C—Br bond.

Baldwin produced a set of rules for which cyclisations are favoured and which disfavoured. Briefly, the disfavoured reactions are the *n-endo-tet* processes with $n < 9$, the *n-endo-trig* processes with $n < 6$, and the *n-exo-dig* processes with $n < 5$. Exceptions to some of the rules are known. Some compounds find different mechanisms, avoiding the disadvantageous features of a more obvious mechanism. Others simply have no alternative paths, and the constraints are not so forbidding as to make the disfavoured path impossible. Pericyclic reactions (Chapter 6) abound with exceptions—Baldwin's rules appear to have little validity there, where there are electrocyclic reactions that are 5-*endo-trig*, 4-*endo-trig* and even 3-*endo-trig* at *both* ends. Nevertheless the rules are a helpful guide in planning a synthesis—if you choose one of the disfavoured reactions for a key step, it is just as well to have a good reason for expecting it to work before you embark on the earlier steps.

The epoxide opening **5.57** giving a product with a four-membered ring is unusual, because five-membered rings are usually formed more rapidly, and there appears at first sight to be a reasonable pathway **5.58** giving a five-membered ring. The explanation lies with Baldwin's rules. The opening **5.57** is uncomplicatedly 4-*exo-trig* at the nucleophilic carbon and 4-*exo-tet* at the electrophilic carbon. In contrast, while the alternative reaction **5.58** is 5-*exo-tet*, if you count only the carbon atoms in the chain, it is also 6-*endo-tet*, if you count the atoms sequentially from 1 to 6 and include the oxygen atom. It is this 6-*endo-tet* aspect that makes the reaction **5.58** less favourable. The problem lies in aligning the nucleophile more or less directly behind the 6-5 C—O bond being broken, which is what $S_N2$ reactions need.

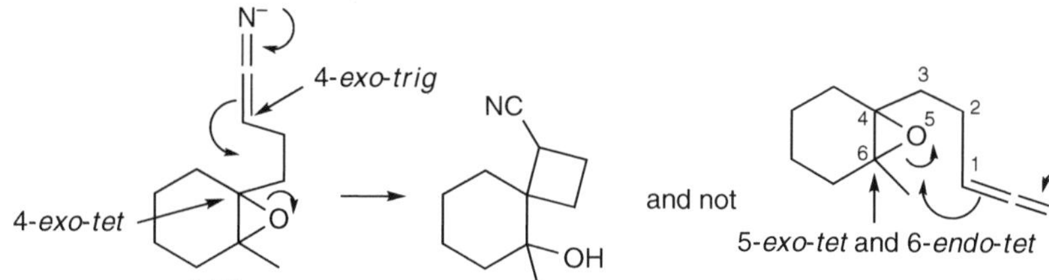

### 5.1.4   The Stereochemistry of Substitution at Trigonal Carbon

Both nucleophilic and electrophilic attack on trigonal carbon can take place by two pathways (Fig. 5.4)—direct attack on one of the $\sigma$ bonds attached to the double bond (path a), or by attack on the $\pi$ bond (path b), with the formation of an ionic intermediate, followed by the loss, respectively, of a nucleofugal group X or an electrofugal group M. There are also formally unimolecular pathways, $S_N1$ and $S_E1$, with ionisation followed by capture by the nucleophile or electrophile, but the former is rare and the latter unknown.

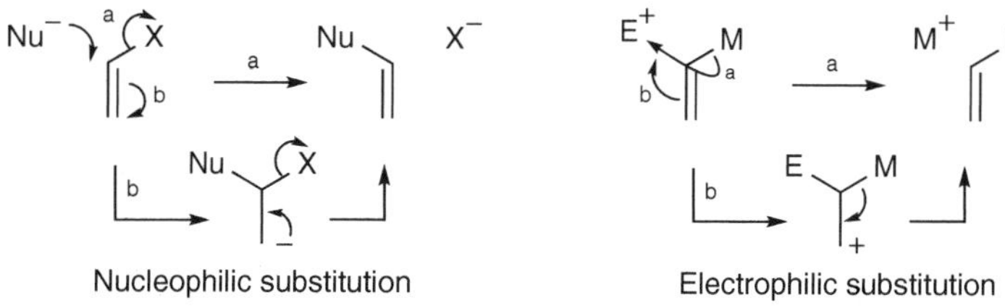

Nucleophilic substitution                    Electrophilic substitution

**Fig. 5.4**   Direct and stepwise substitution reactions at a double bond

The stereochemistry of the direct attack can be expected to resemble the corresponding reactions at a saturated carbon—inversion for nucleophilic substitution, and retention, or perhaps occasionally inversion, for electrophilic substitution. In practice, $S_N2$ reactions at trigonal carbon are rare, and their stereochemistry, where inversion is known, barely established. Electrophilic attack, like the reactions of vinyllithium reagents with protons, aldehydes and carbon dioxide, takes place invariably with retention of configuration. This pattern ties in with the greater difficulty of configurational inversion at trigonal atoms than at tetrahedral atoms. The necessity for inversion in $S_N2$ reactions makes them very difficult at trigonal carbon, and the delicate balance between retention and inversion in $S_E2$ reactions at tetrahedral carbons becomes overwhelmingly in favour of retention at trigonal carbon. The more remarkable stereochemical features are found in those reactions that take the indirect path (path b), with addition followed by elimination.

*5.1.4.1  Nucleophilic  Substitution  by  Addition-Elimination.*  Nucleophilic attack takes place on the $\pi$ bond in the activated alkene **5.59**, creating an intermediate, typically an enolate **5.60**. Rotation about the $\sigma$ bond can take place either clockwise by 60° to give the intermediate **5.61**, or anticlockwise by 120° to give the intermediate **5.62**. Since the C—X bond is lined up with the $\pi$ system of the enolate in each of these intermediates, the loss of the nucleofuge can take place to give, respectively, the products **5.63** of retention of configuration, or **5.64** of inversion of configuration. There is a similar sequence of events for attack from below the $\pi$ bond, which would give the enantiomers of all the intermediates, and is therefore equally probable.

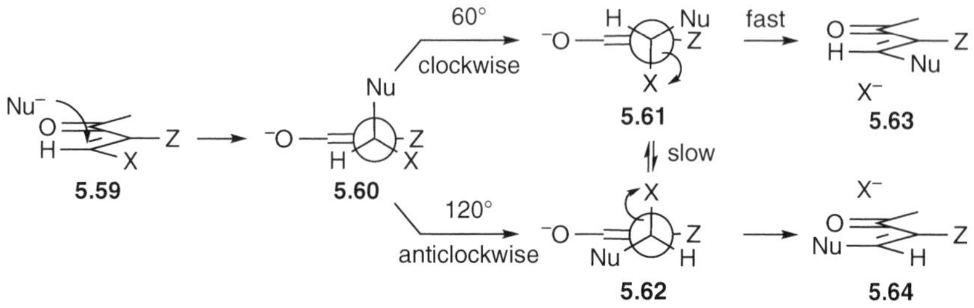

In practice, retention of configuration is commonly observed, as in the stereo-specific reactions of the geometric isomers **5.65** → **5.66**, showing that the 60° rotation, to give the intermediate **5.61**, is understandably more frequent than the 120° rotation. Furthermore, the loss of the nucleofugal group is evidently faster than rotation about the single bond **5.61** → **5.62**. The 60° clockwise rotation causes the stabilising negative hyperconjugation [see (Section 2.2.3) page 77] between the X—C $\sigma$ bond and the $\pi$ system of the enolate steadily to increase as the dihedral angle between the two systems drops from 60 to 0°. This rotation probably occurs in concert with the formation of the X—C bond and the inter-mediate **5.60** may never be formed as such. Only when the intermediate **5.61** has a long lifetime, either because X is a poor nucleofugal group like alkoxide or fluoride, or because the anion-stabilising substituents Z are especially good like nitro, is the stereospecificity lost.

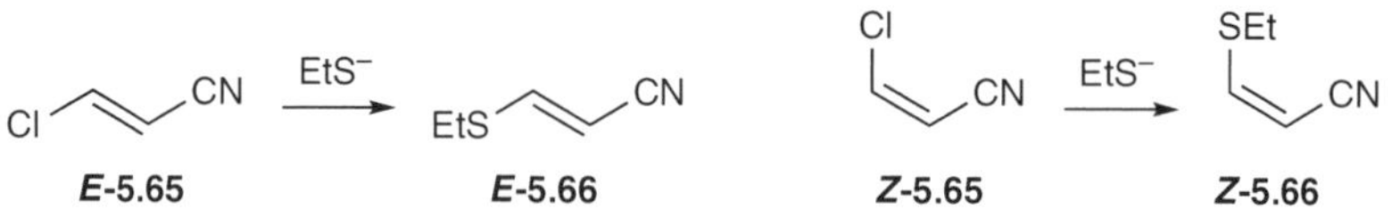

*5.1.4.2  Electrophilic  Substitution  by  Addition-Elimination.*  Electrophilic attack has a parallel series of events, which is best known in the electrophilic substitution of vinylsilanes. The electrophile attacks from above (or below) the $\pi$ bond in the vinylsilane **5.67**, with the creation of an intermediate carbocation **5.68** (with the cationic carbon behind). Rotation about the $\sigma$ bond can take place clockwise by 60° to give the intermediate **5.69**, or anticlockwise by 120° to give the intermediate **5.70**. Since the C—Si bond is lined up with the empty p orbital in both intermediates, it is thermodynamically stabilising, but it is simultaneously kinetically unstable, because it is aligned for removal by a silicophilic nucleophile X⁻ to give, respectively, the products of retention **5.71**, or inversion of configuration **5.72**. The intermediate **5.69** is stabilised by hyperconjugation (see pages 74–76), and rotation **5.69** → **5.70** is slow relative to the ease with which the silyl group is removed. Again, the 60° rotation is easier than the 120° rotation, and the attack and the 60° rotation are probably concerted without formation of an intermediate **5.68** as such.

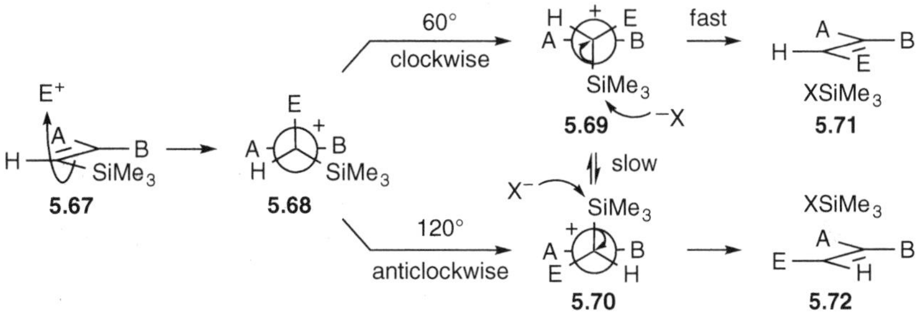

In practice retention of configuration is the normal pattern in protodesilylation and the reactions of vinylsilanes with carbon electrophiles, like the ring closures **5.73** → **5.74**, stereospecifically setting up the exocyclic double bond geometry. The exception to this pattern is bromodesilylation, and a few similar reactions, in which the electrophilic attack **5.75** is followed by nucleophilic opening of the bridged intermediate **5.76**. Rotation **5.77**, followed by anti elimination of the silyl group and bromide ion **5.78**, give the product of inversion of configuration **5.79**.

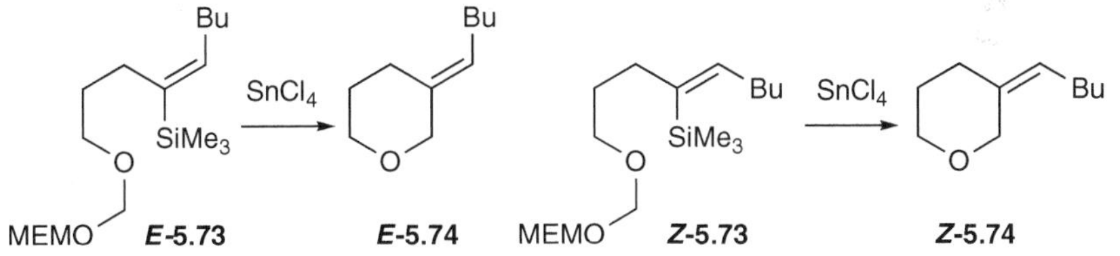

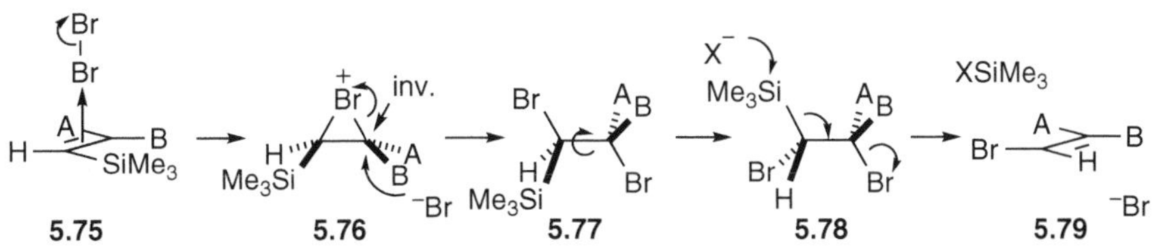

## 5.2   Diastereoselectivity

We have been concerned so far only with stereospecific reactions. Any double bonds involved have the top and bottom surfaces either the same or enantiotopic.

We must now turn to those cases where the attack on one surface gives one diastereoisomer, and attack on the other surface gives a different diastereoisomer, when the surfaces are said to be diastereotopic.

Early success in controlling stereochemistry came by tying a molecule into a more or less rigid ring system, so that one surface of a double bond was more exposed than the other. This approach achieved high levels of stereocontrol, but at the expense of having to build in to the synthetic scheme extra steps to set up the rings, and more to unravel them. Open-chain stereocontrol has, more often than not, been achieved by arranging for the reactions to have cyclic transition structures, for which one conformation is preferred over another—a chair-like six-membered ring, for example, with the larger substituents in equatorial positions.

Understanding how to control genuinely open-chain reactions, those without even a cyclic transition structure, has become possible from a combination of empirical observation and an appreciation of the electronic forces at work. This approach rarely achieves the high levels of selectivity that the constraints in a ring system can impart, whether from the starting material, the product, or the transition structure. The differences between one stereochemistry and another are rarely more than a few kJ mol$^{-1}$. Fortunately for synthetic chemistry, even differences as small as 10–20 kJ mol$^{-1}$ are enough to get very workable levels of selectivity. Understanding in terms of the charge and molecular orbital interactions is hampered by having only a crude tool with which to explain small differences in energy.[17] Nevertheless, whenever a measurable level of control is achieved, it is beholden upon the discoverer to make some attempt to explain it. Steric effects, in which a large substituent hinders the approach of a reagent from one direction, are nearly always a component of such explanations. This is, in one sense, an orbital interaction, since steric effects are the results of the interaction of filled orbitals with filled orbitals, for which there is nearly always an energetic penalty.

There have been several attempts to explain, with more or less rigour, the transmission of electronic effects into a double bond and down a conjugated chain, but none has yet settled in as the accepted way to picture what happens to the orbitals. In the most colourful approach, the electrostatic attraction of a point positive charge to the surface of a starting material is calculated, and then mapped onto the surface with a colour code—red for maximum attraction and blue for minimum attraction. The results are beautiful pictures showing red hot spots on the molecular surface, and they give an immediate and vivid sense of where electrophilic reagents are likely to attack.[18]

The problem is that a p orbital on its own is symmetrical on the top and bottom surfaces. For the top and bottom surfaces to have differently sized lobes some fraction of an s orbital must be mixed in to create a hybrid (Fig. 5.5). An s orbital of one phase will push the lobe up, and an s orbital of the opposite phase

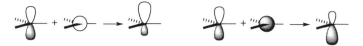

**Fig. 5.5**   Mixing in s orbital character to desymmetrise a p orbital

will push it down. The problem comes in working out what sign the s orbital will have in the frontier orbital in any given situation. The sign can be extracted from calculations, but the sense is far from intuitive when it is induced by a neighbouring stereogenic centre. No method is simple enough to take up and be used with confidence in everyday situations, and the best that can be said is that this problem is still a challenge to theoreticians and physical organic chemists alike.

### 5.2.1 Nucleophilic Attack on a Double Bond with Diastereotopic Faces

The earliest reaction to be studied showing open-chain diastereoselectivity was nucleophilic attack on a carbonyl group, either a carbonyl group with a stereogenic centre adjacent to it or a carbonyl group like that in 4-*tert*-butylcyclohexanone, where the diastereotopic surfaces are distinguished by being axial or equatorial.

*5.2.1.1  The Felkin-Anh Rule.*    The attack of a nucleophile on a carbonyl group adjacent to a stereogenic centre is covered by Cram's rule. When the three substituents differ only in size, and not in electronic nature, there is little difficulty. The explanation suggested by Felkin and Anh is now so widely accepted that it is often referred to as the Felkin-Anh rule. The code used has the groups ranked simply as large L, medium-sized M, and small S. Felkin suggested that the incoming nucleophile would attack the π bond anti to the large substituent, and Anh added that we must also take the Bürgi-Dunitz angle into account. There are therefore two possibilities **5.80**, with the medium-sized group 'inside' (i.e. eclipsing or partly eclipsing the C=O double bond) and **5.81** with the small group 'inside'.

The latter is higher in energy, because the incoming nucleophile is pushed back, close to the medium-sized group, by the forces controlling the Bürgi-Dunitz angle, whereas in the former it is close to the small group. As it happens there is no need for special pleading to account for why the medium-sized group sits inside, since propanal adopts the conformation **2.101a** [see (Section 2.4.3.2) page 94] with the methyl group inside. The sense of attack is therefore covered by **5.80**, and aldehydes and ketones react predominantly in this sense.

Just as the attack angle is not exactly 90°, so none of the angles in the transition structure has to correspond to those drawn schematically in **5.80**, and in particular the large group does not have to be exactly at right angles to the π bond. Angles close to those shown in **5.82** would seem to be near the minimum, although obviously affected in detail by the relative steric requirements of the R, Nu, S, M and L groups.

We need to graft onto this picture consideration of the Flippin-Lodge angle [see (Section 5.1.3.1) page 160]. In aldehydes, with only a hydrogen atom on one side of the carbonyl group, the nucleophile will be tilted away from the stereogenic centre, and hence away from its influence, offering one explanation for why aldehydes show lower selectivity than ketones.

When one of the groups on the stereogenic centre is either more electropositive than carbon or more electronegative, we need to take into account the electronic effect of having a polarised bond next to the reaction site. The lowest-energy conformations of the starting materials are fairly understandable for both cases. A bond to an electropositive element like a silicon atom, will be conjugated to the carbonyl $\pi$ bond **5.83**, because hyperconjugation with an electron deficient group like a carbonyl is stabilising (see pages 74–76). However, a bond to an electronegative element, like a halogen, oxygen or nitrogen, will avoid the conjugation with the carbonyl group, since it is energy-raising and an $\alpha$-haloketone will have the lowest energy when the dipoles are opposed **5.84**. In both cases, the stereochemistry of attack actually observed corresponds to attack from the less hindered side of these conformations, as in the examples **5.85** → **5.86** and **5.87** → **5.88**.

The problem with these explanations is that the lowest energy conformations are in neither case necessarily the most reactive—the Curtin-Hammett principle. The conjugation of the Si—C bond with the carbonyl group in the $\alpha$-silyl aldehyde **5.83** should raise the energy of the LUMO. Both the overall stabilisation and the higher-energy LUMO ought to make this a less reactive conformation than one in which the silyl group is orthogonal to the $\pi$ bond. Nevertheless, the silyl group is the large group, and a transition structure **5.85** with the silyl group anti to the incoming nucleophile might be expected to be preferred, simply using the Felkin-Anh arguments based on steric effects and the probably small electronic effect.

When one of the three groups on the stereogenic centre is an electronegative element, the substituent will be activating—in the higher-energy conformation **5.89**, the overlap of the C—Cl bond with the $\pi$ bond lowers the energy of the LUMO. The sense of the attack in this transition structure is the same as in Cornforth's transition structure **5.87**, and the outcome, whatever the explanation, corresponds to the Felkin-Anh rule if the electronegative element is treated as the large substituent, whatever its actual size relative to the other substituents. The effect of the chlorine atom can be understood as twofold—it activates the carbonyl group to attack, and the fraction of negative charge it carries will repel the incoming nucleophile. Calculations give a transition structure similar to **5.89**, but a good case has been made for reviving the Cornforth explanation.

**5.89**                                                    **5.90**

There is one further complication, but this time easily resolved, well understood, and supported by high-level calculations. When the electronegative element coordinates to a metal that can simultaneously coordinate to, and activate, the carbonyl group, the conformation will be that of a ring **5.90**. The attack from the less hindered side, opposite to the group R, is then relatively easily predicted, and is the opposite of that predicted from the Felkin-Anh rule. It has long been known as chelation control.

*5.2.1.2   Nucleophilic Attack on Cyclohexanones.*   At first sight, attack on a cyclohexanone with a locked conformation **5.91** would appear to resemble the problem covered by the Felkin-Anh arguments. The equatorial hydrogen atoms on C-2 and C-6 are forced to be the inside groups, more or less eclipsing the carbonyl group. The top surface of the carbonyl group, as drawn, is conjugated to the bonds between C-2 and C-3 and between C-5 and C-6, and the bottom surface is conjugated to the axial hydrogen atoms on C-2 and C-6. The steric difference between the two surfaces is therefore clear—the lower surface is less hindered, and equatorial attack should be favoured, giving more of the axial alcohol **5.92** than of the equatorial alcohol **5.93**. This is what happens with large nucleophiles, and with hydride delivered from hindered reagents like selectride (LiBH$^s$Bu$_3$). The problem comes with small nucleophiles like hydride delivered from lithium aluminium hydride, when axial attack is favoured, giving more of the equatorial alcohol **5.93**.

**5.91**                    **5.92**            93:7          **5.93**
                                                10:90

**5.94**

Felkin's explanation draws attention to the fact that the C=O bond is not perfectly eclipsing the equatorial C—H bonds on C-2 and C-6—it is pointing 4–5° lower, as can be seen in the Newman projection **5.94** along the C-1 to C-6 bond of **5.91**. Equatorial attack, therefore, would have to push the oxygen past the two equatorial hydrogen atoms. Axial attack, however, creates a transition structure in which the oxygen can move down into an equatorial position without developing any eclipsing. The energetic cost of the eclipsing caused by equatorial attack is described as *torsional strain*, present in the transition structure for equatorial attack, and is evidently greater than the 1,3-diaxial interaction with the hydrogen atoms on C-3 and C-5 suffered by small nucleophiles during axial attack.

### 5.2.1.3  *Nucleophilic Attack on Cyclic Oxonium and Iminium Ions.*

We have already seen the anomeric effect that makes axial anomeric bonds longer than their equatorial counterparts **2.51**, and that axial nucleofugal groups ionise more easily. The corollary, because the transition structure is low in energy for both the forward and back reactions, is that nucleophiles attack more rapidly from the anomeric direction. The more lone pairs that can be anti to the leaving group or incoming nucleophile, the easier the reaction is in either direction. Thus the orthoester **5.95** reacts with Grignard reagents to give the product **5.96** of axial attack, whereas the equatorial isomer is unreactive. Both the departure of the methoxy group and the attack by the nucleophile are axial, because the anti-periplanar lone pairs weaken the axial bond, and yet stabilise the forming bond. The same pattern is found in imminium ions, where nucleophiles attack from the direction that most easily sets up an anti lone pair, as in the reduction of the imminium ion **5.97** to give monomorine-I **5.98**.

### 5.2.1.4  *The $S_N2'$ Reaction.*

The $S_N2'$ reaction, in which an allyl system equipped with a nucleofugal group undergoes attack at C-3, is the vinylogous version of the $S_N2$ reaction. An allyl halide usually undergoes a straightforward $S_N2$ reaction, with direct attack at C-1 and inversion of configuration at that centre. On the special occasions when the $S_N2'$ reaction does take place, it can, in principle, do so with syn **5.99** → **5.100** or anti stereochemistry **5.99** → **5.101**. Both have been observed, and it is not yet clear what all the factors are that favour the one over the other.

The heart of the problem is to know how the stereogenic centre carrying the nucleofugal group X differentially affects the charge and the orbitals on the top and bottom surfaces of the double bond. We can begin by treating this as a problem of predicting the direction of pyramidalisation, where we extend the idea used to explain axial attack on cyclohexanones by unhindered nucleophiles [see (Section 5.2.1.2) page 171]. The carbon atom and the oxygen atom of the carbonyl group and the two substituents C-2 and C-6 do not lie perfectly in the same plane—C-1 is a little above the plane defined by C-2, C-6 and the oxygen atom in **5.94**. The carbon is said to be pyramidalised. There is therefore the possibility that we can predict the stereochemistry of attack on a trigonal carbon from the sense of pyramidalisation, and the stereochemistry of many reactions is indeed well correlated to the sense of quite small degrees of pyramidalisation as measured by X-ray crystal structure determinations.

The direction of pyramidalisation in open-chain systems is also fairly easy to predict from first principles—it takes place away from the substituent held most nearly at right angles to the plane of the carbonyl group, because it leads to a greater degree of staggering **5.102** in that direction. It is independent of whether the substituent is an electron donor or an acceptor, as calculations for 2-silylace-taldehyde **5.103** and 2-fluoroacetaldehyde **5.104** show. The former, with an Si—C—C—O dihedral angle of 92.4° is the global minimum, but the latter has to be constrained to the same angle, otherwise a structure with the fluorine in the plane of the carbonyl group is obtained. The sense of pyramidalisation in these two cases matches the sense of nucleophilic attack discussed earlier[see (Section 5.2.1.1) page 170].

If pyramidalisation is induced by increasing the degree of staggering, the trigonal carbon atoms in a conjugated chain ought to be pyramidalised in alter-nating directions. Thus a C=C double bond **5.105** pyramidalised downwards at C-2 by the anti-periplanar R group on C-1 ought, other factors being equal, to be pyramidalised upwards at C-3, and so on down a longer chain.

Applying this idea to the $S_N2'$ reaction, with the advantage that we can be confident that the nucleofugal group in a concerted reaction will be conjugated to the double bond at the time of reaction, and will therefore be the prime influence inducing the pyramidalisation, the structure we obtain is **5.106**, implying that the preferred reaction ought to be on the upper surface of C-3, syn to the leaving group X. We come to the same conclusion with tau bonds—the attack on the upper surface of C-3 is anti to the lower tau bond **5.107**, showing inversion of configuration at that atom, and that in turn is anti to the leaving group X, which is also effectively displaced with inversion.

**5.106**                                             **5.107**

However primitive these treatments are, the conclusion has a certain appeal, for it implies that having two more electrons in the transition structure changes the stereochemistry, from inversion in the $S_N2$ reaction to syn in the $S_N2'$ reaction. We saw the same change in substitution reactions, retention for $S_E2$ to inversion for $S_N2$, and for elimination reactions, anti for E2 to syn for E2′ [see (Section 5.1.2.2) page 157], but in neither of those cases, nor in the $S_N2'$ reaction, is it reliable.

Experimentally, there is a trend in favour of this pattern for the $S_N2'$ reaction, but it is not fixed immutably. In the first examples to which stereochemistry was assigned, Stork found that **5.108** → **5.109** and **5.110** → **5.111** were stereospecifically syn, but changing to a sulfur nucleophile reduced the degree of stereospecificity from almost completely syn to variably syn or anti, depending upon the solvent. In an open-chain system **5.112** having only the one stereocentre, designed to minimise the problem that the reference stereocentre in Stork's examples might impart some conformational preferences, and with only the difference between hydrogen and deuterium at C-3, the reaction **5.112** → **5.113** was highly syn stereospecific (> 96:4). There are also a few reactions showing more or less preference for an anti $S_N2'$ reaction.

**5.108**                 **5.109**                **5.110**                **5.111**

**5.112**                          **5.113**

The reactions of alkyl cuprates with allylic acetates are always stereospecifically anti **5.114** → **5.115**, although the formation of racemic product shows that regiocontrol has been lost. These reactions are not *mechanistically* $S_N2'$ reactions, since they involve in the first step coordination of the copper on the lower surface, followed by the formation of a $\pi$-allyl system at the same time as the acetate leaves. The delivery of the methyl group to the same surface, and equally to both ends of the allylic system, is a reductive elimination. It seems likely that the decisive step determining the stereoselectivity is that coordination of the copper anti to the nucleofugal group is needed before it will leave.

| **5.114** | Me₂CuLi → | *anti* $S_N2'$ | + | **5.115** | *inversion* $S_N2$ |

### 5.2.2  Nucleophilic and Electrophilic Attack on Cycloalkenes

We now turn to the stereochemistry governed by a ring system, and we shall look at both nucleophilic and electrophilic attack, since usually they have similar stereochemical preferences rather than contrasting preferences. In addition to several reactions that are straightforwardly electrophilic attack, we shall see several which can be described as electrophilic in nature, like the reactions of alkenes with osmium tetroxide, with peracids, with some 1,3-dipoles, and with boranes, and of dienes with dienophiles in Diels-Alder reactions. Some of these reactions are pericyclic, the pericyclic nature of which we shall meet in Chapter 6. For now, it is only their diastereoselectivity that will concern us.

#### 5.2.2.1  *Monocyclic alkenes.*  Cycloalkenes have a preferred conformation, which may or may not influence the stereochemistry of attack upon the double bond. The attack is always more or less along the line of the p orbitals, as discussed earlier [see (Section 5.1.3) page 158], but there may be steric or electronic effects operating to affect which of the two surfaces of the double bond is best presented to an incoming reagent. At its most simple, a single substituent on a four- or five-membered ring usually causes electrophiles (and nucleophiles) to attack anti to it. The uncomplicated explanation is that the large group hinders approach from the side it occupies. The relatively small degree of kinking away from a flat conformation in these small rings can usually be ignored, but it cannot be ignored with six-membered and larger rings. In a six-membered ring a large group will be equatorial in the most abundant conformation, and the two surfaces of a cyclohexene are sterically not very different. The alkene **5.116** is attacked by peracid with little selectivity, and what little there is happens to be in favour of attack syn to the resident group.

**5.116**

**5.117**    **5.118** 70:30    **5.119**    **5.120** 83:17

In the epoxidation reaction, electrophilic attack is taking place concertedly at both carbons of the double bond. In contrast, when the attack is only at one of the two carbons, the degree of stereocontrol in a six-membered ring becomes high. One simple way to appreciate this is to see the pyramidalisation as taking place to move the cyclohexene conformation from a half chair closer to a full chair. The enolate methylations **5.117** and **5.119** are selective for axial attack, leading to chair conformations **5.118** and **5.120** in the product. In the former, the attack is anti to the *tert*-butyl group and in the latter it is syn, in contrast to what happens in five-membered rings. The reason can be seen on both sides of the reaction coordinate—in the presumed pyramidalisation of the starting material, and in the greater ease of forming a chair conformation rather than a boat in the product. There are similar explanations for the stereochemistry of nucleophilic attack in six-membered rings such as 5-substituted cyclohexenones, which again allow stereochemistry to be transmitted effectively from one stereogenic centre to another three atoms away.

Medium-sized rings have a different feature allowing high levels of diastereo-selectivity. In these rings, a double bond does not lie flat in the average plane of the ring. Instead it presents one face towards reagents, while the other is obscured by the ring wrapped around behind it. If the conformation can be controlled, high levels of stereocontrol can be achieved, whether attack is by an electrophile **5.121** or a nucleophile **5.122**. The problem with medium-sized rings is to predict their conformation. Only with simple cases is it easy to see why conformations like **5.121** and **5.122** are the most populated, with the rings kinked in the direction that makes the methyl group equatorial, and why attack from the front surface as drawn is most favourable.

**5.121**    86:14    **5.122**    99:1

*5.2.2.2 Bicyclic alkenes.* Bicyclic systems like the alkene **5.123** are well known to be attacked from the exo direction, on the less hindered convex face of the bicyclic system. Similarly high levels of stereocontrol are found for nucleophilic attack on bicyclic systems, as in the reduction of the ketone **5.124**, in which the preference for exo attack overwhelms the steric hindrance offered by the adjacent methyl group. Bicyclic systems, especially like these with a zero bridge, are often used in synthesis to give reliably high levels of stereocontrol, with the penalty that there may be many steps needed to open them up to reveal a target structure.

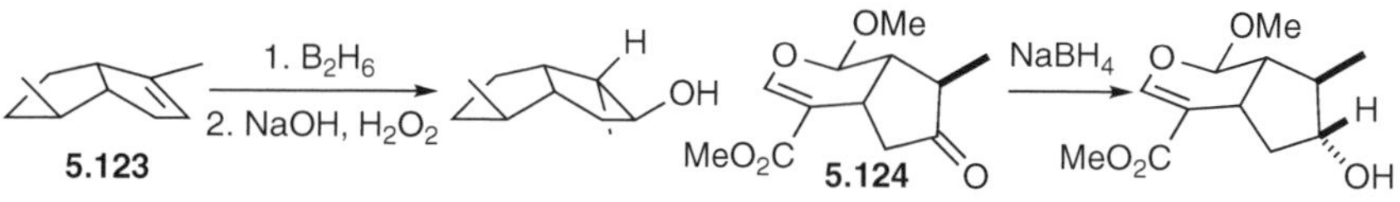

However, there are anomalies, where a steric effect is clearly not enough to explain the observed stereoselectivity. The steric argument, although commonly invoked, is weak for norbornene **5.125** and for the bicyclo[2.2.2]octadiene **5.126**, but the selectivity for exo attack in the former and endo in the latter is strong. The steric argument is nearly nonexistent for the diene **5.127**, but it shows high levels of diastereoselectivity, for attack in the endo direction, which is, if anything, the more hindered.

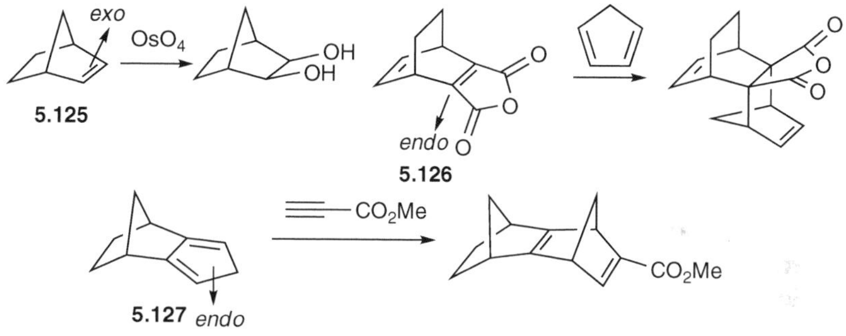

One possible explanation for these results is based on the sense of pyramidalisation at the reacting trigonal carbons. In norbornene they are pyramidalised with the p orbitals bulging in the exo direction **5.128**, anti to the two-carbon bridge, since the $\sigma$ bonds leading to them, marked bold, are better aligned for overlap with the orbitals of the $\pi$ bond **5.129** than the $\sigma$ bonds of the one-carbon bridge. Even though the degree of pyramidalisation is small, electrophilic attack from the exo direction will induce less torsional strain, just as axial attack by nucleophiles does on cyclohexanones. The double bond undergoing attack in the bicyclooctadiene **5.126** pyramidalises with the p orbitals bulging downwards **5.130**, because the $\sigma$ bonds from the tetrahedral carbons, marked in bold, are more effective donor substituents than the $\sigma$ bonds from the trigonal carbons. The latter have more $\sigma$ bonding involving the 2s orbitals than the former, making the tetrahedral carbons effectively more electropositive than the trigonal carbons. In the diene **5.127**, the alternation of pyramidalisation along a chain of double bonds [see (Section 5.2.1.4) page 173] implies that the termini of the double bonds would be pyramidalised downwards, exaggerated and shown for the HOMO in **5.131**, and that is the direction from which the electrophilic dienophile, attacks.

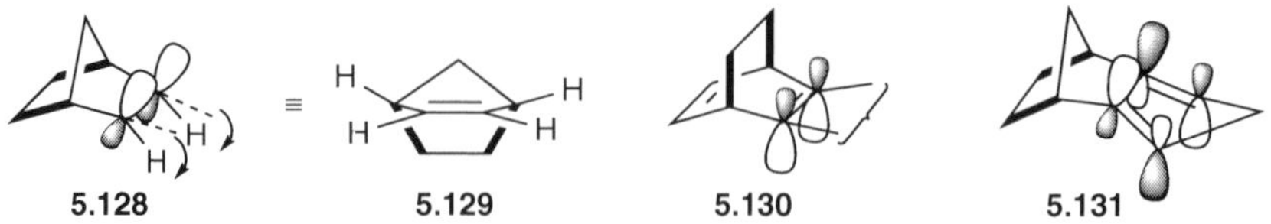

| 5.128 | 5.129 | 5.130 | 5.131 |

### 5.2.3   Electrophilic Attack on Open-Chain Double Bonds with Diastereotopic Faces

#### 5.2.3.1   The Houk Rule for Steric Effects in Electrophilic Attack on Open-Chain Alkenes.
The factors affecting which side of an open-chain alkene with a neighbouring stereogenic centre is attacked depends upon its conformation, which is affected by the fact that it is a C=C double bond and not a C=O. In contrast to the Felkin-Anh picture, the *small* group is likely to be inside **5.132**, more or less eclipsing the double bond, but electrophilic attack is still, in the absence of stereoelectronic effects, likely to be anti to the large group.

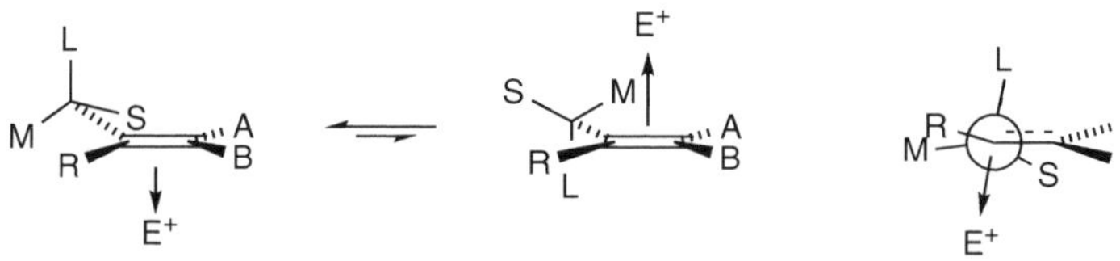

| 5.132 | 5.133 | 5.134 |

An exception to this pattern is when the medium-sized group M is small, as with a methyl group, and when, at the same time, the substituent A on the double bond, *cis* to the stereogenic centre, is a hydrogen atom. In this case, the conformation **5.133**, with the medium-sized group inside, although not usually the lowest in energy, is populated, and attack from the less hindered side in this conformation becomes plausible, especially as it is now taking place syn to the small group instead of syn to a medium-sized group. Furthermore, the size of the group R will affect the outcome—if it is large, the conformation **5.133** can actually be lower in energy than conformation **5.132**. As with nucleophilic attack on a carbonyl group, the dihedral angles will not have the large group precisely at right angles, and the transition structure may well be more like that shown in **5.134**, which applies to a nonbridging electrophile attacking with an obtuse angle outside the double bond. Electrophilic attack is often exothermic, and the structure of the starting material is therefore likely to be a good guide to the transition structure. A straightforward example illustrating this sense of attack is the methylation of the lithium enolate **5.135** to give mainly the ketone

**5.137**, and the protonation of the enolate **5.136** in the same sense to give mainly the diastereoisomeric ketone **5.138**.

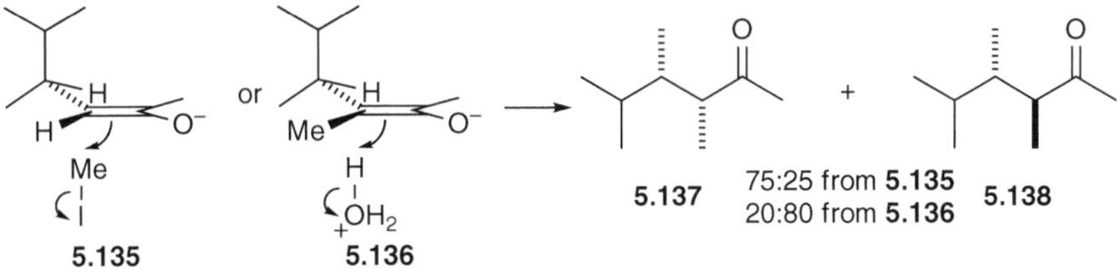

A reaction in the alternative sense **5.133** is the cycloaddition of a nitrile oxide to a terminal alkene, which gives mainly the diastereoisomer **5.141** by way of the transition structure **5.139**. Nitrile oxide cycloadditions are among those dipolar cycloadditions which are electrophilic in nature. The substituent A is a hydrogen atom, and the medium-sized group is only a methyl group, so it fits the criteria that make this pathway plausible.

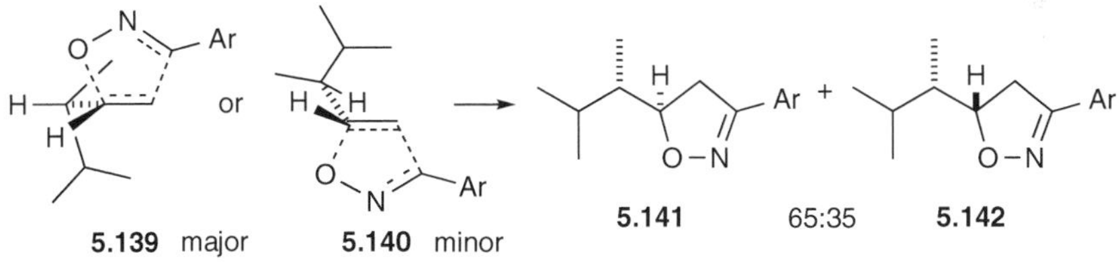

Bridging electrophiles, as in epoxidation, are fairly well behaved in the sense **5.132**, possibly because the acute approach angle does not leave room for the medium-sized group to sit inside. Bromination, however, is reversible in the first step, and the stereochemistry actually observed, although often in the sense **5.132**, is partly governed by the relative ease with which each of the diastereoisomeric epibromonium ions is opened. As a consequence, the ratio of diastereoisomers is not reliably a measure of the relative rates of attack on the diastereotopic faces of the alkene.

### 5.2.3.2 The Influence of Electropositive Substituents.

The story remains the same in the presence of the stereoelectronic effect of a donor substituent. A silyl group on the stereogenic centre is the best studied among these, where it is notable for imparting a strikingly high level of open-chain stereocontrol in the sense **5.132** with the silyl group as the large substituent. Any electronic component will encourage attack anti to the silyl group, since the bond to an incoming *electrophile* is electron deficient, and the transition structure will be especially well stabilised by an anti-periplanar donor substituent like silyl.

Open-chain $S_E2'$ reactions are special cases of electrophilic attack on a C=C double bond in the sense **5.132**, with attack specifically at C-3, and followed by the loss of an electrofugal group from the stereogenic centre. The most studied of these are the electrophilic substitution reactions of stereodefined allylsilanes, where the silyl group is the large substituent, which take place overall in the anti sense **5.143** → **5.144**, especially when the group R is large enough, as with R = Ph, to experience severe $A^{1,3}$ interactions in the conformation **5.133** with R as the medium-sized group.

E = DO$_2$CCF$_3$ and
$^t$BuCl, RCHO and
RCOCl + Lewis acid

There are a number of $S_E2'$ reactions which are not open-chain reactions. The electrophile is typically an aldehyde coordinated at the time of reaction to an electropositive atom like boron, tin or zinc on the stereogenic centre. These reactions usually use cyclic, chair-like transition structures, are called metallo-ene reactions, and are inherently syn overall.

### 5.2.3.3   *The Influence of Electronegative Substituents.*   The real problem explaining the diastereoselectivity of electrophilic attack comes when one of the bonds on the stereogenic centre is to an electronegative atom, making it an electron-withdrawing group. We shall leave out of consideration those cases where an oxygen atom delivers the reagent by hydrogen bonding or Lewis acid-base coordination—reactions like epoxidation and the Simmons-Smith reaction on allylic alcohols. These reactions use cyclic transition structures, and the diastereoselectivity is determined by the conformation of the ring—molecular orbital considerations are very secondary.

A C—X bond conjugated with a π bond will lower the energy of the HOMO, and make the alkene less reactive towards electrophiles. Consequently, when it is not delivering the reagent, an electronegative substituent often adopts a confor-mation in which the C—X bond is not in conjugation with the π bond. For example, in the cyclic alkene **5.145** dipolar cycloaddition takes place syn to the chlorine atoms, and the Diels-Alder reaction on acetoxycyclopentadiene **5.146** takes place syn to the acetoxy substituent. These results are in striking contrast to the reactions on small rings, which usually take place anti to the resident sub-stituents whatever they are. Relatively mild distortions allow the C—H bonds to be lined up to overlap with the π bond, with the C—Cl and C—O bonds mostly bent out of conjugation, and the incoming reagent will approach the pyramidalised p orbitals bulging anti to the better conjugated C—H bonds.

In open chain alkenes there is more flexibility for the bond to the electronegative substituent to avoid conjugation with the double bond. When the substituent A (in **5.132**) is larger than a hydrogen atom, we might expect most reactions to take place with a transition structure like **5.132**, with the electronegative atom as the medium-sized group oriented to be as little in conjugation with the double bond as possible. In practice this is rarely the case, and reaction in this sense appears to be most likely when the group A is significantly larger than a methylene group and the group R larger than a hydrogen atom. An example is the dihydroxylation of the Z-alkene **5.148** with osmium tetroxide, in which the $A^{1,3}$ interaction between the inside hydroxyl group and the ethoxycarbonyl group is too severe for the conformation **5.147** to be significantly populated. The ethyl group counts as the large group, and the hydroxyl as the medium-sized group, and dihydroxylation gives only the lactone **5.149** by attack on the lower surface.

When the substituent A is a hydrogen atom, it appears that the lowest energy conformation in the transition structure is often that with the electronegative substituent inside. Reactions therefore take place in the sense **5.133** with the electronegative element treated as the medium-sized group, which it often is. An example is the dihydroxylation of the E-alkene **5.150**, which reacts in a conformation with the hydroxyl group inside, and the product is the lactone **5.151** with the opposite stereochemistry at C-3 from that of the product **5.149** derived from the Z-isomer. Reactions often take place with the medium-sized oxygen substituent inside, even when the substituent A is larger than a hydrogen atom. This propensity is known as the 'inside alkoxy' effect, and only when the substituent A is larger, as it is with the ethoxycarbonyl group in the alkene **5.147**, is the alkoxy group pushed into an outside position. Other reactions cleanly showing this stereochemistry are nitrile oxide cycloadditions and Diels-Alder reactions with acetylenic dienophiles.

In contrast to the dihydroxylations, the hydroboration of allylic alcohols takes place without an 'inside alkoxy' effect in the sense **5.132** with the oxygen atom treated as the medium-sized group. It is probably significant that the 'inside alkoxy' effect is most noticeable with reagents which are relatively electrophilic in nature, and not with boranes, which are only mildly electrophilic.

### 5.2.4  Diastereoselective Nucleophilic and Electrophilic Attack on Double Bonds Free of Steric Effects

On account of the delicate balance between steric and electronic effects, and the difficulty of teasing them apart, a number of reactions have been carried out on substrates designed as far as possible to remove the steric component, and to leave an electronic one. Substrates which have been used in this kind of experiment include the adamantanones **5.152** (Y = O), the norbornanones **5.153** (Y = O), and the cyclopentanones **5.154**, for nucleophilic attack, and the corresponding alkenes **5.152** (Y = CH$_2$), **5.153** (Y = CH$_2$) and **5.155** for electrophilic attack by such electrophiles as peracid, dichlorocarbene and dichloroketene. These ketones and alkenes are designed to have the possibility of purely electronic effects trans-mitted rationally through the $\sigma$ framework, and they do all show diastereoselec-tivity, occasionally to a high degree.

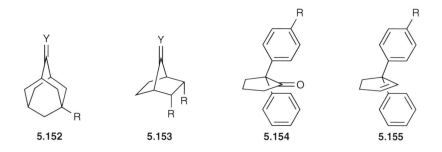

5.152     5.153     5.154     5.155

As an example of the kind of experiment that is carried out in this area, borohydride attack on the ketone **5.152** (Y = O, R = F) gave mainly (62:38) the alcohol derived from attack syn to the electron-withdrawing substituent, whereas the ketone **5.152** (Y = O, R = SnMe$_3$) gave mainly (52:48) the alcohol derived from attack anti to the electron-donating substituent. The degree of stereoselectivity is unimpressive, but the electron-withdrawing fluorine and the electron-donating trimethylstannyl group exert their effects in opposite senses.[19]

In trying to explain these results, we no longer have to consider which bond is conjugated to the carbonyl group at the time of reaction—the bonds are fixed in their orientation, and attack is always axial in one ring and equatorial in the other. One explanation is that the C—F bond in the fluoroadamantone is conjugated to the C—C bonds, as emphasised with the intervening bold bonds in **5.156**. This conjugation makes the emboldened C—C bonds less electron-donating than the C—C bonds on the other side of the carbonyl group, emphasised in **5.157**. Following an argument advanced by Cieplak, the nucleophile attacks the side opposite the better donor, since the bond that is being formed is inherently electron-deficient. With the electron-donating trimethylstannyl group, it is the bonds on the side of the stannyl group, emphasised in **5.158**, that are made the better donors, and the nucleophile attacks anti to them.

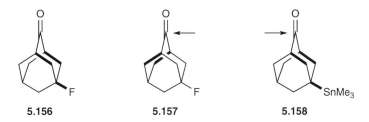

5.156         5.157         5.158

Similarly, since Cieplak's argument suggests that electrophiles will also attack anti to the better donor, the corresponding electrophilic attack on the alkenes **5.152** ($Y = CH_2$) by borane is also syn selective for the fluoroadamantane system ($R = F$), and anti selective for the trimethylsilyladamantane system ($R = SiMe_3$). Although these experiments, and several others similar to them, support Cieplak's contention, there are enough exceptions and unexplained features that this account of the selectivity is not completely accepted. The most frequently cited alternative explanation for the effect of remote substituents is electrostatic effects through space rather than relayed through the bonds.

## 5.3 Exercises

1. In $S_N2$ reactions the LUMO is the antibonding C—Halogen orbital, not the antibonding C—H orbital. In $S_E2$ reactions the HOMO is the bonding C—Metal orbital, not the bonding C—H orbital. Show by an interaction diagram that this is justified: that the frontier orbitals for a C—Halogen bond are lower than those of a C—H (or C—C) bond, and that those of a C—Li bond are higher. It is probably simpler to use hybridised orbitals.

2. Suggest explanations for why aldehydes are less selective in the Felkin-Anh sense than ketones.

3. The dioxan-5-one **5.159** is attacked by nucleophiles with a much higher degree of axial selectivity than cyclohexanones, whereas the dithia analogue **5.160** is attacked from the equatorial direction. Consider the arguments that might account for these two effects.

Ph—⟨O—O⟩—MeMgI → Ph—⟨O—O⟩—OH    Ph—⟨S—S⟩—MeMgI → Ph—⟨S—S⟩—OH

**5.159**         >98:2         **5.160**         92:8

4. The protonation of the steroidal enol ethers ($E$)-**5.161** and ($Z$)-**5.161** give diastereoisomeric products **5.162** and **5.163**, respectively. Identify the features that explain why the geometry of the double bond appears to control the diastereoselectivity.

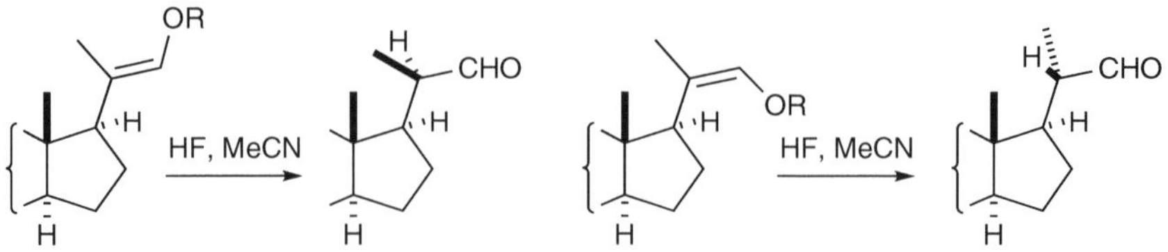

(E)-**5.161**    **5.162** 80:20    (Z)-**5.161**    **5.163** 80:20

5. Predict what stereochemistry you might expect for the borohydride reduction of 5-azaadamantanone **5.164**.

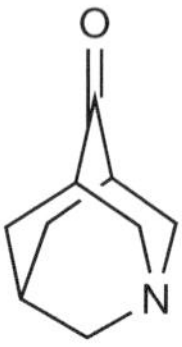

**5.164**

# 6 Thermal Pericyclic Reactions

Pericyclic reactions[20,21] are the second distinct class of the three, more or less exclusive categories of organic reactions—ionic (Chapters 4 and 5), pericyclic (this Chapter) and radical (Chapter 7). Their distinctive features are that they have cyclic transition structures with all the bond-making and bond-breaking taking place in concert, without the formation of an intermediate, and they are highly stereospecific, adding to the examples in Chapter 5. The Diels-Alder reaction and the Alder 'ene' reaction are venerable examples.

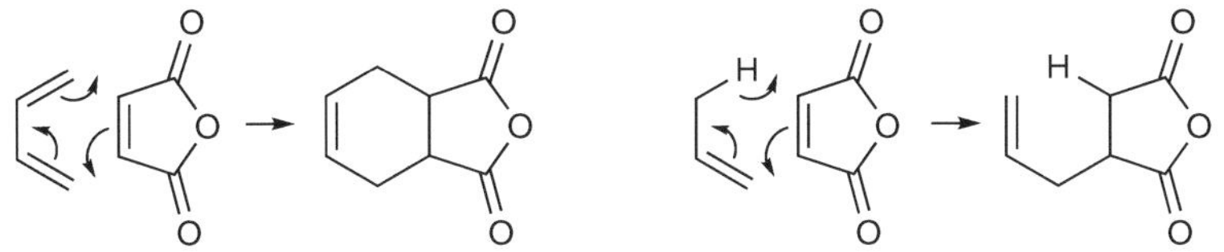

A Diels-Alder reaction                      An Alder ene reaction

The curly arrows are drawn clockwise, but they could equally well have been drawn anticlockwise. Thus, there is no absolute sense in which the hydrogen atom that moves from one carbon atom to the other in the ene reaction is a hydride shift, as seems to be implied by the clockwise curly arrow, or a proton shift, as it would seem to be if the arrows were to have been drawn in the opposite direction. In other words, neither component can be associated with the supply of electrons to any of the new bonds. The curly arrows therefore have a somewhat different meaning from those used in ionic reactions. They share with all curly arrows the function of showing where to *draw* the new bonds and which ones not to draw in the resulting structure. They are related to the arrows used to illustrate resonance in benzene, in having no sense of direction, but the Diels-Alder reaction has starting materials and a product, and aromatic resonance in benzene does not.

Within the overall category of pericyclic reactions, it is convenient to divide them into four main classes. These are *cycloadditions*, *electrocyclic reactions*, *sigmatropic rearrangements*, and *group transfer reactions*, each of which possesses special features not shared by the others, and some of which employ a terminology that cannot be used without confusion if applied to a reaction belonging to one of the other classes. It is a good idea to be clear about which class of reactions you are dealing with, in order to avoid using inappropriate terminology.

## 6.1　The Four Classes of Pericyclic Reactions

*Cycloadditions* are the largest class. They are characterised by two components coming together to form two new $\sigma$ bonds, one at each end of both components, joining them together to form a ring, with a reduction in the length of the conjugated system of orbitals in each component (Fig. 6.1a), with the Diels-Alder reaction and [1,3]-dipolar cycloadditions the most important, featureful and useful of all pericyclic reactions. They are reversible, and the reverse reaction is called a *retro-cycloaddition* or a *cycloreversion.*

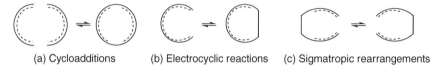

(a) Cycloadditions　　　(b) Electrocyclic reactions　　　(c) Sigmatropic rearrangements

**Fig. 6.1**　The three most important classes of pericyclic reaction

*Cheletropic reactions* are a special group of cycloadditions in which the two $\sigma$ bonds are made or broken to the *same* atom. Thus sulfur dioxide adds to butadiene to give an adduct, for which the sulfur has provided a lone pair to one of the $\sigma$ bonds and has received electrons in the formation of the other. It is an oxidative addition to the sulfur dioxide, changing it from $S^{IV}$ to $S^{VI}$. The reaction is readily reversible on heating.

both $\sigma$ bonds
made to the same
atom

*Electrocyclic reactions* are unimolecular, unlike cycloadditions, and are characterised by the creation of a ring from an open-chain conjugated system. A $\sigma$ bond is formed across the ends of a conjugated system, with the conjugated system becoming shorter by one p orbital at each end (Fig. 6.1b). The reactions are inherently reversible, with the direction they take being determined by thermodynamics. Most electrocyclic reactions are ring-closings, since a $\sigma$ bond is created at the expense of a $\pi$ bond, but a few are ring-openings, because of ring strain. Representative electrocyclic reactions are the ring-opening of cyclobutene to give butadiene, and the ring-closing of hexatriene to give cyclohexadiene.

*Sigmatropic rearrangements* are unimolecular isomerisations, and formally involve overall the movement of a σ bond from one position to another, with a concomitant movement of the conjugated systems in order to accommodate the new bond (Fig. 6.1c). The oldest known example is the first step in the Claisen rearrangement, when a phenyl allyl ether is heated. The first step is the sigmatropic rearrangement in which the single bond, drawn in bold, moves to its new position in the intermediate. It has effectively moved three atoms along the carbon chain (from C-1 to C-3), and three atoms along the chain of the oxygen atom and two carbon atoms (O-1' to C-3'). This type of rearrangement is called a [3,3]-shift, with the numbers identifying the number of atoms along the chain that each end of the bond has moved. The second step forming *o*-allylphenol is an ordinary ionic reaction—the enolisation of a ketone. It is perhaps a timely reminder that ionic reactions often precede or follow a pericyclic reaction, sometimes disguising the pericyclic event.

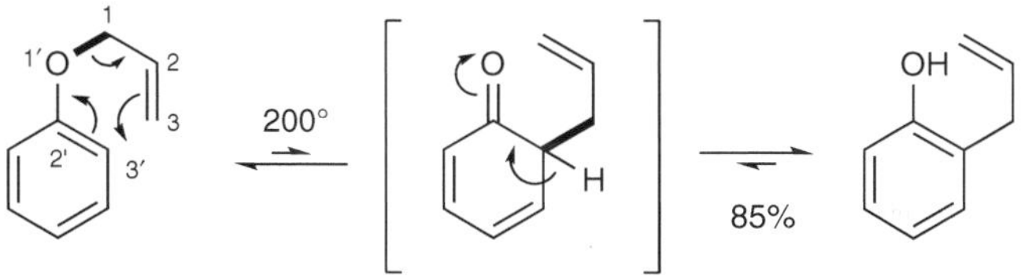

A quite different looking sigmatropic rearrangement is the hydrogen atom shift long known from the chemistry of vitamin D. In this case, the end of the H—C bond attached to the hydrogen atom (H-1') necessarily remains attached to the hydrogen, but the other end has moved seven atoms (C-1 to C-7) along the conjugated carbon chain. This reaction is therefore called a [1,7]-shift.

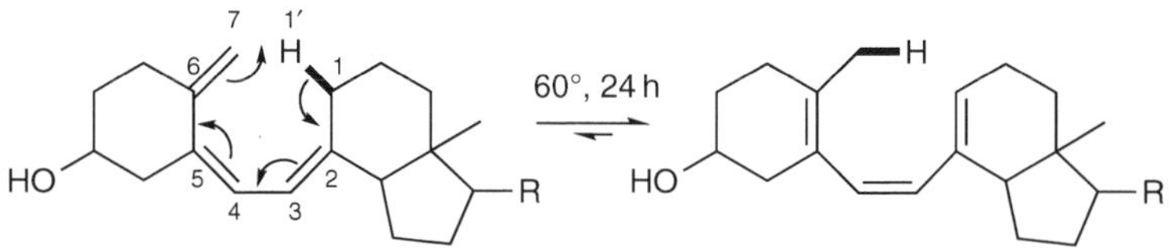

Another sigmatropic reaction is the Mislow rearrangement of allyl sulfoxides, which is invisible because it is thermodynamically unfavourable, but the ease with which it takes place explains why they racemise so much more easily than other sulfoxides. One end of the C—S bond moves from the sulfur (S-1') to the oxygen atom (O-2') and the other end moves from C-1 to C-3. This is therefore called a [2,3]-sigmatropic shift, the bond marked in bold moving two atoms at one end and three at the other.

*Group transfer reactions* make up the fourth category; they have few representatives, with ene reactions by far the most common. The ene reaction usually needs to have electron-withdrawing groups attached to the enophile, and they usually take place from left to right as shown at the start of this Chapter, since overall a $\pi$ bond is replaced by a $\sigma$ bond. They resemble [1,5]-sigmatropic rearrangements, since a $\sigma$ bond moves, and they also resemble cycloadditions like Diels-Alder reactions, since one of the $\pi$ bonds of the diene has been replaced by a $\sigma$ bond in the ene component. Nevertheless, since the reaction is bimolecular and no ring is formed, they are neither sigmatropic rearrangements nor cycloadditions.

Ene reactions have a hydrogen atom moving from the ene component to the enophile, but other atoms can in principle move. In practice, the only elements other than hydrogen that are commonly found in this kind of reaction are metals, when the reaction is called a metalla-ene reaction. The carbon chains may also have one or more heteroatoms in place of the carbons. Thus if the atom carrying the hydrogen is an oxygen atom and the atom to which it is moving is also an oxygen atom, it becomes an aldol reaction. Aldol reactions are usually carried out with acid or base catalysis, and most are not significantly pericyclic in nature. The other well known type of group transfer is represented by the concerted syn delivery of two hydrogen atoms from diimide to an alkene or alkyne.

## 6.2   Evidence for the Concertedness of Bond Making and Breaking

The characteristic feature of all pericyclic reactions is the concertedness of all the bond making and bond breaking, and hence the absence of any intermediates.

Naturally, organic chemists have worked hard, and devised many ingenious experiments, to prove that this is true, concentrating especially on Diels-Alder reactions and Cope rearrangements. The following is a brief description of some of the more telling experiments.

The Arrhenius parameters for Diels-Alder reactions show that there is an exceptionally high negative entropy of activation, typically between $-150$ J K$^{-1}$ mol$^{-1}$ and $-200$ J K$^{-1}$ mol$^{-1}$, with a low enthalpy of activation reflecting the exothermic nature of the reaction. Bimolecular reactions inherently have high negative entropies of activation, but the extra organisation for the two components to approach favourably aligned, so that both bonds can form at the same time, accounts for the exceptionally high value in cycloadditions. The compact transition structure is also in agreement with the negative volumes of activation measured by carrying out the reaction under pressure. The rates of Diels-Alder reactions are little affected by the polarity of the solvent. Typically, a large change of dipole moment in the solvent, from 2.3 to 39, causes an increase in rate by a factor of only 10. If a zwitterionic intermediate were involved, the intermediate would be more polar than either of the starting materials, and polar solvents would solvate it more thoroughly. Stepwise ionic cycloadditions take place with increases in rate of several orders of magnitude in polar solvents. This single piece of evidence rules out stepwise ionic pathways for most Diels-Alder reactions, and the only stepwise mechanism left is that involving a diradical.

Deuterium substitution on the four carbon atoms changing from trigonal to tetrahedral as the reaction proceeds, gives rise to inverse secondary kinetic isotope effects, small, but measurable both for the diene and the dienophile. If both bonds are forming at the same time, the isotope effect when both ends are deuterated would be geometrically related to the isotope effects at each end. If the bonds are being formed one at a time, the isotope effects are arithmetically related. It is a close call, but the experimental results, both for cycloadditions and for cyclo-reversions, suggest that they are concerted. Similar isotope effects in Cope and Claisen rearrangements, and in the 'ene' reaction, come even more firmly to the conclusion that these are concerted reactions.

Another way of testing how one end of the dienophile affects the other end is to load up the dienophile with up to four electron-withdrawing groups, and see how each additional group affects the rate. A stepwise reaction between butadiene and tetracyanoethylene ought not to take place much more than statistically faster than a similar reaction with 1,1-dicyanoethylene, but a concerted reaction ought to, and does, take place much faster.

Finally the single most impressive piece of evidence comes from the very fact that pericyclic reactions obey the rules that we are about to expand upon. These rules only apply if the reactions are concerted. To have a few reactions accidentally obeying the rules would be reasonable, but to have a very large number of reactions seemingly fall over themselves to obey strict stereochemical rules, sometimes in what look like the most improbable circumstances, is overwhelmingly strong evidence about the general picture. Of course, no single reaction can be proved to be pericyclic, just because it obeys the rules—obedience to the rules is merely a necessary condition for a reaction to be considered as pericyclic. It seems more than likely that most reactions thought to be pericyclic actually are.

## 6.3  Symmetry-Allowed and Symmetry-Forbidden Reactions

Between 1965 and 1969 Woodward and Hoffmann presented rules for each of the different classes of pericyclic reaction. They showed that the allowedness or otherwise of reactions depended critically upon their stereochemistry. We shall go through the rules twice: first the rules class-by-class, and then again using the generalised rule that applies to all classes of pericyclic reactions.

### 6.3.1  The Woodward-Hoffmann Rules—Class by Class

**6.3.1.1  Cycloadditions.**    A cyclic movement of electrons can be drawn for any number of cycloadditions, but not all of them take place. Thus butadiene undergoes a Diels-Alder reaction with maleic anhydride, but ethylene and maleic anhydride do not give a cyclobutane when they are heated together.

It is not that this cycloaddition is energetically unprofitable—the cyclobutane is lower in energy than the two alkenes—so there must be a high kinetic barrier to the cycloaddition of one alkene to another. This is a deeply important point, and it is just as well that it is true—if alkenes and other double-bonded compounds could readily dimerise to form four-membered rings, there would be few stable organic molecules, and life would be impossible.

Diels-Alder reactions are classified as [4 + 2] cycloadditions, and the reaction giving the cyclobutane would be a [2 + 2] cycloaddition. This classification is based on the number of *electrons* involved. Diels-Alder reactions are not the only [4 + 2] cycloadditions. Conjugated ions like allyl cations, allyl anions and pentadienyl cations are all capable of cycloadditions. Thus, an allyl cation can be a 2-electron component in a [4 + 2] cycloaddition, as in the reaction of the methallyl cation **6.2** derived from its iodide **6.1**, with cyclopentadiene giving a seven-membered ring cation **6.3**. The diene is the 4-electron component. The product eventually isolated is the alkene **6.4**, as the result of the loss of the neighbouring proton, the usual fate of a tertiary cation. This cycloaddition is also called a [4 + 3] cycloaddition if you were to count the atoms, but this is a structural feature not an electronic feature. In this chapter it is the number of electrons that counts.

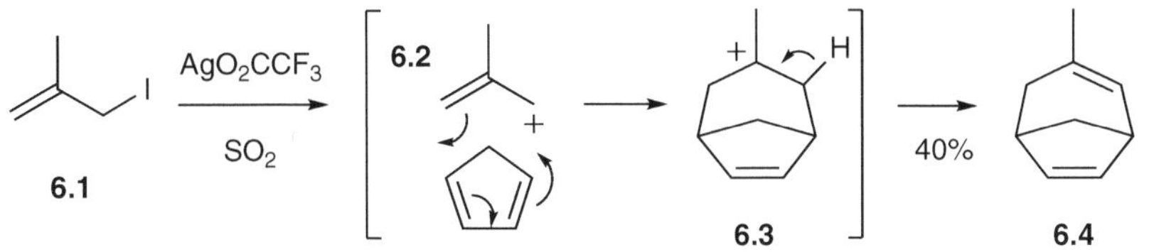

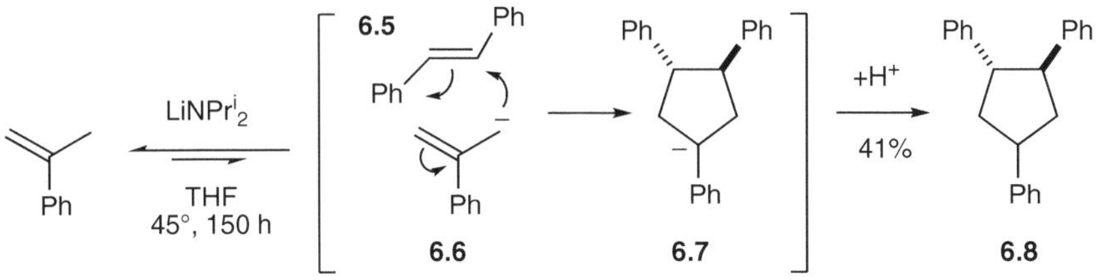

An allyl anion such as the 2-phenylallyl anion **6.6**, prepared in an unfavourable equilibrium by treating α-methylstyrene with base, undergoes a cycloaddition to an alkene such as stilbene **6.5**, present *in situ*, to give the cyclopentyl anion **6.7**, and hence the cyclopentane **6.8** after protonation. It is not always legitimate to think of conjugated anions simply as symmetrical conjugated systems of p orbitals, since a metal is usually bonded to the system. Allyl anion + alkene cycloadditions are rare, and calculations suggest that the few that are known are actually stepwise. It is evidently a penalty of the fact that allyl anions are not usually simple conjugated systems of p orbitals, making it difficult for the overlap to develop at both ends simultaneously. Nevertheless, the pericyclic pathway has the energetic benefit of forming both new σ bonds in the same step, and so some of these reactions may be pericyclic. Structurally it is a [3 + 2] cycloaddition, but electronically it is a [4 + 2] cycloaddition, just like the Diels-Alder and the allyl cation + diene reactions.

Yet another [4 + 2] cycloaddition, rather rare, is that between a pentadienyl cation and an alkene. The best known example is the perezone-pipitzol transformation **6.9 → 6.11**, where it is heavily disguised. It can be understood as beginning with an intramolecular proton transfer to give the intermediate **6.10**, which can then undergo an intramolecular [4 + 2] cycloaddition with the pentadienyl cation, emphasised in bold, acting as the 4-electron component and the pendant alkene, also bold, as the 2-electron component.

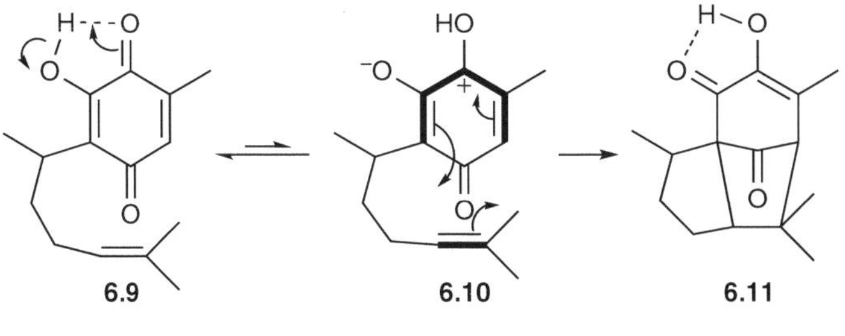

1,3-Dipolar cycloadditions **6.12** + **6.13** → **6.14**, however, are a large group of [4 + 2] cycloadditions isoelectronic with the allyl anion + alkene reaction. There is much evidence that these reactions are usually concerted cycloadditions. They have a conjugated system of three p orbitals with four electrons in the conjugated system, but the three atoms, X, Y, and Z in the dipole **6.12** and the two atoms A and B in the dipolarophile **6.13**, are not restricted to carbon atoms. The range of possible structures is large, with X, Y, Z, A and B able to be almost any combination of C, N, O and S, and with a double **6.12** or, in those combinations that can support it, a triple bond **6.15** between two of them.

All the reactions described so far have mobilised six electrons, but other numbers are possible, notably a few [8 + 2] and [6 + 4] cycloadditions involving 10 electrons in the cyclic transition structure. A conjugated system of eight electrons would normally have the two ends of the conjugated system far apart, but there are a few molecules in which the two ends are held close enough to participate in cycloadditions to a double or triple bond. Thus, the tetraene **6.17** reacts with dimethyl azodicarboxylate **6.18** to give the [8 + 2] adduct **6.19**, and tropone **6.20** adds as a 6-electron component to the 4-electron component cyclopentadiene to give the adduct **6.21**.

Adequately for most purposes, we can state a rule for which cycloadditions can take place and which not: thermal pericyclic cycloadditions are allowed if the total number of electrons involved can be expressed in the form $(4n + 2)$, where $n$ is an integer.

This rule needs to be qualified, because it applies to those reactions taking place in the sense shown in Fig 6.2a, in which the orbital overlap that is developing to

suprafacial component

suprafacial component

antarafacial component

suprafacial component

(a) Suprafacial overlap developing in both components

(b) Suprafacial overlap developing in one component and antarafacial overlap developing in the other

**Fig. 6.2**   Suprafacial and antarafacial defined for cycloaddition reactions

form the new $\sigma$ bonds takes place on the same surface of each of the conjugated systems, represented here by a curved line, but implying a continuous set of overlapping p orbitals from one end to the other. The dashed lines represent the two developing $\sigma$ bonds. Most cycloadditions have this stereochemistry, but an alternative possibility is that one of the two components might develop overlap with one bond forming on the top surface and the other on the bottom surface in the sense shown in the component on the left in Fig. 6.2b. Obviously considerable twisting in the conjugated systems has to take place before this kind of overlap can develop, and reactions showing this feature are exceedingly rare.

When both new bonds are formed on the same surface of the conjugated system, that component is described as undergoing *suprafacial* attack. When one bond forms to one surface and the other bond forms to the other surface, that component is described as undergoing *antarafacial* attack. The $(4n+2)$ rule applies to the common, indeed almost invariable, cases where both components are attacking suprafacially on each other. It does not apply to the case where one component is suprafacial and the other antarafacial—these are allowed when the total number of electrons is a $(4n)$ number. They are exceedingly rare, but one example may be the $[14+2]$ cycloaddition of tetracyanoethylene **6.22** to heptafulvene, where the heptafulvalene is attacked in an antarafacial manner **6.23**, one of the dashed lines, on the left, showing overlap developing to the bottom surface of the conjugated system, and the other to the top surface, presumably helped by some twisting in the conjugated system. This reaction may not be pericyclic, but it is striking that the two hydrogens at the point of attachment in the product **6.24** are *trans* to each other, revealing that the heptafulvalene behaved as an antarafacial component.

**6.22**                    **6.23**                    **6.24**

*6.3.1.2 Electrocyclic Reactions.* The parent members of the most simple electrocyclic reactions, are the equilibria between the allyl cation **6.25** and the cyclopropyl cation **6.26**, the pentadienyl cation **6.27** and the cyclopentenyl cation **6.28**, the heptatrienyl cation **6.29** and the cycloheptadienyl cation **6.30**, the allyl anion **6.31** and the cyclopropyl anion **6.32**, the pentadienyl anion **6.33** and the cyclopentenyl anion **6.34**, the heptatrienyl anion **6.35** and the cycloheptadienyl anion **6.36**, butadiene **6.37** and cyclobutene **6.38**, hexatriene **6.39** and cyclohexadiene **6.40**, and octatetraene **6.41** and cyclooctatriene **6.42**. These reactions are rarely seen in their unadorned state, and the direction in which they go is determined by such factors as ring strain, the gain or loss of aromaticity, and the substituents or heteroatoms stabilising one side of the equilibrium or the other. There are, of course, heteroatom-containing analogues, with nitrogen or oxygen in the chain of atoms, and lone pairs of electrons on the heteroatoms can take the place of the carbanion centre.

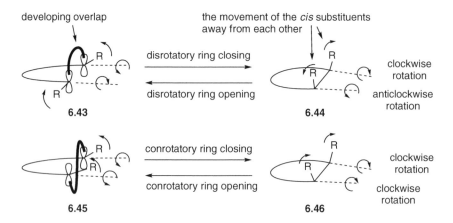

In contrast to cycloadditions, which almost invariably take place with a total of $(4n + 2)$ electrons, there are many examples of electrocyclic reactions taking place when the total number of electrons is a $(4n)$ number. However, those electrocyclic reactions with $(4n)$ electrons differ strikingly in their stereochemistry from those reactions mobilising $(4n + 2)$ electrons, as revealed when the parent systems are stereochemically labelled with substituents. The stereochemistry is not dependent upon the direction in which the reaction takes place, but it does depend upon whether there are $(4n)$ or $(4n + 2)$ electrons.

There are two possible stereochemistries for the ring-closing and ring-opening reactions. They are called *disrotatory* and *conrotatory*, and are illustrated for the general cases in Fig. 6.3. In the ring-closing disrotatory reaction **6.43 → 6.44**, the two outer substituents R move upwards, so that the top lobes of the p orbitals turn towards each other to form the new σ bond. The word disrotatory reflects the fact that the rotation about the terminal double bonds is taking place clockwise at one end but anticlockwise at the other. In the corresponding ring opening, **6.44 → 6.43**, there is similarly a clockwise and anticlockwise rotation as the

**Fig. 6.3** Disrotatory and conrotatory defined

$\sigma$ bond breaks, and the two upper substituents R move apart to become the outer substituents in the open-chain conjugated system. There is an equally probable disrotatory ring closure, in which both R groups fall, with the lower lobes of the p orbitals forming the new $\sigma$ bond, and there is a possible alternative disrotatory ring opening, in which both R groups move towards each other.

In contrast, in conrotatory ring closing, **6.45** → **6.46**, one of the outer substituents and one of the inner substituents, both labelled R, rise to become *cis*, so that the top lobe of the p orbital at the far end forms a $\sigma$ bond by overlap with the bottom lobe of the p orbital at the near end. The rotations are now in the same sense, either both clockwise or both anticlockwise. In the ring opening, **6.46** → **6.45**, the two substituents that are *cis* to each other move in the same direction, one to an outer position and the other to an inner position by clockwise rotations, or they could both move by anticlockwise rotations.

The rules are that thermal electrocyclic reactions involving a total number of electrons that can be expressed in the form $(4n+2)$ are disrotatory, and thermal electrocyclic reactions in which the total number of electrons can be expressed in the form $(4n)$ are conrotatory.

Examples showing stereochemistry in agreement with these rules are the cyclobutene openings **6.47** → **6.48** and **6.49** → **6.50**, and the hexatriene closings **6.51** → **6.52** and **6.53** → **6.54**. The conrotatory opening **6.47** → **6.48** and the disrotatory closing **6.51** → **6.52** are stereochemically contrathermodynamic, with the products the less stable stereoisomer.

Among ions, the opening of a cyclopropyl anion is exemplified by the reactions of the *trans* and *cis* aziridines **6.55** and **6.58**, which are isoelectronic with the cyclopropyl anion. They open in the conrotatory sense to give the W- and sickle-shaped ylids **6.56** and **6.59**, respectively, which are isoelectronic with the corresponding allyl anions. This step is an unfavourable equilibrium, which can be detected by the 1,3-dipolar cycloaddition of the ylids to dimethyl acetylenedicarboxylate, which takes place suprafacially on both components to give the *cis* and *trans* dihydropyrroles **6.57** and **6.60**. The conrotatory closing of a pentadienyl cation can be followed in the NMR spectra of the ions **6.61** and **6.62**, and the disrotatory closing of a pentadienyl anion can be seen in what is probably the oldest known pericyclic reaction, the formation of amarine **6.64** from the anion **6.63**.

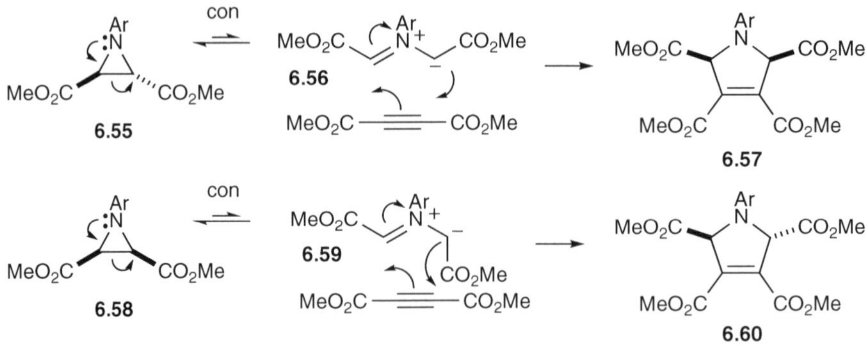

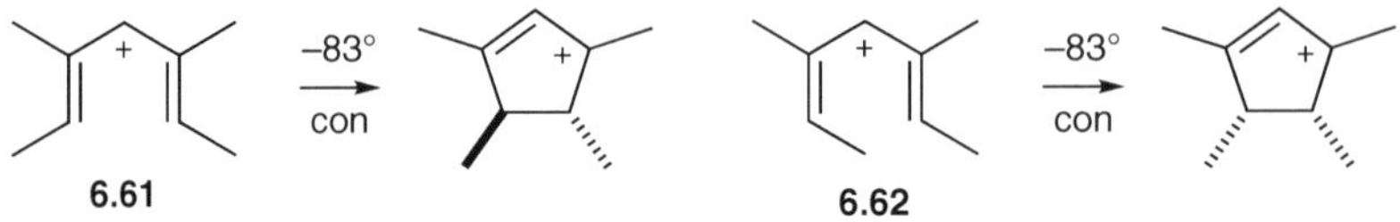

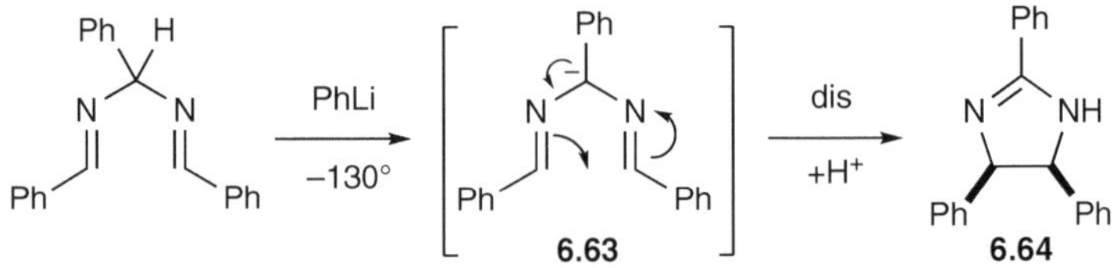

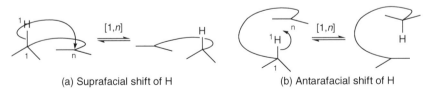

**6.3.1.3  Sigmatropic Rearrangements.**    Sigmatropic rearrangements in which a hydrogen atom moves are all [1,*n*] sigmatropic rearrangements. Those involving a total of (4*n*+2) electrons take place with the hydrogen atom moving from one surface of the conjugated system to the same surface at the other end in the general sense of Fig. 6.4a. This is called a *suprafacial* shift. Those involving a total of (4*n*) electrons show the alternative stereochemistry in which the hydrogen atom leaves one surface of the conjugated system and arrives at the other end on the opposite surface in the general sense of Fig. 6.4b, with the hydrogen atom leaving the upper surface, above C-1, but arriving on the lower surface, below C-*n*. This is called an *antarafacial* shift. Note that, in the section on cycloadditions, the words suprafacial and antarafacial refer to the stereochemical sense of the developing overlap, whereas they are used somewhat differently here to describe a structural change.

(a) Suprafacial shift of H                          (b) Antarafacial shift of H

**Fig. 6.4**    Suprafacial and antarafacial defined for a [1,*n*] migration of hydrogen

The [1,5]-suprafacial shift is found in open-chain systems and in rings, where it is striking that the shift in the cyclopentadiene **6.65** equilibrates the three isomers **6.65–6.67** at room temperature, whereas the cycloheptatriene **6.68** does not undergo the analogous but forbidden suprafacial [1,7]-shift. Instead it undergoes the geometrically more contorted, but allowed, suprafacial [1,5]-shift, **6.68** (arrows) at a much higher temperature. Both reactions **6.65** and **6.68** are described as [1,5]-shifts, because they are made possible by the overlap along the numbered set of atoms from C-1 to C-5. The former is *structurally* a [1,2]-shift, since the hydrogen atom moves to the adjacent carbon, but it is not mechanistically a [1,2]-shift. The C-1 to C-5 bond plays no electronic part in the mechanism—it merely serves to hold the two atoms close to each other, speeding up the reaction. The same reaction can take place when the C-1 to C-5 bond is not present, but it is then much slower.

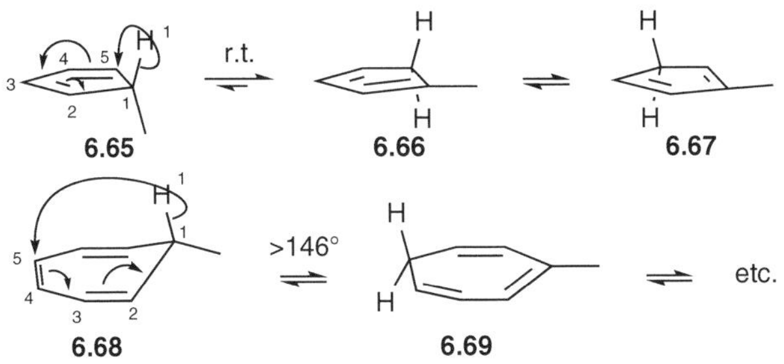

[1,7]-Antarafacial shifts can only occur in open chain systems, as in the vitamin D reaction shown earlier [see (Section 6.1) page 187], because it is sterically impossible in a ring like that in a cycloheptatriene **6.68** for a hydrogen atom leaving one surface of the ring to develop overlap onto the other surface. The antarafacial stereochemistry of the vitamin D reaction has been supported in the model system **6.70**, which gives a mixture of the 10S **6.71** and 10R **6.72** diastereoisomers, the former by antarafacial deuterium shift from C-15 (arrows), and the latter by antarafacial hydrogen shift from the top surface of C-15 to the bottom surface of C-10.

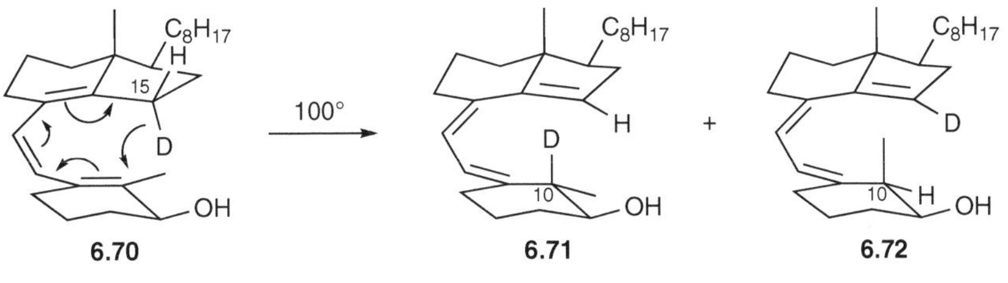

When the migrating group in a [1,n]-shift is a carbon atom, two more possibilities arise. In addition to moving either suprafacially or antarafacially, the migrating group can migrate with retention of configuration or with inversion of configuration. Suprafacial migrations are by far the most common, and they take place with retention of configuration when the total number of electrons is a $(4n + 2)$ number, and with inversion of configuration when the total number of electrons is a $(4n)$ number.

Thus the [1,2]-shift of an alkyl group towards an electron deficient atom always takes place with retention of configuration, whether it be towards carbon in Wagner-Meerwein rearrangements **6.73**, towards nitrogen in Beckmann or Curtius rearrangements, or towards oxygen in Baeyer-Villiger rearrangements. These are two-electron reactions, and the allowed suprafacial migration is geometrically reasonable. However, in the corresponding anion, there are four electrons, and neither of the allowed pathways, suprafacial with inversion **6.74** nor antarafacial with retention **6.75**, is reasonable. Accordingly anions do not normally rearrange, and when they do they use a stepwise pathway.

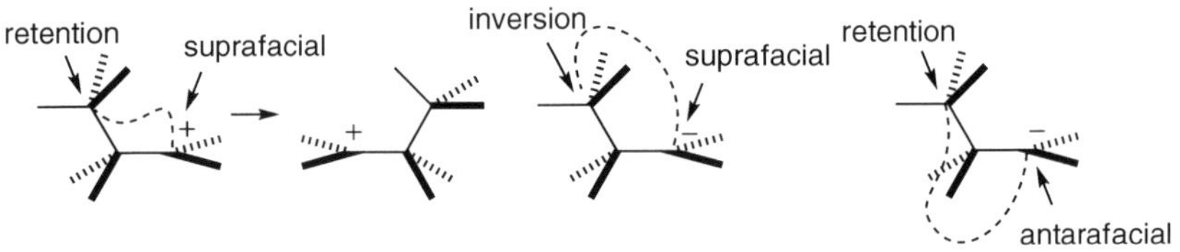

**6.73**                    **6.74**                    **6.75**

Examples for which inversion of configuration in the migrating group has been proved are the suprafacial [1,3]-shift of the bridge in the bicyclo[3.2.0]heptene **6.76** and the [1,4]-shift in the bicyclo[3.1.0]hexenyl cation **6.77**. The stereochemistry in the latter case is proved by the NMR spectrum: as the rearrangement takes place all the methyl groups attached to the five-membered ring become equivalent on the NMR timescale, but the signals of the two methyl groups labelled i (for inside) and o (for outside) remain distinct. Note that the inside methyl group remaining inside corresponds to an inversion of configuration—the bond between C-1 and C-1' is on the front face of C-1', and the bond that is forming from C-4 to C-1' is on the back side.

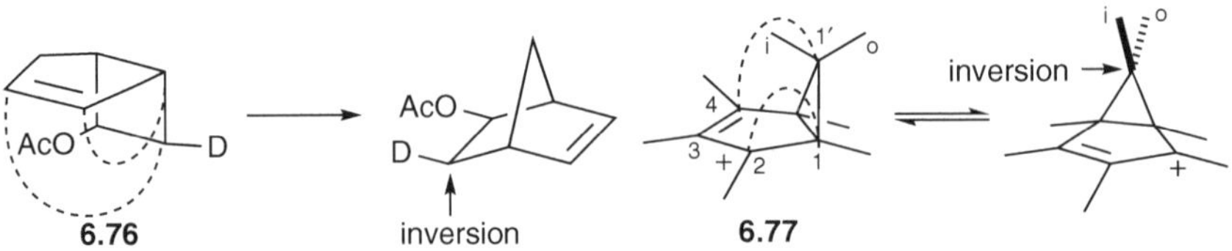

**6.76**                    inversion                    **6.77**

The rules for [$m,n$]-sigmatropic rearrangements, where $m \neq 1$ and $n \neq 1$, are more complex still. The bond can migrate suprafacially or antarafacially on either component, with the great majority of known reactions being

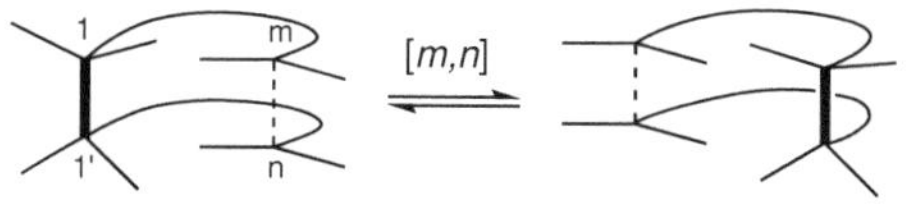

**Fig. 6.5**  Suprafacial-suprafacial defined for [*m,n*]-shifts in general

suprafacial on both components, as in Fig. 6.5. In principle, however, one or both migrations could be antarafacial, but it is difficult to maintain continuous overlap in such systems, and [*m,n*]-sigmatropic rearrangements with antarafacial components are correspondingly rare. When the total number of electrons is a (4*n* + 2) number, [*m,n*]-sigmatropic rearrangements are allowed if both migrations are suprafacial (or, unlikely but allowed, both antarafacial). When the total number of electrons is a (4*n*) number, [*m,n*]-sigmatropic rearrangements are allowed if one end of the bond migrates suprafacially and the other antarafacially.

The great majority of [*m,n*]-sigmatropic rearrangements involve the all-suprafacial participation of (4*n*+2) electrons. Much the most common are the various [3,3]-sigmatropic rearrangements, such as the Claisen rearrangement [see (Section 6.1) page 187], the Ireland-Claisen rearrangement **6.78 → 6.79**, and the all-carbon version, which is called the Cope rearrangement, as in the reaction **6.80 → 6.81**. Common [2,3]-sigmatropic rearrangements are the Mislow reaction [see (Section 6.1) page 187], and the anionic [2,3]-Wittig rearrangement **6.83 → 6.84** of allyl ethers **6.82** in the presence of strong base.

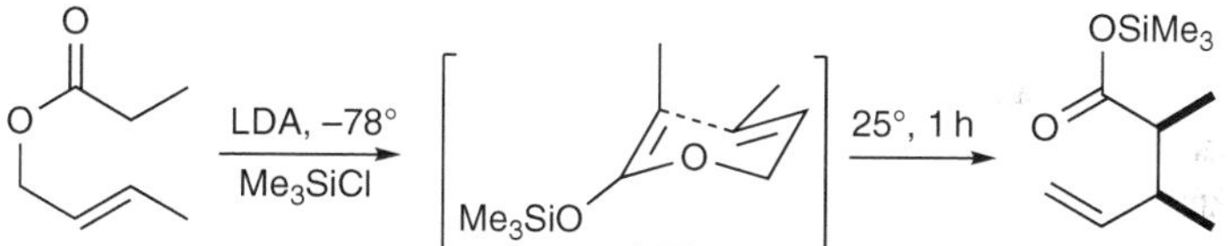

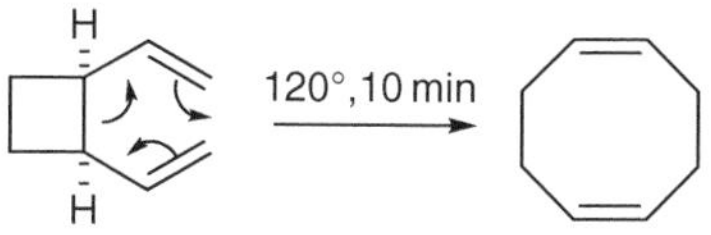

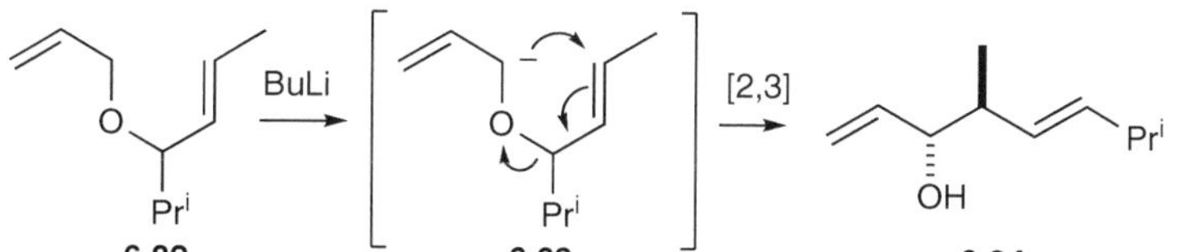

There are a few sigmatropic rearrangements with more than six electrons, such as the 10-electron doubly vinylogous Stevens rearrangement **6.86 → 6.87** of the unsaturated ammonium salt **6.85**, and the 10-electron benzidine rearrangement **6.88 → 6.89**.

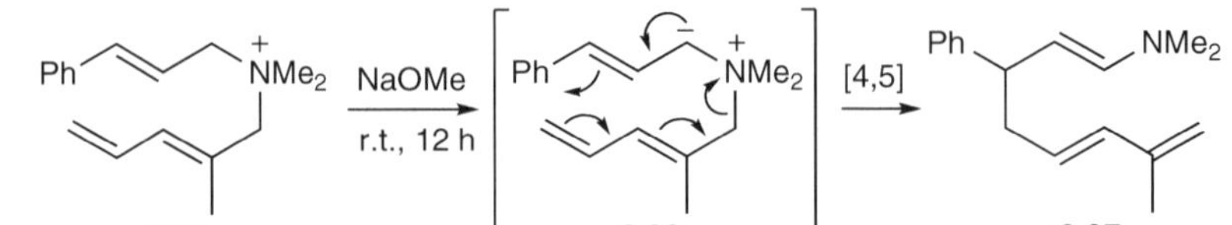

6.85    6.86    6.87

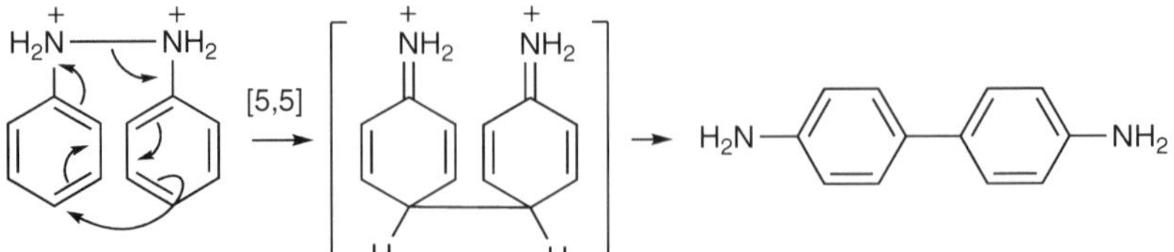

6.88    6.89

#### 6.3.1.4 Group Transfer Reactions.

There are so few of these reactions that a fully general rule for them can wait until the next section, where we see the final form of the Woodward-Hoffmann rules. For now, we can content ourselves with a simplified rule which covers almost all known group transfer reactions. When the total number of electrons is a $(4n+2)$ number, group transfer reactions are allowed with all-suprafacial stereochemistry.

The stereochemistry of the ene reaction $6.90 + 6.91 \rightarrow 6.93$ is such that the hydrogen atom delivered to the enophile **6.91** leaves *from* the same surface of the ene **6.90** as the surface to which the C—C bond is forming, and the hydrogen atom is delivered *to* the same surface of the enophile as the forming C—C bond **6.92**, so that both components are reacting suprafacially. The full stereochemistry is not proved in this example, because neither the methyl group, C-1, nor the $\beta$ carbon has any stereochemical label, but the all-suprafacial pathway provides a reasonable explanation for the relative stereochemistry set up between the $\alpha$ carbon and C-3.

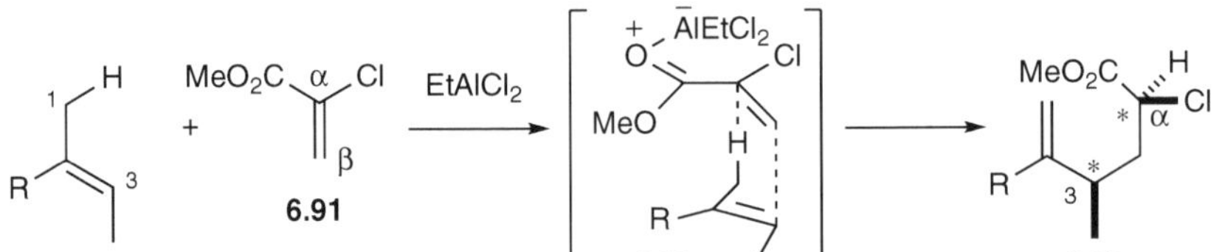

6.90    6.91    6.92    6.93

### 6.3.2 The Generalised Woodward-Hoffmann Rule

We have now seen a large number of rules, expressed differently for each kind of pericyclic reaction. Learning them would seem to impose a considerable burden, but Woodward and Hoffmann saved us from this effort by rewriting them in one all-encompassing rule that applies to all thermal pericyclic reactions:[20]

> A ground-state pericyclic change is symmetry allowed when the total number of $(4q+2)_s$ and $(4r)_a$ components is odd.

This admirably concise statement is compelling, but we must now see what it means, and learn how to apply it to each of the classes of pericyclic reaction.

*6.3.2.1   Cycloadditions.*   Let us begin with the bare bones of the Diels-Alder reaction in Fig. 6.6. The *components* of a cycloaddition are obvious enough—we have been using the word already to refer to the core electronic systems undergoing change. For a Diels–Alder reaction the components are the $\pi$ orbitals of the diene, containing four electrons, and the $\pi$ bond of the dienophile, containing two electrons. We ignore all substituents not directly involved, treating them only, for the purposes of following the rule, as stereochemical labels. We ask ourselves two questions: (1) which of these components is acting in a suprafacial manner and which in an antarafacial manner; and (2) in which of these components can the number of electrons be expressed in the form $(4q + 2)$ and which in the form $(4r)$, where $q$ and $r$ are integers? For the Diels-Alder reaction both components are undergoing bond formation in a suprafacial sense, as shown by the dashed lines in Fig. 6.6, and so the answer to the first question is: both components.

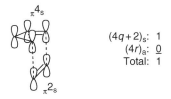

$(4q+2)_s$: 1
$(4r)_a$: 0
Total: 1

**Fig. 6.6**   The Diels-Alder reaction as a $[_\pi 4_s + _\pi 2_s]$ reaction

The diene has four electrons, a number that can be expressed in the form $(4r)$, with $r = 1$. Since the new bonds are forming on the diene in a suprafacial manner, both lines coming to the lower surface, the diene is a $(4r)_s$ component. The dienophile has two electrons, a number that can be expressed in the form $(4q + 2)$, with $q = 0$. Since the new bonds are forming on the dienophile in a suprafacial manner, both lines coming to the upper surface, the dienophile is a $(4q + 2)_s$ component. Thus, the Diels-Alder reaction has one $(4q + 2)_s$ component and no $(4r)_a$ components. We ignore $(4q + 2)_a$ and $(4r)_s$ components, when there are any. The total number of $(4q + 2)_s$ and $(4r)_a$ components is therefore 1, and, since this is an odd number, the reaction is symmetry-allowed. The Diels-Alder reaction is, as we have been calling it all along, a $[4 + 2]$ cycloaddition. Since it takes place suprafacially on both components, it is more informatively described as a $[4_s+2_s]$ cycloaddition, and finally, because both components are $\pi$ systems, it is fully described as a $[_\pi 4_s + _\pi 2_s]$ cycloaddition.

The description $[_\pi 4_s + _\pi 2_s]$ for a Diels-Alder reaction does not supplant the older name—it is not the only reaction that is $[_\pi 4_s + _\pi 2_s]$. 1,3-Dipolar cycloadditions **6.12** + **6.13** are equally easily drawn as $[_\pi 4_s + _\pi 2_s]$, and so are the combinations: allyl cation **6.2** and diene, allyl anion **6.6** and alkene, and pentadienyl cation **6.10** and alkene. Furthermore, $[_\pi 4_s + _\pi 2_s]$ is not the only way of describing a Diels-Alder reaction. It would be easy to overlook the fact that the diene can be treated as one component, and to see it instead as two independent $\pi$ bonds. Although it makes extra work to see it this way, it does not cause the rule to break down. For example, the drawing on the left of Fig. 6.7 might have been used instead of the one in Fig. 6.6. The dashed line representing the developing overlap for the formation of the $\pi$ bond is from the lower lobe on C-2 to the lower lobe on C-3. This makes all three components suprafacial—the $\pi$ bond between C-1 and C-2 has both dashed lines to the lower lobes, and the $\pi$ bond between C-3 and C-4 also has both dashed lines to the lower lobes. In other words both are suffering suprafacial development of overlap. The same is true for the $\pi$ bond of the dienophile. Overall the sum is changed to having three $(4q + 2)_s$ components, which is still an odd number, and so the reaction remains allowed. It is now described as a $[_\pi 2_s + _\pi 2_s + _\pi 2_s]$ cycloaddition, but it is of course the same reaction.

**Fig. 6.7**   The Diels-Alder reaction as a $[_\pi 2_s + _\pi 2_s + _\pi 2_s]$ and as a $[_\pi 2_s + _\pi 2_a + _\pi 2_a]$ reaction

Another drawing, on the right of Fig. 6.7, still representing the same reaction, places the dashed line between the upper lobes on C-2 and C-3. This changes each of the $\pi$ bonds of the diene to suffering notional antarafacial development of overlap. It is just as valid a representation as either of the earlier versions, and the sum still comes out with an odd number of $(4q + 2)_s$ components and no $(4r)_a$ components. The two $_\pi 2_a$ components do not have to be counted, because they are $(4q + 2)_a$ and not $(4r)_a$. The reaction is now a $[_\pi 2_s + _\pi 2_a + _\pi 2_a]$ cycloaddition. Clearly, the three designations $[_\pi 4_s + _\pi 2_s]$, $[_\pi 2_s + _\pi 2_s + _\pi 2_s]$, and $[_\pi 2_s + _\pi 2_a + _\pi 2_a]$ are all the same reaction, and none of them defines a Diels-Alder reaction. The three designations, in fact, describe the three *drawings* in Figs 6.6 and 6.7, and no reaction should be described in this way in the absence of a drawing like these.

It is easy enough to see how to extend the labelling of cycloaddition reactions to those involving larger conjugated systems. As just one example, the cycloaddition of heptafulvalene to tetracyanoethylene has overlap developing on opposite sides of the 14-electron component, which is therefore a $(4q + 2)_a$

component, and does not count towards the sum. The overlap on the 2-electron component, although not proved, is probably suprafacial, and, as a $(4q + 2)_s$ component, does count.

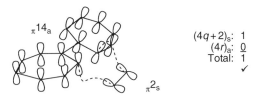

$$(4q+2)_s: 1$$
$$(4r)_a: \underline{0}$$
$$\text{Total: } 1$$
$$\checkmark$$

### 6.3.2.2 Electrocyclic Reactions.

For electrocyclic reactions, we need to see how the words suprafacial and antarafacial are defined for $\sigma$ bonds, for which we use orbitals made from overlapping $sp^n$ hybrids. Just as a suprafacial event on a $\pi$ bond has overlap developing to the two overlapping lobes that contribute to bonding, so with $\sigma$ bonds, overlap that develops to the two large lobes of the $sp^n$ hybrids is suprafacial. Less obviously, overlap that develops to the two small lobes is also suprafacial, because it is the counterpart to overlap developing to the other two lobes in a $\pi$ bond. Antarafacial overlap is when one bond is forming to an inside lobe and one to an outside lobe, either way round.

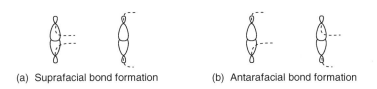

(a) Suprafacial bond formation    (b) Antarafacial bond formation

The electrocyclic interconversion of the cyclobutene **6.47** and the *cis, trans*-butadiene dicarboxylic ester **6.48** is shown in Fig. 6.8a. The components

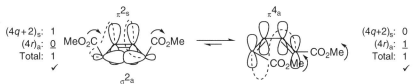

$$(4q+2)_s: 1 \qquad\qquad (4q+2)_s: 0$$
$$(4r)_a: \underline{0} \qquad\qquad (4r)_a: \underline{1}$$
$$\text{Total: } 1 \qquad\qquad \text{Total: } 1$$
$$\checkmark \qquad\qquad \checkmark$$

(a) The allowed conrotatory interconversion of a cyclobutene and a butadiene

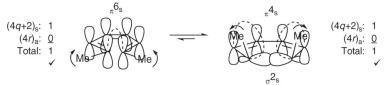

$$(4q+2)_s: 1 \qquad\qquad (4q+2)_s: 1$$
$$(4r)_a: \underline{0} \qquad\qquad (4r)_a: \underline{0}$$
$$\text{Total: } 1 \qquad\qquad \text{Total: } 1$$
$$\checkmark \qquad\qquad \checkmark$$

(b) The allowed disrotatory interconversion of a hexatriene and a cyclohexadiene

**Fig. 6.8** The allowed electrocyclic reactions

for the ring opening are the $\sigma$ bond made from two sp$^3$ hybrids drawn in front and the $\pi$ bond drawn at the back, and the conrotatory ring opening is shown as a $[_\sigma2_a + _\pi2_s]$ process. In the ring-closing direction there is only one component, the $\pi$ system of the diene, and the conrotatory ring closing is shown as a $[_\pi4_a]$ process. The small sums show that they are allowed by the Woodward-Hoffmann rule. Notice how the dashed lines and the curved arrows correspond to the clockwise direction in which the substituents move in the ring opening and anticlockwise in the ring closing, and that the geometry shown in the products matches these movements in both directions. As usual, there is an equally allowed conrotatory process with anticlockwise rotations in the ring opening and clockwise in the ring closing.

The symmetry-allowed disrotatory ring closing of the *trans,cis,trans*-dimethyl-hexatriene **6.51** and the disrotatory ring opening of the *cis*-dimethylcyclohexa-diene **6.52** are shown in Fig. 6.8b as $[_\pi6_s]$ and $[_\sigma2_s + _\pi4_s]$ processes. There is a similar reaction in both directions interconverting the *cis,cis,cis*-triene and the *cis*-dimethylcyclohexadiene, in which the methyl substituents move inwards in the ring opening, instead of outwards, following an energetically less favourable pathway, but one that is equally allowed by symmetry.

In order to describe the ring opening of the aziridine **6.55**, we need to define what suprafacial and antarafacial mean when applied to a p orbital. This is shown in Fig. 6.9, and applied there to the conrotatory aziridine opening. When both lines are drawn into the same lobe it is suprafacial, and when there is one line drawn into the top lobe and one into the bottom, it is antarafacial. Since this is neither a $\pi$ nor a $\sigma$ orbital, it is given the Greek letter $\omega$. The same designations apply whether the orbital is filled (on the left) or unfilled (on the right), and whether it is a p orbital or any of the sp$^n$ hybrids.

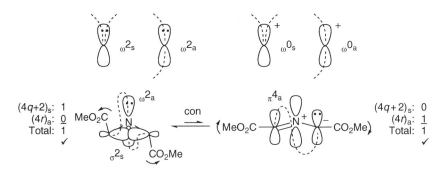

**Fig. 6.9**   Suprafacial and antarafacial defined for a p orbital, and the allowed conrotatory interconversion of an aziridine with an azomethine ylid

In the aziridine opening shown in Fig. 6.9, the aryl group behind the nitrogen atom is left out for clarity. The dashed lines are drawn from the large lobes of the $\sigma$ bond, making this a $[_\sigma2_s]$ component. Both substituents move anticlockwise in this conrotatory mode, so the dashed line on the left goes up, to overlap with the upper lobe of the p orbital on the nitrogen atom, and the dashed line on the right goes down to overlap with the lower lobe. With one overlap drawn as developing to the top and one to the bottom, the p orbital is an $[_\omega 2_a]$ component, making the overall reaction drawn in this way a $[_\sigma 2_s + _\omega 2_a]$ process. In the opposite direction, the clockwise conrotatory ring closing of the azomethine ylid **6.56** is simply a $[_\pi 4_a]$ process. The alternative conrotatory process taking place in the clockwise direction would place both methoxycarbonyl groups inside in a U-shaped ylid; this would be thermodynamically less favourable but just as allowed by symmetry. In the corresponding *cis*-disubstituted aziridine, the stereochemistry is still conrotatory, but now the geometry of the ylid is sickle-shaped, with one of the methoxycarbonyl groups outside and the other inside.

*6.3.2.3  Sigmatropic Rearrangements.*   The overlap developing in a suprafacial [1,5]-hydrogen shift in a diene is drawn in Fig. 6.10 both as a $[_\sigma 2_s + _\pi 4_s]$ process and as a $[_\sigma 2_a + _\pi 4_a]$ process. In both cases, the [1,5]-shift is suprafacial in the structural sense, but the overlap developing is selected in different ways in the two drawings—all-suprafacial on the left and all-antarafacial on the right. Thus the word suprafacial does not have the same meaning in the two contexts in which it is used here, although there is an obvious relationship. Both drawings, of course, are equally valid, and both would show that the reaction is symmetry-allowed if we were to complete the little sum that has accompanied all the drawings up to this point, but which we shall leave out from now on. The antarafacial [1,7]-hydrogen shift is similarly drawn in two ways in Fig. 6.11. One component is suprafacial and one antarafacial in each, but in both the hydrogen atom shifts in a structurally antarafacial sense, and the reaction is the same.

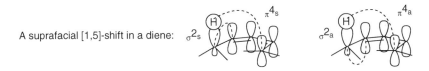

A suprafacial [1,5]-shift in a diene:

**Fig. 6.10**   The [1,5]-suprafacial shift of an H atom drawn as $[_\sigma 2_s + _\pi 4_s]$ and $[_\sigma 2_a + _\pi 4_a]$ processes

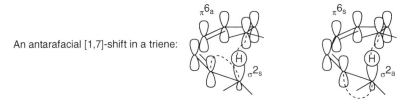

An antarafacial [1,7]-shift in a triene:

**Fig. 6.11**   The [1,7]-antarafacial shift of an H atom drawn as $[_\sigma 2_s + _\pi 6_a]$ and $[_\sigma 2_a + _\pi 6_s]$ processes

The [1,2]-shift of an alkyl group with retention of configuration in the migrating group, shown for the carbocation **6.73** [see (Section 6.3.1.3) page 198] has a dashed line identifying it as a suprafacial migration. A second dashed line connecting the lower end of the $\sigma$ bond to the same lobe of the empty p orbital would make it $[_{\sigma}2_s + _{\omega}0_s]$, but the same reaction could equally be described with a different line coming from the lower lobe of the empty p orbital to the lower side of the carbon atom of the $\sigma$ bond, making it $[_{\sigma}2_a + _{\omega}0_a]$. Similarly, the [1,3]-shift of an alkyl group with inversion of configuration in the migrating group, shown for the bicyclo[3.2.0]heptene **6.76**, already has the dashed lines identifying it as a $[_{\sigma}2_a + _{\pi}2_s]$ process, and the [1,4]-shift of an alkyl group with inversion of configuration, shown for the bicyclo[3.1.0]hexyl cation **6.77**, has the dashed lines for a $[_{\sigma}2_a + _{\pi}2_s]$ process, both of which are allowed by the unified rule.

The [3,3]-sigmatropic rearrangement, found in the Claisen and Cope rearrangements, is drawn for the chair transition structure **6.94** as an all-suprafacial process in the middle of Fig. 6.12, and as one suprafacial and two antarafacial processes on the right. The bold bond marked 1,1' is leaving the lower surface of the conjugated system from C-1 and arriving on the lower surface at C-3; the same bond is leaving the upper surface at C-1' and arriving on the upper surface at C-3'. Thus the reaction, however it is described for the purposes of the Woodward-Hoffmann rule, involves a structurally suprafacial migration of the $\sigma$ bond across each of the surfaces as defined for the general case in Fig. 6.5.

[2,3]-Sigmatropic rearrangements, like the [2,3]-Wittig rearrangement **6.83** and its sulfur and aza analogues, are drawn for the general case in Fig. 6.13, where X=O, S, SR$^+$, or NR$_2$$^+$. An envelope-shaped transition structure is almost always involved, because this allows the smooth development of head-on overlap

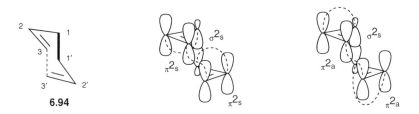

**Fig. 6.12** A [3,3]-sigmatropic rearrangement drawn as $[_{\sigma}2_s + _{\pi}2_s + _{\pi}2_s]$ and $[_{\sigma}2_s + _{\pi}2_a + _{\pi}2_a]$ processes

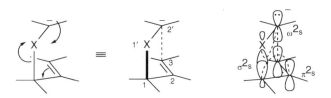

**Fig. 6.13** A [2,3]-sigmatropic rearrangement drawn as a $[_{\sigma}2_s + _{\omega}2_s + _{\pi}2_s]$ process

in the formation of the $\sigma$ bond between C-2′ and C-3 at the same time that the $\sigma$ bond between C-1′ and C-1 is conjugated with the $\pi$ bond. These reactions all have an $\omega$ component in the form of a lone pair or filled p orbital, and can be described in the all-suprafacial mode drawn in Fig. 6.13 as $[_\sigma2_s + _\omega2_s + _\pi2_s]$.

*6.3.2.4 Group Transfer Reactions.* The ene reaction **6.92** is drawn again on the left of Fig. 6.14, showing that it can be described as a $[_\sigma2_s + _\pi2_s + _\pi2_s]$ process, and a dihydrogen transfer similar to that in diimide reduction is redrawn on the right of Fig. 6.14, showing that it can be described as a $[_\sigma2_s + _\sigma2_s + _\pi2_s]$ process.

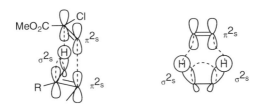

**Fig. 6.14**  An ene reaction drawn as $[_\sigma2_s + _\pi2_s + _\pi2_s]$ and a dihydrogen transfer drawn as $[_\sigma2_s + _\sigma2_s + _\pi2_s]$

*6.3.2.5 Some Hints about Drawing Diagrams for the Woodward-Hoffmann Rule.*  The first requisite for a good understanding of a pericyclic reaction is to have a good drawing of the transition structure. Begin with the flat, curly arrow-based representation, because this helps to identify the *components*—they are the lone pairs and the bonds that the curly arrows apply to—the bonds that are broken and the bonds that are made, and the lone pairs that are mobilised or localised. Then try to draw a three-dimensional view, in order to assess how reasonable the reaction is. Boat-like, chair-like and envelope transition structures are common, easily drawn, and are likely to be a good starting point. A good drawing will show the component orbitals lined up to develop overlap with the right geometry—head-on if it is creating a $\sigma$ bond or sideways-on if it is creating a $\pi$ bond—as drawn for the ene reaction in Fig. 6.14. Sometimes this is not possible, especially with electrocyclic ring-closing reactions. Any attempt to bring the orbitals at the ends of the diene in Fig. 6.8a and the triene in Fig. 6.8b into a position to show the developing $\sigma$ overlap will so distort the conjugated systems that the drawing will be hard to read. The reactions take this path, but it is probably wise to avoid drawings close to the transition structures in cases like these.

Then there is the problem of assessing whether the reaction is symmetry-allowed or not using the Woodward-Hoffmann rule. All reactions using $(4n + 2)$ electrons (an odd number of curly arrows) are allowed in the all-suprafacial mode, and so it is helpful to draw the dashed or solid lines (or better still use a line with a distinctive colour) to show the developing overlap with only suprafacial components. The $(4q + 2)_s$ components will then add up to an odd number, and the task is done.

If the dashed, bold or coloured lines are not those for an all-suprafacial reaction, as on the right-hand side of Fig. 6.7, for example, all is not lost—simply do the sum to find out whether the drawing corresponds to an allowed reaction or not. The all-suprafacial drawing is no better than the other representations, but it is a quick way to arrive at a drawing showing that a $(4n + 2)$ reaction is reasonable and allowed, and it works for a very high proportion of pericyclic reactions.

Reactions involving a total of $(4n)$ electrons (an even number of curly arrows) are allowed if there is one antarafacial component and all the others are suprafacial. In this case, if the dashed, bold or coloured lines include only one antarafacial component, the number of $(4q + 2)_s$ and $(4r)_a$ components will add up to an odd number, and the drawing will show the geometry of a symmetry-allowed reaction.

When working out whether or not a reaction is obeying the rule, it is inappropriate to shade the orbitals—what we have been doing in the whole of Section 6.3 is not a frontier orbital treatment, which we shall come to later. No particular orbital is being considered when analyses like those in Figs 6.6–6.14 are carried out. There is no real need to draw the lobes at all, as long as the perspective used to make the drawings and the placing of the dashed, bold or coloured lines clearly identify the surfaces of the conjugated systems on which the new bonds are developing and the old bonds breaking.

### 6.3.2.6 Some Symmetry-Allowed but Geometrically Unreasonable Reactions.

Some reactions are symmetry-allowed, but they do not take place, because they cannot attain a geometry that allows the continuous development of overlap. The $[2 + 2]$ reaction shown in Fig. 6.15a is a $[_\pi 2_s + _\pi 2_a]$ reaction, fully allowed by the Woodward-Hoffmann rule, but it does not take place, because the molecule is not flexible enough for the overlap developing on the left-hand side to take place at the same time as the overlap developing on the right-hand side. A longer conjugated system might have the necessary flexibility, and this could be the case in the reaction between heptafulvene and tetracyanoethylene [see (Section 6.3.2.1) page 203].

(a) A symmetry-allowed but unreasonable $[_\pi 2_s + _\pi 2_a]$ cycloaddition

(b) A symmetry-allowed but unreasonable $[_\sigma 2_a + _\pi 2_s]$ [1,3]-sigmatropic rearrangement

(c) A symmetry-allowed but unreasonable $[_\sigma 2_s + _\pi 2_a]$ [1,3]-sigmatropic rearrangement

**Fig. 6.15** Symmetry-allowed but geometrically inaccessible reactions

Another example of an allowed but essentially impossible pathway is the [1,2]-shift in an anion, whether it is suprafacial with inversion of configuration in the migrating group **6.74** or antarafacial with retention of configuration in the migrating group **6.75**. Slightly less unlikely are [1,3]-sigmatropic shifts, which are also allowed to be either suprafacial with inversion ($[_\sigma2_a + _\pi2_s]$ in Fig. 6.15b) or antarafacial with retention ($[_\sigma2_s + _\pi2_a]$ in Fig. 6.15c). Neither looks geometrically reasonable, but the former may just possibly explain the stereochemistry of the reaction of the bicyclo[3.2.0]heptene **6.76**. Perhaps most convincingly it is seen in the [1,3]-shift of a silyl group, taking place with inversion of configuration at the silicon atom, when an allylsilane is heated to a high temperature. There is no good example of an antarafacial [1,3]-shift with retention of configuration, a geometrically even less probable reaction.

### 6.3.2.7 Some Geometrically Reasonable but Symmetry-Forbidden

**Reactions.**  [2 + 2] Cycloadditions, for which the only reasonable transition structures are for suprafacial attack on both components, are symmetry-forbidden, but they are only forbidden if they are concerted—there is nothing forbidden about the formation of *one* bond, provided that the other is not forming at the same time. If only one bond forms, it will create either a zwitterionic or a diradical intermediate. The ionic pathway becomes reasonable if either or both of the ionic centres in the zwitterionic intermediate is equipped with stabilising groups: C or Z for anions and C or X for cations. Thus the enamine **6.95** reacts with the $\alpha,\beta$-unsaturated ester **6.96** to give the cyclobutane **6.98** in a stepwise ionic reaction, because the intermediate zwitterion **6.97** has the cationic centre stabilised as an iminium ion by the lone pair of the amino group, and the anionic centre stabilised as an enolate by the adjacent carbonyl group. The reaction is still a cycloaddition, but it is not pericyclic.

Equally, not all [4 + 2] cycloadditions are concerted. If a zwitterionic or a diradical intermediate is well enough stabilised, one bond can form ahead of the other, as in the reaction between the same enamine **6.95** and the $\alpha,\beta$-unsaturated ketone **6.99** giving the dihydropyran **6.101**. This is formally a hetero-Diels-Alder reaction, but it is almost certainly stepwise, taking place by way of the zwitterion **6.100**. A stepwise reaction forming only one bond does not suffer from the same high negative entropy of activation that forming both together does, whereas a concerted Diels-Alder reaction, although it suffers from a high negative entropy of activation, does not demand the high degree of stabilisation found in intermediates like **6.97** and **6.100**.

Stepwise reactions by way of diradical intermediates are also possible, as in the coupling of the halogenated alkene **6.102** with butadiene **6.103**. As we saw in Chapter 2 (see pages 67–68), any group, C, Z or X, stabilises a radical. Both radical centres in the intermediate **6.104** are stabilised, the one at the top by the $\alpha$-chlorines and the $\beta$-fluorines, and the one below because it is allylic. These combine rapidly **6.104** (arrow on the upper drawing) to give the cyclobutane **6.105**.

Normally a six-membered ring is formed more rapidly than a four-membered ring, so why does the diradical **6.104** (arrows in the lower drawing) not give a six-membered ring? For the first step in a stepwise pathway, in contrast to the Diels-Alder reaction, the diene can remain in the more populated s-*trans* conformation. The allyl radical produced from the s-*trans* diene has half of a W-configuration [see (Section 2.3.1.2) page 83], in which rotation about the bond between C-2 and C-3 is more restricted than it was in the diene. The chlorine-stabilised radical does not attack the allylic radical at C-1, because, were it to do so, it would give a *trans*-cyclohexene **6.106**. Rotation about the bond between C-2 and C-3 is evidently too slow to compete with the radical combination **6.104** (the arrow in the upper drawing).

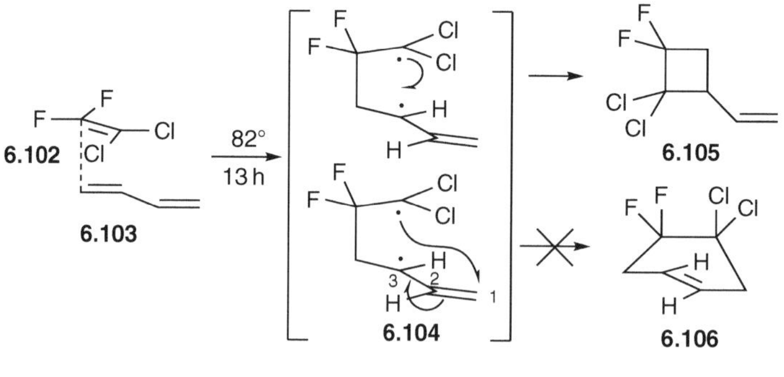

6.102    6.103    82° 13 h    6.104    6.105    6.106

Some other stepwise reactions seeming to disobey the rules, are the 1,2-shifts of ylids, like the Stevens rearrangement **6.107** → **6.109**. The symmetry-allowed reactions, suprafacial-with-inversion **6.74** or antarafacial-with-retention **6.75**, are unreasonable—there is no flexibility for migration across only two atoms, and yet reactions like this take place easily. It is now clear that these reactions are stepwise, taking place by homolytic cleavage **6.107** → **6.108**, followed by rapid radical recombination **6.108** → **6.109**.

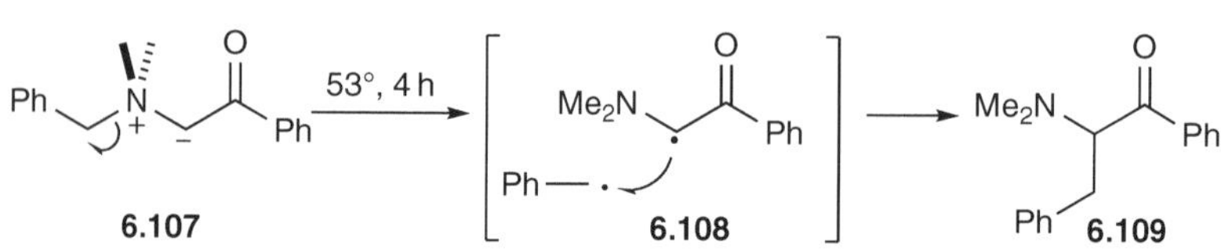

6.107    53°, 4 h    6.108    6.109

### 6.3.2.8 Reactions of Ketenes, Allenes and Carbenes which Appear to be Forbidden.

Some [2+2] cycloadditions only appear to be forbidden. One of these is the cycloaddition of ketenes to alkenes. These reactions have some of the characteristics of pericyclic cycloadditions, such as being stereospecifically syn with respect to the double bond geometry, and hence suprafacial at least on the one component, as in the reactions of the stereoisomeric cyclooctenes **6.110** and **6.112** giving the diastereoisomeric cyclobutanones **6.111** and **6.113**. However, stereospecificity is not always complete, and many ketene cycloadditions take place only when there is a strong donor substituent on the alkene. An ionic stepwise pathway by way of an intermediate zwitterion is therefore entirely reasonable in accounting for many ketene cycloadditions.

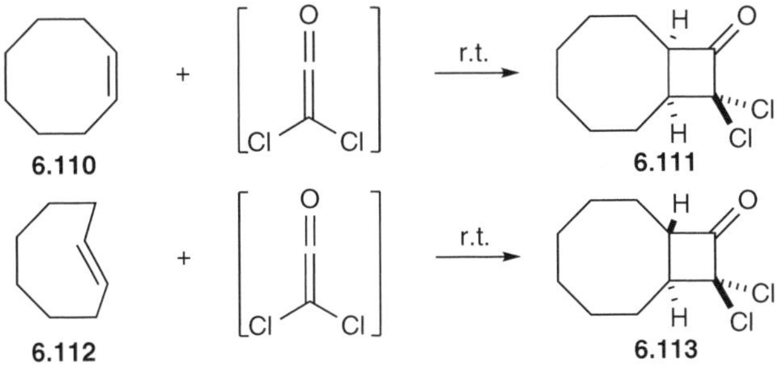

Somewhat similarly, dimethylallene undergoes a cycloaddition to dimethyl fumarate and dimethyl maleate giving mainly the cyclobutanes **6.114** and **6.116**, respectively, together with a little of the regioisomers **6.115** and **6.117**, but with a high level of stereospecificity, implying either that the reaction is concerted and suprafacial or, less probably, that any intermediate diradical or zwitterion has not had time to lose configurational information. Allenes also undergo cyclodimerisation, with enantiomerically enriched allenes leading to enantiomerically enriched products, with the details in agreement with the possibility that the reactions are concerted cycloadditions.

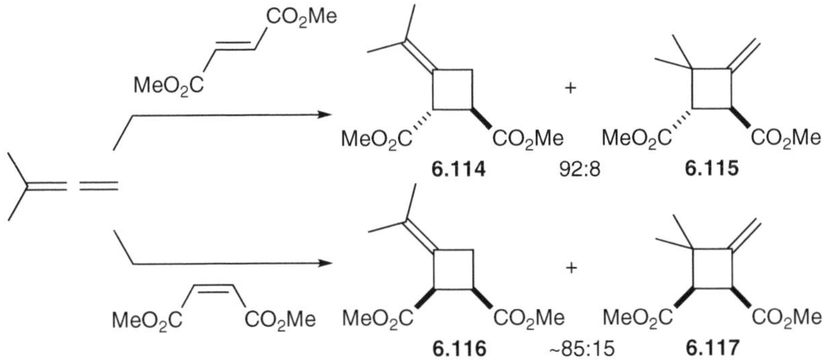

It seems likely that some of these ketene and allene cycloadditions are peri-cyclic and some not, with the possibility of there being a rather blurred border-line between the two mechanisms, with one bond forming so far ahead of the other that any symmetry-control from the orbitals is essentially lost. But if it is pericyclic, how does it overcome the symmetry-imposed barrier? One sugges-tion is that the two molecules approach each other at right angles, with overlap developing in an antarafacial sense on the ketene or allene like that in Fig. 6.15a, making the reaction the allowed $[_\pi 2_s + _\pi 2_a]$ cycloaddition that we have dismissed as being unreasonable. This is the simplest explanation, but it is unsatisfactory.

The probability is that some $[2 + 2]$ cycloadditions of ketenes and allenes are concerted by virtue of the fact that ketenes and allenes have two sets of $\pi$ orbitals at right angles to each other. Overlap can develop to orthogonal orbitals **6.118** and **6.119** (solid lines), and in addition there may be some transmission of information from one orbital to its orthogonal neighbour (dashed lines). In the case of the allene there is an implied direction of rotation **6.119** (arrow) of the terminal groups on the double bond not involved in forming the ring, a detail which becomes important later when we consider regio- and stereochemistry.

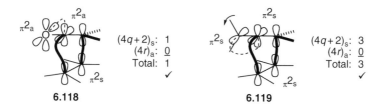

|  | $(4q+2)_s$: | 1 |
|---|---|---|
|  | $(4r)_a$: | 0 |
|  | Total: | 1 ✓ |

**6.118**

|  | $(4q+2)_s$: | 3 |
|---|---|---|
|  | $(4r)_a$: | 0 |
|  | Total: | 3 ✓ |

**6.119**

This is a legitimate but somewhat contrived way of making the electronic con-nection cyclic and hence pericyclic. This version identifies the reactions as allowed $[_\pi 2_s + (_\pi 2_a + _\pi 2_a)]$ or $[_\pi 2_s + (_\pi 2_s + _\pi 2_s)]$ cycloadditions. In essence the ketene or allene is able to take up the role of antarafacial component by using an orbital that has turned through 90° towards the alkene component. Several calculations support this picture, giving a transition structure with substantial C—C bonding to the carbonyl carbon (1.71–1.78 Å) and much less (2.43–2.47 Å) at the other C—C bond, and with a severely twisted four-membered ring. A variant of the approach, perhaps the simplest way of thinking about these reactions, is to omit the overlap drawn with dashed lines in **6.118** and **6.119**. This removes the symmetry-imposed barrier, because the reaction is no longer being thought of as strictly pericyclic. The two bonds are still being formed more or less in concert, but independently, without symmetry information being transmitted from one orbital to the other.

Related to ketene cycloadditions are the group of cycloadditions with vinyl cation intermediates. The reaction between 2-butyne **6.120** and chlorine giving the dichlorocyclobutene **6.122** is the Smirnov-Zamkow reaction, and there is a similar reaction between allene **6.123** and hydrogen chloride giving the dichlorocyclobutane **6.125**. Both reactions take place by cycloaddition of a vinyl cation **6.121** and **6.124** to another molecule of the starting material. Vinyl cations, like ketenes, have two p orbitals at right angles to each other, and overlap can develop to each simultaneously. In a sense, a ketene is merely a special case of a vinyl cation, with the carbonyl group a highly stabilised carbocation.

Several reactions in organometallic chemistry also appear to contravene the rule, but which can be explained in a somewhat similar way. Hydrometallation [**5.45**, see (Section 5.1.3.4) page 162], carbometallation, metallo-metallation, and olefin metathesis reactions are all stereospecifically suprafacial [2 + 2] additions to an alkene or alkyne, for which the all-suprafacial pathway is forbidden. Hydroboration, for example, begins with electrophilic attack by the boron atom, but it is not fully stepwise, because electron-donating substituents on the alkene do not speed up the reaction as much as they do when alkenes are attacked by electrophiles. Nevertheless, the reaction is stereospecifically syn—there must be some hydride delivery more or less concerted with the electrophilic attack. The empty p orbital on the boron is the electrophilic site and the s orbital of the hydrogen atom is the nucleophilic site. These orbitals are orthogonal, and so the addition **6.126** is not pericyclic.

An anomalous cycloaddition is the insertion of a carbene into an alkene. Some cheletropic reactions are straightforwardly allowed pericyclic reactions, which we can illustrate with the drawing **6.127** for the suprafacial addition of sulfur dioxide to a diene, and with the drawing **6.128** for the 8-electron antarafacial addition of sulfur dioxide to a triene. The problem comes with the insertion of a carbene into a double bond, which is well known to be stereospecifically suprafacial on the alkene with singlet electrophilic carbenes [see (Section 4.6.2) page 149]. This is clearly a forbidden pericyclic reaction if it takes place in the sense **6.129**.

**6.127**

$(4q+2)_s$: 1
$(4r)_a$: 0
Total: 1 ✓

**6.128**

$(4q+2)_s$: 1
$(4r)_a$: 0
Total: 1 ✓

**6.129**

$(4q+2)_s$: 2
$(4r)_a$: 0
Total: 2 ✗

This is known as the linear approach, in which the carbene, with its two sub-stituents already lined up where they will be in the product, comes straight down into the middle of the double bond. The two sulfur dioxide reactions above, **6.127** and **6.128**, are also linear approaches, but these are both allowed, the former because the total number of electrons (6) is a $(4n+2)$ number, and the latter because the triene is flexible enough to take up the role of antarafacial component. The alternative for a carbene is a nonlinear approach **6.130**, in which the carbene approaches the double bond on its side, and then has the two substituents tilt upwards as the reaction proceeds, in order to arrive in their proper orientation in the product **6.131**. The carbene is effectively able to take up the role of the antarafacial component; as with ketenes, it is possible to connect up the orthogo-nal orbitals, as in **6.132** (dashed line), to make the nonlinear approach classifiably pericyclic and allowed. This avoids any problem there might be with reactions like **6.127** and **6.128** being pericyclic and the clearly related reaction **6.130** → **6.131** seeming not to be. Similar considerations apply to the insertion of carbenes into σ bonds.

**6.130**          **6.131**          **6.132**

$(4q+2)_s$: 2
$(4r)a$: 1
Total: 3 ✓

## 6.4   Explanations for the Woodward-Hoffmann Rules

Three levels of explanation have been advanced to account for the patterns of reactivity encompassed by the Woodward-Hoffmann rules. The first draws atten-tion to the frequency with which pericyclic reactions have a transition structure with $(4n+2)$ electrons in a cyclic conjugated system, which can be seen as being aromatic. The second makes the point that the interaction of the appropriate frontier orbitals matches the observed stereochemistry. The third is to use orbital and state correlation diagrams in a compellingly satisfying treatment for those cases with identifiable elements of symmetry. Molecular orbital theory is the basis for all these related explanations.

### 6.4.1 The Aromatic Transition Structure

We saw earlier that the all-suprafacial $[4+2]$, $[8+2]$, and $[6+4]$ thermal cycloadditions are common, and that $[2+2]$, $[4+4]$, and $[6+6]$ cycloadditions are almost certainly stepwise or, as we shall see in Chapter 8, photochemically induced. The total number of electrons in the former are $(4n+2)$ numbers, analogous to the number of electrons in aromatic rings. This wonderfully simple idea was the first explanation for the patterns of allowed and forbidden pericyclic reactions. At first sight, it is a bit more difficult to explain those pericyclic reactions that take place smoothly in spite of their having a total of $4n$ electrons. They all show stereochemistry involving an antarafacial component, but it is possible to include this very feature in the aromatic transition structure model. If the p orbitals that make up a cyclic conjugated system have a single twist, like a Möbius strip, then the appropriate number of electrons for an aromatic system becomes $4n$ rather than $(4n+2)$. The antarafacial component in a conrotatory electrocyclic closure, for example, with overlap developing from the top lobe at one end to the bottom lobe at the other (**6.45** in Fig. 6.3), is equivalent to the twist in a Möbius conjugated system.

### 6.4.2 Frontier Orbitals

The easiest explanation is based on the frontier orbitals—the highest occupied molecular orbital (HOMO) of one component and the lowest unoccupied orbital (LUMO) of the other. Thus if we compare a $[2+2]$ cycloaddition **6.133** with a $[4+2]$ cycloaddition **6.134** and **6.135**, we see that the former has frontier orbitals that do not match in sign at both ends, whereas the latter do, whichever way round, **6.134** or **6.135**, we take the frontier orbitals. In the $[2+2]$ reaction **6.133**, the lobes on C-2 and C-2′ are opposite in sign and represent a repulsion—an anti-bonding interaction. There is no barrier to formation of the bond between C-1 and C-1′, making stepwise reactions possible; the barrier is only there if both bonds are trying to form at the same time. The $[4+4]$ and $[6+6]$ cycloadditions have the same problem, but the $[4+2]$, $[8+2]$ and $[6+4]$ do not. Frontier orbitals also explain why the rules change so completely for photochemical reactions, as we shall see in Chapter 8.

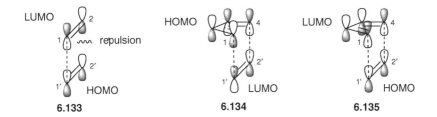

Applying frontier orbital theory to unimolecular reactions like electrocyclic ring closures and sigmatropic rearrangements is inherently contrived, since we are looking at only one orbital. To set up an interaction between frontier orbitals, we have artificially to treat a single molecule as having separate components. To take one of the less dubious examples, since the component orbitals are at least orthogonal, the electrocyclic conrotatory opening of a cyclobutene can be treated as the addition of the HOMO of the single bond $\sigma$ to the LUMO of the double bond $\pi^*$ **6.136**, where the dashed lines connect the lobes of the atomic orbitals of the same sign. For the ring-closing direction, which is more dubious, since the component orbitals are conjugated, we can treat the double bonds as separate components **6.137**, one bond providing the HOMO, $\pi$ on the left, and the other the LUMO, $\pi^*$ on the right. Alternatively, we can look only at the HOMO of the diene, $\psi_2$ in **6.138**, where the development of bonding from C-1 to C-4 corresponding to conrotatory ring-closing does not have a sign change. This is hardly compelling, since it is not obvious why we should take the HOMO.

**6.136**                    **6.137**                         **6.138**

### 6.4.3   Correlation Diagrams

Correlation diagrams provide a compelling explanation, at least for those reactions that have well defined elements of symmetry preserved throughout the reaction. The idea is to identify the symmetry elements maintained throughout the reaction, classify the orbitals undergoing change with respect to those symmetry elements, and then see how the orbitals of the starting materials connect with those of the product. The assumption is that an orbital in the starting material must feed into an orbital of the same symmetry in the product, preserving the symmetry throughout the reaction. Substituents, whether they technically break the symmetry or not, are treated as insignificant perturbations on the orbitals actually undergoing change.

*6.4.3.1   Orbital Correlation Diagrams.*   We shall begin with an allowed reaction, the ubiquitous Diels-Alder.

*Step 1*. Draw the bare bones of the reaction **6.139**, and draw the curly arrows for the forward and backward reactions. Any substituents, even if they make the diene or dienophile unsymmetrical, do not fundamentally disturb the symmetry of the orbitals directly involved.

a plane of
symmetry
intersects the
page here

**6.139**

**6.140**

*Step 2.* Identify the molecular orbitals undergoing change. The curly arrows help you to focus on the components of the reaction—what we want now is the molecular orbitals of those components. For the starting materials, they are the $\pi$ orbitals ($\psi_1$-$\psi_4^*$) of the diene unit and the $\pi$ orbitals ($\pi$ and $\pi^*$) of the C=C double bond of the dienophile. For the product, they are the $\pi$ bond ($\pi$ and $\pi^*$) and the two newly formed $\sigma$ bonds ($\sigma$ and $\sigma^*$ for each).

*Step 3.* Identify any symmetry elements maintained throughout the course of the reaction. There may be more than one. For a Diels-Alder reaction, which we know to be suprafacial on both components, there is only the one **6.140**, a plane of symmetry bisecting the bond between C-2 and C-3 of the diene and the $\pi$ bond of the dienophile.

*Step 4.* Rank the orbitals by their energy, and draw them as energy levels, one above the other, with the starting material on the left and the product on the right (Fig. 6.16).

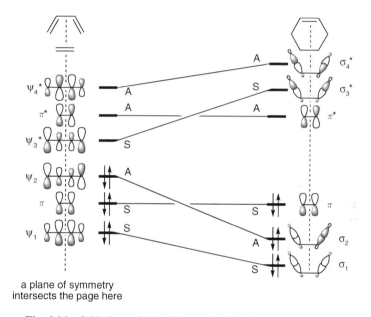

**Fig. 6.16**   Orbital correlation diagram for the Diels-Alder reaction

*Step 5.* Beside each energy level, draw the orbitals, showing the signs of the coefficients of the atomic orbitals. All the $\pi$ bonds are straightforward, but we meet a problem with the two $\sigma$ bonds in the product, which appear to be independent entities. In the next step we have to identify the symmetry these orbitals have with respect to the plane of symmetry maintained through the reaction, and it is not possible to do this for a pair of independent orbitals. The answer is to combine them; they are held one bond apart, and they must interact in a $\pi$ sense. The interaction of the two bonding $\sigma$ orbitals (Fig. 6.17a) and the two antibonding $\sigma^*$

orbitals (Fig. 6.17b) leads to a new set of four molecular orbitals $\sigma_1$, $\sigma_2$, $\sigma_3^*$ and $\sigma_4^*$, one pair ($\sigma_1$ and $\sigma_3^*$) lowered in energy by the extra $\pi$ bonding, and the other pair ($\sigma_2$ and $\sigma_4^*$) raised in energy by the extra $\pi$ antibonding.

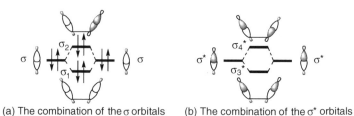

(a) The combination of the $\sigma$ orbitals          (b) The combination of the $\sigma^*$ orbitals

**Fig. 6.17**   Molecular orbitals from a pair of interacting $\sigma$ orbitals

*Step 6.* Classify each of the orbitals with respect to the symmetry element. Starting at the bottom left of Fig. 6.16, the lowest-energy orbital is $\psi_1$ of the diene, with all-positive coefficients in the atomic orbitals, in other words with unshaded orbitals across the top surface of the conjugated system. The atomic orbitals on C-1 and C-2 are reflected in the mirror plane, intersecting the page at the dashed line, by the atomic orbitals on C-3 and C-4, and $\psi_1$ is therefore classified as symmetric (S). Moving up the left-hand column, the next orbital is the $\pi$ bond of the dienophile, which is also symmetric with respect to reflection in the plane. The next orbital is $\psi_2$ of the diene, in which the atomic orbitals on C-1 and C-2 have positive coefficients, and those on C-3 and C-4 have negative coefficients, because of the node halfway between C-2 and C-3. The atomic orbitals on C-1 and C-2 are not reflected in the mirror plane by the orbitals on C-3 and C-4, and this orbital is antisymmetric (A). It is unnecessary to be any more sophisticated in the description of symmetry than this. The remaining orbitals can all be classified similarly as symmetric or antisymmetric. Likewise with the orbitals of the product on the right, $\sigma_1$ is symmetric, $\sigma_2$ antisymmetric, and so on.

*Step 7.* Fill in the orbital correlation (Fig. 6.16). Following the assumption that an orbital in the starting material must feed into an orbital of the same symmetry in the product, draw lines connecting the orbitals of the starting materials to those of the products nearest in energy and of the same symmetry. Thus, $\psi_1$ (S) connects to $\sigma_1$ (S), $\pi$ (S) connects to $\pi$ (S), and $\psi_2$ (A) connects to $\sigma_2$ (A), and similarly, with the unoccupied orbitals, $\psi_3^*$ (S) connects to $\sigma_3^*$ (S), $\pi^*$ (A) connects to $\pi^*$ (A), and $\psi_4^*$ (A) connects to $\sigma_4^*$ (A).

Let us go through the same steps for a symmetry-forbidden reaction, the $[_\pi2_s+_\pi2_s]$ cycloaddition **6.141**. We first draw the reaction and put in the curly arrows—the orbitals are evidently the $\pi$ and $\pi^*$ of each of the $\pi$ bonds. There are two symmetry elements maintained this time—a plane like that in the Diels-Alder reaction, bisecting the $\pi$ bonds, but also another between the two reagents, which reflect each other through that plane.

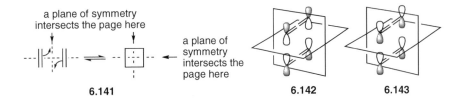

**6.141**                                            **6.142**          **6.143**

a plane of symmetry intersects the page here

a plane of symmetry intersects the page here

In order to classify the symmetry of the orbitals with respect to that plane, we have to take the approaching $\pi$ bonds and pair them up in a lower energy symmetric **6.142** and a higher energy antisymmetric combination **6.143**. These are the molecular orbitals developing as the two molecules approach each other. Pairing the orbitals like this is essentially the same device as pairing the $\sigma$ bonds in setting up $\sigma_1$-$\sigma_4^*$ in Fig. 6.17. We shall also have to repeat that exercise in this case, to deal with the two $\sigma$ bonds in the cyclobutane product.

We are ready to construct the orbital correlation diagram Fig. 6.18, but we must classify the symmetry of the orbitals twice over, once for the plane bisecting the $\pi$ bonds, represented by the vertical dashed line in Fig. 6.18, and then for the plane between the two reagents, the horizontal dashed lines. Thus the lowest-energy orbital in the starting materials is the bonding combination $\pi_1$ of the two bonding $\pi$ orbitals. This orbital is reflected through both planes and is classified as symmetric with respect to both (SS). The next orbital up is the antibonding combination $\pi_2$ of the two bonding $\pi$ orbitals. This orbital is reflected through the first plane, but not in the second, so it is classified as symmetric with respect to one and antisymmetric with respect to the other (SA). Working up through the two

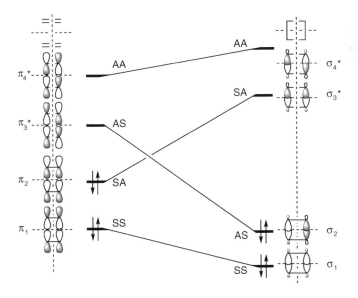

**Fig. 6.18**  Orbital correlation diagram for a $[_\pi 2_s + _\pi 2_s]$ cycloaddition

antibonding $\pi$ orbitals reveals that $\pi_3{}^*$ and $\pi_4{}^*$ are AS and AA, respectively. The product side is similar—except for the addition of the second symmetry classification, it reproduces the pattern for the $\sigma$ bonds that we saw in Fig. 6.16.

We can now complete Fig. 6.18 by correlating the energy levels, feeding the orbitals in the starting materials into orbitals of the same symmetry in the product, SS to SS, SA to SA, AS to AS, and AA to AA. This time, the filled, bonding orbitals of the starting materials, $\pi_1$ and $\pi_2$, do not lead to the ground-state orbitals of the product—one of them, $\pi_1$, leads to the lower bonding orbital $\sigma_1$, but the other, $\pi_2$, leads to one of the antibonding orbitals $\sigma_3{}^*$.

It is common practice to stop here, since we can already see a difference between the allowed and the forbidden reactions. However, an important feature is revealed if we complete the analysis by constructing state correlation diagrams.

### 6.4.3.2   State Correlation Diagrams.
Going back to the Diels-Alder reaction in Fig. 6.16, the ground state of the starting materials is designated $(\psi_1{}^2\pi{}^2\psi_2{}^2)$. Because all the terms are squared (each of the orbitals is doubly occupied), it is described overall as symmetric (S). Similarly the ground state of the product is $(\sigma_1{}^2\sigma_2{}^2\pi{}^2)$, and it too is symmetric. The lines in Fig. 6.16 connect the filled orbitals of the ground state on the left with the filled orbitals of the product on the right, and the state correlation diagram is correspondingly easy. As the individual orbitals of the ground state in the starting material correlate with the individual orbitals of the ground state of the product, the important part of the state correlation diagram (Fig. 6.19) consists simply of a line joining the ground state with the ground state.

**Fig. 6.19**   State correlation diagram for the Diels-Alder reaction

In contrast, the state correlation diagram for the forbidden cycloaddition (Fig. 6.20) is not so simple. The ground state of the starting materials on the left, $\pi_1{}^2\pi_2{}^2$, is overall symmetric, because both terms are squared. Following the

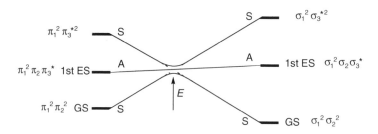

**Fig. 6.20**   State correlation diagram for a $[_\pi2_s + _\pi2_s]$ cycloaddition

lines across Fig. 6.18, we see that this state feeds into a doubly excited state, $\sigma_1{}^2\sigma_3{}^{*2}$, in the product, which is also symmetric because both terms are squared. If we now start at the ground state of the product, $\sigma_1{}^2\sigma_2{}^2$, and follow the lines (SS and AS) in Fig. 6.18 back to the orbitals of the starting material from which they are derived, we find another doubly excited state $\pi_1{}^2\pi_3{}^{*2}$. Both of these states, with both terms squared, are again symmetric. Any hypothetical attempt by the molecules to follow these paths in either direction, supposing they had the very large amounts of energy necessary to do so, would be thwarted because states of the same symmetry cannot cross. The hypothetical reaction would in fact lead from ground state to ground state, but it would have to traverse a very substantial barrier, represented in Fig. 6.20 by the line $E$, which leads up to the avoided crossing. This barrier provides, at last, a convincing explanation of why the forbidden [2 + 2] cycloaddition is so difficult—the energy needed to surmount it is far above that available in most thermal reactions.

We should look now at the first excited state in the starting materials, $\pi_1{}^2\pi_2\pi_3{}^*$, which is produced by promoting one electron from $\pi_2$ to $\pi_3{}^*$. Following the lines in Fig. 6.18 from the occupied and the two half-occupied orbitals on the left (SS, SA and AS), we are led to the orbitals of the first excited state of the product on the right, $\sigma_1{}^2\sigma_2\sigma_3{}^*$. In the state correlation diagram, Fig. 6.20, both of these states are antisymmetric, and there is a line joining them, passing close to the avoided crossing in the ground-state correlation. The value of $E$ is approaching the energy of electronic excitation. It also explains why the photochemical [2 + 2] reaction is allowed—the electrons in the orbitals of the first excited state move smoothly over into the orbitals of the first excited state of the product. This does not mean that the reaction ends there, for the electron in $\sigma_3{}^*$ must somehow drop into $\sigma_2$ to give the ground state, disposing of a large amount of energy—by no means a simple event. All we need to understand in the present context is that the photochemical reaction does not meet a symmetry-imposed barrier like that for the ground-state reaction.

Correlation diagrams have given us a convincing sense of where the barriers come from for those reactions that we have been calling forbidden. In principle, of course, no reaction is forbidden—what these reactions have is a formidable symmetry-imposed barrier, and something very unusual is needed if barriers of this magnitude are to be crossed.

Correlation diagrams take quite a bit of thought, and there are some pitfalls in their construction—however satisfying they may be, they are not for everyday use, and it was for this reason that Woodward introduced the simple rule that we covered earlier [see (Section 6.3.2) page 201].

## 6.5   Secondary Effects

The Woodward-Hoffmann rules arise fundamentally from the conservation of orbital symmetry seen in the correlation diagrams. These powerful constraints govern which pericyclic reactions can take place and with what stereochemistry. As we have seen, frontier orbital interactions are consistent with these features, but they are not the best way of explaining them. In contrast, there are many secondary effects for which the frontier orbitals do provide the most immediately telling explanation. These are the substituent effects on rates and regioselectivity; secondary stereochemical effects like the endo rule for Diels-Alder reactions; periselectivity; and torquoselectivity. We are still on weak ground, for all the usual reasons undermining frontier orbital theory when it is applied too ruthlessly [see (Section 3.7) page 110], but for the organic chemist seeking some kind of explanation for all these phenomena, it is nearly indispensable.

### 6.5.1 The Energies and Coefficients of the Frontier Orbitals of Alkenes and Dienes

In order to apply frontier orbital arguments to these phenomena, we need to know the effect of C-, Z- and X-substituents on the frontier orbitals of alkenes. In Section 2.1.2 (see pages 60–65), we deduced, without carrying out any calculations, that all three kinds of substituents, C, Z and X, lowered the *overall* energy. Using the same arguments, we also deduced the relative energies of the frontier orbitals of C-, Z- and X-substituted alkenes.

The effect of a C-substituent (vinyl and phenyl) poses no problem, because it is seen in the orbitals of a simple alkene and a diene—the HOMO is raised in energy in going from ethylene to butadiene (or to styrene), and the LUMO is lowered in energy (Figs 1.32 and 1.35). For a Z-substituted alkene like acrolein, we saw in Fig. 2.3 that the HOMO energy is close to that of a simple alkene at $1\beta$ below the $\alpha$ level. It lies somewhere between the HOMO of an allyl cation ($\psi_1$) and the HOMO of a diene ($\psi_2$). However, the LUMO energy of a Z-substituted alkene is well below that of a simple alkene, because it lies somewhere between the LUMO of an allyl cation ($\psi_2$ at $0\beta$) and the LUMO of butadiene ($\psi_3^*$ at $0.62\beta$), both of which are lower in energy than $\pi^*$ of a simple alkene at $1\beta$ above the $\alpha$ level. The argument for an X-substituted alkene was even easier: we saw in Fig. 2.5 that it simply mixes in a bit of allyl anion-like character to the unsubstituted alkene, raising the energy of the HOMO relative to the energy of the HOMO of the simple alkene on the left, and, to a smaller extent, raising the energy of the LUMO relative to the energy of the LUMO of the simple alkene.

The same arguments can be used for dienes with a substituent on C-1. A C-substituent raises the energy of the HOMO and lowers the energy of the LUMO, in going from butadiene to hexatriene (Fig. 1.35). For a Z-substituent, the comparison would be between a pentadienyl cation on the one hand and hexatriene on the other, and for an X-substituent, the comparison would be between a pentadienyl anion and the unsubstituted diene. The orbitals for these systems can be found in Fig. 1.35, and estimating an average between the extremes will show that the HOMO of a Z-substituted diene is either unaffected or lowered slightly in energy relative to the HOMO of butadiene, and that the LUMO is distinctly lowered in energy relative to the LUMO of butadiene. Similarly, the HOMO of a 1-X-substituted diene is distinctly raised in energy relative to the HOMO of butadiene, and the LUMO is slightly raised in energy.

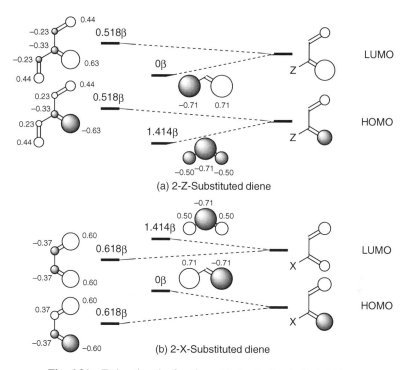

**Fig. 6.21** Estimating the frontier orbitals of a 2-substituted diene

For dienes with substituents at C-2, similar arguments can be used and similar results obtained, as seen in Fig. 6.21. For the Z-substituent, some allyl cation character is mixed into the orbitals of 2-vinylbutadiene, and, for an X-substituent, some allyl anion character is mixed into the orbitals of butadiene.

In summary, the effects of the three kinds of substituent on the energies of the frontier orbitals are:

> C- (extra conjugation)
>   raises the energy of the HOMO
>   lowers the energy of the LUMO
>
> Z- (an electron-withdrawing group)
>   slightly lowers the energy of the HOMO
>   substantially lowers the energy of the LUMO
>
> X- (an electron-donating group)
>   substantially raises the energy of the HOMO
>   slightly raises the energy of the LUMO

The energies of the frontier orbitals are used to explain the effect of substituents on the *rates* of cycloadditions, whereas the relative sizes of the coefficients are used to account for *regioselectivity*. We therefore need to return to the arguments that were used earlier [see (Section 2.1.2) pages 60–65] to estimate the relative magnitude of the coefficients of the atomic orbitals in the frontier orbitals of substituted alkenes and dienes.

We saw in Fig. 1.31 that a C-substituted alkene has higher coefficients at the unsubstituted terminus than at the atom carrying the substituent, both in the HOMO and the LUMO. We saw in Fig. 2.4 that a Z-substituent has only a small effect on the HOMO, with the coefficient at the unsubstituted terminus probably the larger. We also saw that the LUMO was strongly polarised, with the coefficient at the terminal carbon substantially the larger. Finally, we saw in Fig. 2.6 that an X-substituent increases the coefficient in the HOMO at the terminal carbon and reduces it in the LUMO. Similar arguments can be carried over to 1- and 2-substituted dienes.

Putting all these arguments onto a firmer base, Houk estimated energies for the HOMOs and LUMOs of alkenes and 1- and 2-substituted dienes with representative C-, Z- and X-substituents, using experimental measurements of photoelectron spectra for the occupied orbitals, and a combination of electron affinity measurements, charge transfer spectra and polarographic reduction potentials for the unoccupied orbitals.[22] They are summarised in Fig. 6.22, to which we shall frequently refer in discussing the rates and regioselectivity of Diels-Alder reactions and 1,3-dipolar cycloadditions. The circles represent a cross-section of the lobes of the p orbitals looked at from above the plane of the paper, and the shaded and unshaded circles are of opposite sign in the usual way. They are not s orbitals.

### 6.5.2 Diels-Alder Reactions

*6.5.2.1 The Rates of Diels-Alder Reactions.* Most Diels-Alder reactions require that the dienophile carries a Z-substituent before they take place at a reasonable rate. Butadiene **6.144** will react with ethylene, but it needs a temperature of 165 °C and high pressure, whereas the reaction with acrolein is faster, taking less time at a lower temperature. An X-substituent on the diene, on C-1 or C-2, increases the rate further, with 1-methoxybutadiene **6.145** and 2-methoxybutadiene **6.146** reacting with acrolein at lower temperatures. Times and temperatures are not a reliable way of measuring relative rates, but these four reactions were taken to the point where the yields of isolated product are close to 80%.

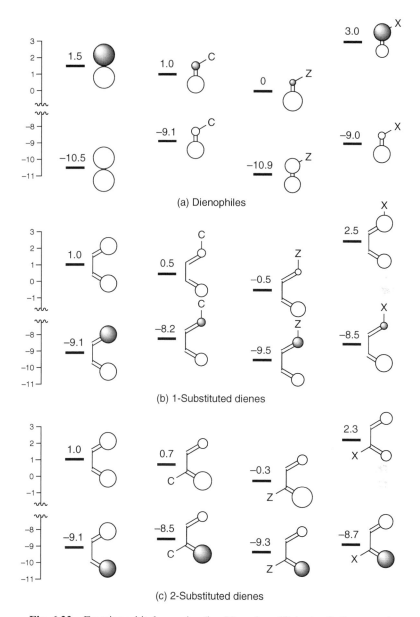

(a) Dienophiles

(b) 1-Substituted dienes

(c) 2-Substituted dienes

**Fig. 6.22** Frontier orbital energies (in eV) and coefficients of alkenes and dienes. The energies are representative values for each class of alkene and diene (1 eV = 23 kcal = 96.5 kJ)

The simplest explanation for how the substituents affect the rates is provided by the relative energies of the frontier orbitals: a Z-substituent lowers the energy of the LUMO, an X-substituent raises the energy of the HOMO, and the smaller the energy gap the faster the reaction. In more detail, taking energy values from

Fig. 6.22, the energy separation between the HOMO of butadiene (at −9.1 eV) and the LUMO of a Z-substituted dienophile (at 0 eV), emphasised with a bold, double-headed arrow in Fig. 6.23b, is less than that between the HOMO of butadiene (at −9.1 eV) and the LUMO of ethylene (at 1.5 eV). Similarly, the energy separation between the HOMO of an X-substituted diene (at −8.5 eV) and the LUMO of the Z-substituted dienophile (at 0 eV), is even smaller (Fig. 6.23c), and the reaction is faster still. To be a good dienophile in a normal Diels-Alder reaction, the most important factor is a low-lying LUMO. Thus, the more electron-withdrawing groups we have on the double bond, the lower the energy of the LUMO, the smaller the separation of the HOMO$_{(diene)}$ and the LUMO$_{(dienophile)}$, and the faster the reaction. Tetracyanoethylene is a very good dienophile.

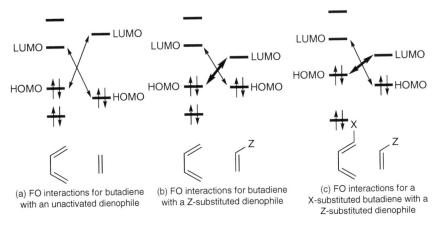

(a) FO interactions for butadiene with an unactivated dienophile

(b) FO interactions for butadiene with a Z-substituted dienophile

(c) FO interactions for a X-substituted butadiene with a Z-substituted dienophile

**Fig. 6.23**   Frontier orbital interactions for Diels-Alder reactions

Another way of producing a low-lying LUMO is to have an oxygen or nitrogen atom in the π bond. Because p orbitals on these atoms lie so much lower in energy than those on carbon, the π molecular orbitals that they make will inevitably have a lower-energy HOMO and LUMO. This is what happens with O=O and —N=N— double bonds, which is one reason why singlet oxygen, and azadienophiles like dimethyl azodicarboxylate, are such good dienophiles.

   It is also possible, but usually less effective, to increase the rate of a Diels-Alder reaction by having electron-donating (X-) groups on the dienophile, raising the energy of the HOMO, and electron-withdrawing (Z-) substituents on the diene, lowering the energy of the LUMO. These Diels-Alder reactions are described as taking place with *inverse electron demand*. The energy separation in Houk's diagram (Fig. 6.22) between the HOMO of an X-substituted dienophile (at −9.0 eV) and the LUMO of a Z-substituted diene (at −0.5 eV) is actually the same as that between the HOMO of an X-substituted diene (at −8.5 eV) and the LUMO of a Z-substituted dienophile (at 0 eV), but it is less effective to carry out Diels-Alder reactions with inverse electron demand. It is not obvious why.

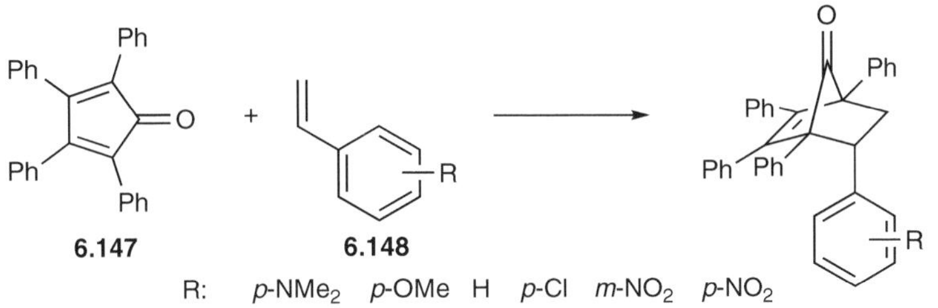

| R: | p-NMe₂ | p-OMe | H | p-Cl | m-NO₂ | p-NO₂ |
|---|---|---|---|---|---|---|
| $k_{rel}$ | 4.6 | 1.4 | 1 | 1.07 | 1.08 | 1.2 |

By adjusting the energies of the HOMO and LUMO of both the diene and the dienophile, we can set up reactions to achieve both normal and inverse electron demand in the same system. For example, Konovalov found that tetracyclone **6.147** reacts with the unsubstituted styrene **6.148** (R = H) slower than with the substituted styrenes, whether the substituent is electron-donating or -withdrawing. The energies of the HOMO and LUMO of this diene are evidently so placed, with respect to those of the styrenes, that when R is H, neither HOMO/LUMO interaction is much stronger than the other. When R is an electron-withdrawing substituent, the LUMO of the dienophile is lowered, bringing it closer in energy to the HOMO of the diene, and when R is an electron-donating substituent, the HOMO of the dienophile is raised, bringing it closer in energy to the LUMO of the diene.

*6.5.2.2 The Regioselectivity of Diels-Alder Reactions.* Regioselectivity refers to the orientation of a cycloaddition: for example, methoxybutadiene **6.145** reacts with acrolein to give more of the 'ortho' adduct **6.149** than of the 'meta' adduct **6.150**. To explain regioselectivity we look at the coefficients of the atomic orbitals in the more important pair of frontier orbitals. We should perhaps remind ourselves that the *sign* of the lobe that is overlapping with another lobe is a much more important factor in determining the energy change than is the second-order effect of its *size*, but from now on we shall ignore the sign because we shall only be looking at allowed (and hence observed) reactions.

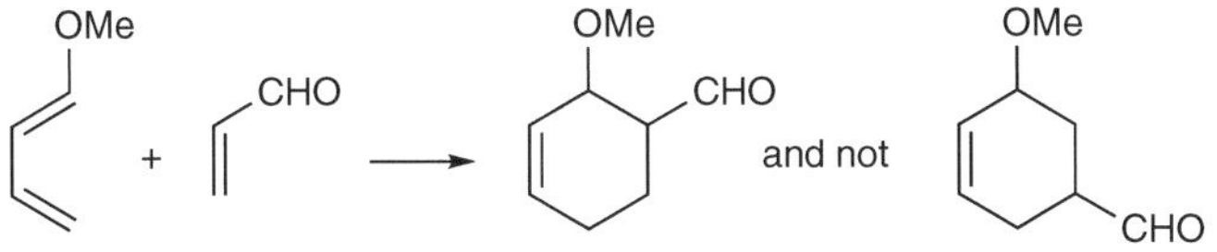

We already know from the data in Fig. 6.22, and from the arguments used to create Fig. 6.23, that the important interaction in a case like this with normal electron demand will be between the HOMO of the diene and the LUMO of the dienophile, shown on the left in Fig. 6.24 ($E_{LUMO} - E_{HOMO} = 8.5$ eV), not the other way round shown on the right ($E_{LUMO} - E_{HOMO} = 13.4$ eV). We see that the two larger atomic orbitals overlap (dashed line) in forming the observed product **6.149**.

$$E_{LUMO} - E_{HOMO} = 8.5 \text{ eV} \qquad E_{LUMO} - E_{HOMO} = 13.4 \text{ eV}$$

**Fig. 6.24**   Coefficients of the frontier orbitals of methoxybutadiene and acrolein

It is not self evident that the choice of the large-large interaction in Fig. 6.24 is better than two large-small interactions. Here is a simple proof that it is right. Consider two interacting molecules X and Y in Fig. 6.25: let the square of the terminal coefficients on X be $x$ and $x + n$, and let the square of the coefficients on Y be $y$ and $y + m$. For the large-large/small-small interaction (Fig. 6.25a), the contribution to the numerator of the third term of the Salem-Klopman equation (Equation 3.4) will be: $xy + (x + n)(y + m)$. For the large-small/ small-large case (Fig. 6.25b), the contribution will be: $x(y + m) + (x + n)y$. Subtracting the latter from the former gives $nm$. In other words, the former interaction is greater so long as $n$ and $m$ are of the same sign; that is, $x + n$ and $y + m$ are either the two large (as shown) or the two small lobes. The implication of a large-large interaction leading over the small-small interaction, as in Fig. 6.25c, is that the transition structure of an unsymmetrical Diels-Alder reaction is itself unsymmetrical, having the two $\sigma$ bonds being formed to different extents. The reaction is still *concerted*, with both bonds forming at the same time, but it is not *synchronous*, with them both formed to an equal extent in the transition structure. Secondary deuterium isotope effects, which are so successful in confirming that Diels-Alder reactions are concerted cycloadditions have also been applied to asynchronous reactions, where they match the calculated values allowing for the asynchronicity.

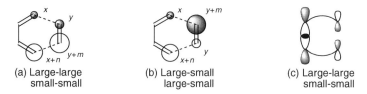

| (a) Large-large | (b) Large-small | (c) Large-large |
| small-small | large-small | small-small |

**Fig. 6.25**   Large-large/small-small pairing of frontier orbitals compared with large-small/ large-small

You may feel that we have laboured hard to justify an example of regioselectivity which any experienced organic chemists would have predicted would go this way round. They would have drawn curly arrows **6.151** creating the canonical structures **6.152** to express an electron distribution making C-4 of the diene nucleophilic and C-3′ of the acrolein electrophilic. They would acknowledge that the cycloaddition is concerted, but would say that the bonding between C-4 and C-3′ was advanced, as in Fig. 6.25c, over that which develops between C-1 and C-2′. This reasoning is fine, but it cannot be applied to all cases.

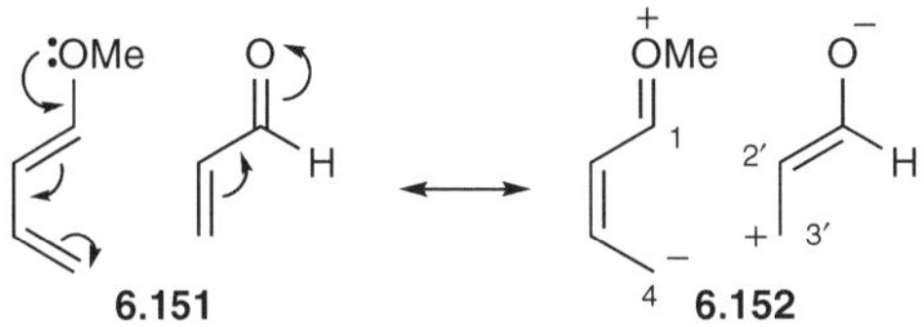

**6.151**                          **6.152**

For example, it does not work in the case of the reaction between butadienecarboxylic acid **6.154** and acrylic acid **6.155**. Curly arrows **6.153** establish that C-4 of the dienecarboxylic acid can be expected to carry a small positive charge, just like C-3′ of the acrylic acid, and the charges on C-4 and C-3′ ought to repel each other. The adduct expected would be the 'meta' adduct **6.157**, whereas the reaction gives mainly the 'ortho' adduct **6.156**.

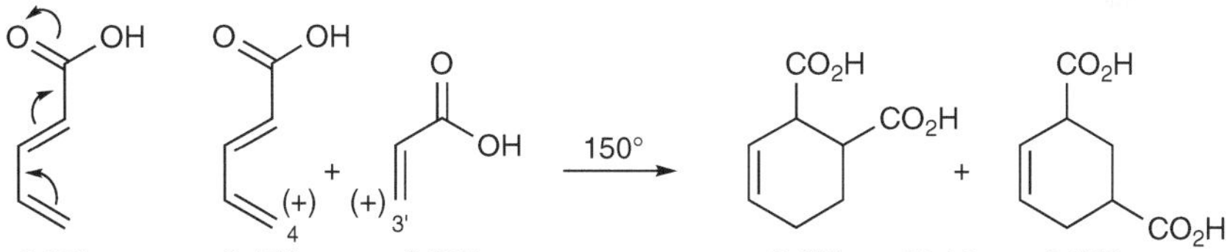

**6.153**      **6.154**      **6.155**      150°      **6.156**   90:10   **6.157**

Clearly this valence bond argument is not good enough. Taking the orbitals and the energies from Fig. 6.22, we get Fig. 6.26, in which the usual $HOMO_{(diene)}/LUMO_{(dienophile)}$ combination has the smaller energy gap. These orbitals are polarised with the marginally higher coefficient in the HOMO of the diene on C-4, which stems from the hexatriene-like character of the conjugated system. As a result, the counter-intuitive combination with bonding between C-4 and C-3′ wins.

The anions of these acids also undergo a Diels-Alder reaction. The contribution of a carboxylate ion group ($CO_2^-$) to the frontier orbitals will be even more like that of a simple C-substituent and less like that of a Z-substituent. The prediction from the frontier orbitals is therefore the same—an 'ortho'

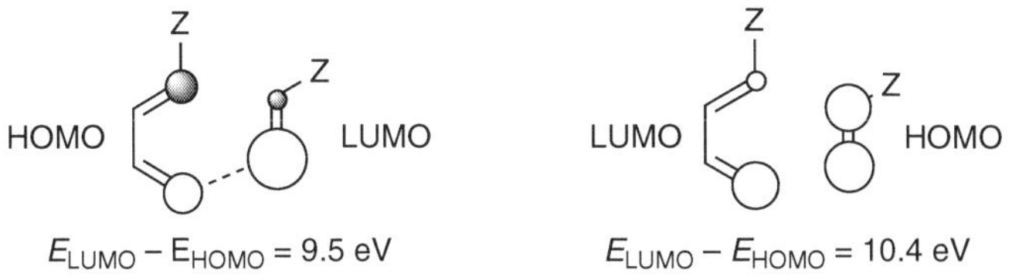

HOMO                          LUMO          LUMO                          HOMO

$E_{LUMO} - E_{HOMO} = 9.5$ eV                    $E_{LUMO} - E_{HOMO} = 10.4$ eV

**Fig. 6.26**   Frontier orbitals for a Z-substituted diene and a Z-substituted dienophile

adduct—but this time the negative charges will strongly repel each other, favouring the 'meta' adduct. The observation of a 50:50 mixture of 'ortho' and 'meta' adducts shows how powerful a directing effect the orbital contribution must be.

There are 18 possible combinations of C-, Z- and X-substituted dienes and C-, Z- and X-substituted dienophiles. The 'ortho' adduct is predicted (and found) to be the major product for almost all of the nine possible combinations with 1-substituted dienes, and the 'para' adduct predicted (and found) for almost all of the nine possible combinations of 2-substituted dienes. Here are some examples to add to those as discussed above.

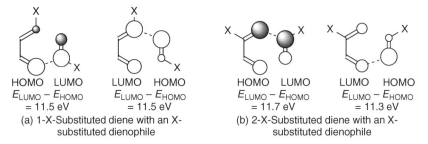

The two exceptions to the 'ortho-para' rule are the reactions between an X-substituted diene and an X-substituted dienophile. The frontier orbitals in Fig. 6.27, taken from Fig. 6.22, indicate that the preferred combination, whichever pair of frontier orbitals and whichever diene, 1-substituted or 2-substituted, is taken, will lead to the 'meta' adduct. Neither frontier orbital interaction is between orbitals close in energy ($>11$ eV in all combinations), with the result that the reaction can be expected to be very slow, and is in practice exceedingly rare. One example is the combination of the (vinylogous) 2-X-substituted diene **6.158**, derived from the benzcyclobutene by electrocyclic ring-opening, and ethyl vinyl ether **6.159**. Although not usefully regioselective, the reaction does take place, assisted by the aromatisation of the diene, and the 'meta' adduct is the major product.

HOMO  LUMO          LUMO  HOMO              HOMO  LUMO          LUMO  HOMO
$E_{LUMO} - E_{HOMO}$    $E_{LUMO} - E_{HOMO}$        $E_{LUMO} - E_{HOMO}$    $E_{LUMO} - E_{HOMO}$
= 11.5 eV              = 11.5 eV                  = 11.7 eV              = 11.3 eV
(a) 1-X-Substituted diene with an X-                    (b) 2-X-Substituted diene with an X-
    substituted dienophile                                  substituted dienophile

**Fig. 6.27**   Frontier orbitals for the combination of X-substituted dienes with an X-substituted dienophile

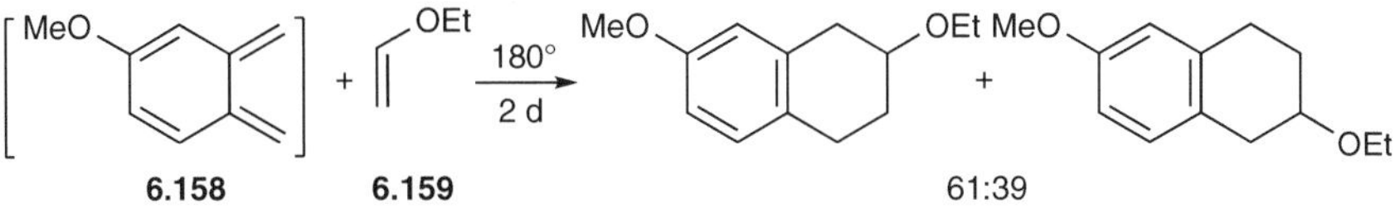

**6.158**     **6.159**                                    61:39

An alternative explanation for regioselectivity has been the diradical theory, in which the major adducts correspond to those which would be obtained if a diradical intermediate were to be involved. Although originally put forward as a mechanism for the Diels-Alder reaction, the idea still works if there is radical character in the transition structure for a concerted cycloaddition. Thus, taking a 2-substituted diene, the intermediate diradical **6.160** leading to the 'para' adduct **6.161** can be expected to be better stabilised than either of the diradicals **6.162** or **6.163** leading to the 'meta' adduct **6.164**, given that any substituent, C, Z or X, stabilises a radical (see pages 67–68). This explanation works to explain the regioselectivity in the majority of cases, but not for the combination of X-substituted diene and X-substituted dienophile, which is therefore support for an orbital-based theory.

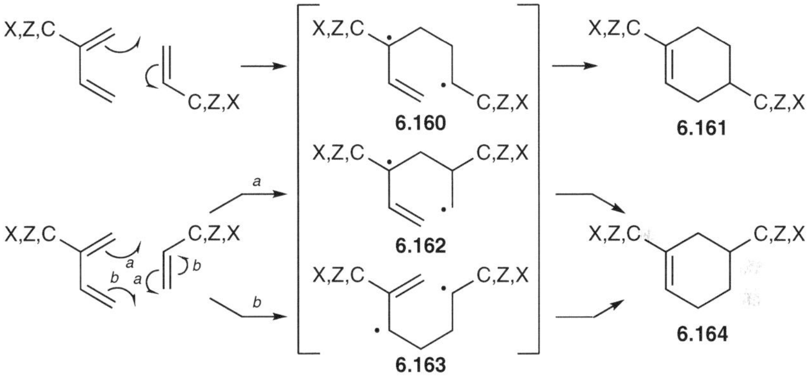

A special case is the effect of a boron substituent. Since it is electropositive and has an empty p orbital, it will be a $\sigma$ donor and a $\pi$ acceptor. If the effect in the $\pi$ system overrides that in the $\sigma$ system, it should be like a Z-substituent activating a dienophile by lowering the LUMO energy. It almost certainly does, as shown by the easy reaction of butadiene with vinyl-9-BBN **6.166**, which takes place at room temperature 200 times faster than the reaction with methyl acrylate. Vinyl boronic acids and esters, in which the empty p orbital is conjugated to oxygen lone pairs, are much less reactive. Vinylboranes show anomalous regioselectivity, giving mainly (>98:2) the 'meta' adduct **6.167** with *trans*-piperylene **6.165**. This appears to be a steric effect, since the reaction with piperylene is significantly slower than the reaction with butadiene, which is not the normal pattern, and the less sterically hindered vinyldimethylborane is no longer selective for the 'meta' isomer.

A number of other examples of regioselectivity in Diels-Alder reactions are less straightforwardly categorised. Thus, citraconic anhydride and 1-phenylbutadiene react to give the 'ortho' adduct **6.168**. One simple-minded way of looking at it is to say that the conjugation **6.169** of the methyl group, through the double bond, with the carbonyl group attached to C-3 will reduce the electron-withdrawing effect of that carbonyl group. The result is that the carbonyl group attached to C-2 is more powerful than the C-3 carbonyl in polarising the double bond, and the reaction becomes another example of a C-substituted diene with a Z-substituted dienophile [See also (Section 4.5.2.3), pages 142 and 143].

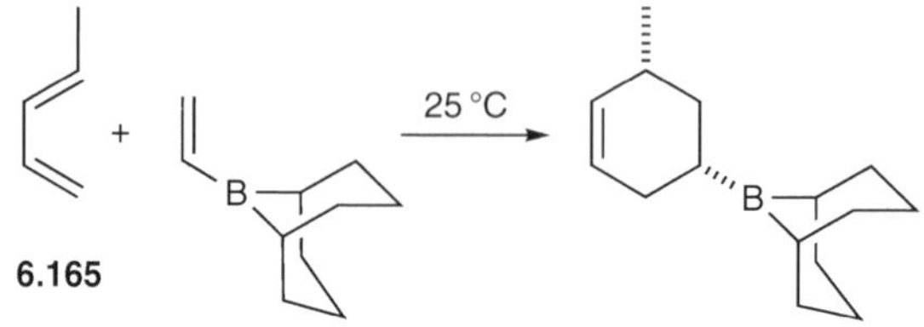

6.165        6.166        6.167

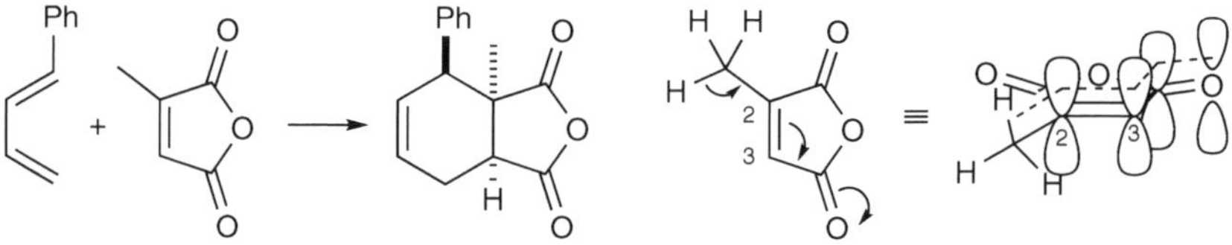

6.168        6.169

An unfortunate consequence of this regiochemistry was a setback to a steroid synthesis. 2,6-Xyloquinone **6.171** reacted with the diene **6.170** to give the adduct **6.172**, and not the adduct which would have been useful for a steroid synthesis. The polarisation of the LUMO of citraconic anhydride deduced above, and the HOMO of a 1-substituted diene, explain the observed regioselectivity. Evidently the 2-aryl substituent has not changed the relative sizes of the coefficients, although it might have been expected to boost the coefficient at C-1. Substituents at C-2 are usually less effective in polarising a frontier orbital than those at C-1. For a sequel to this setback, see (Section 6.5.2.6) page 240.

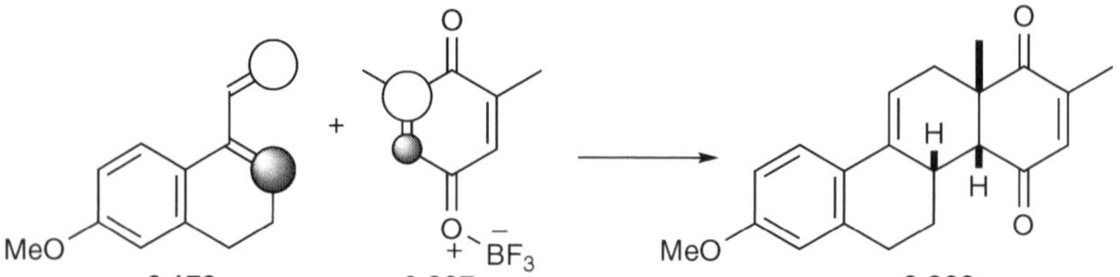

6.170        6.207        6.208

Another special case, in which the unsymmetrical diene is part of a conjugated system which cannot easily be placed in any of the categories, C-, Z- or X-substituted, is tropone **6.173** when it reacts as a diene. On account of its symmetry, we have to work out the coefficients of the atomic orbitals by some other means than by the simple arguments used above and in Chapter 2 (pp. 60–65). The coefficients of the HOMO and LUMO are shown in Fig. 6.28. The numbers for the HOMO are not easily guessed at, and we must be content, in this more complicated situation, to accept the calculation which led to them.

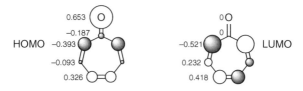

**Fig. 6.28**    The frontier orbital coefficients of tropone

With this pattern in mind, we can see that the regioselectivity shown when tropone reacts as a diene with styrene to give the adduct **6.174** and with acrylonitrile to give the adduct **6.175** is readily explained, whichever pair of frontier orbitals we take in the second case.

**6.5.2.3    The Regioselectivity of Hetero Diels-Alder Reactions.**    In a few cases, carbonyl, nitrosyl, cyano, and other double bonds with one or more electronegative heteroatoms have acted as dienophiles in Diels-Alder reactions. The carbonyl group has a HOMO and a LUMO as shown in Fig. 1.51. The energies of both orbitals are relatively low, and most of their Diels-Alder reactions will therefore be guided by the interaction between the HOMO of the diene and the LUMO of the carbonyl compound. This explains the regioselectivity in the cycloaddition of dimethylbutadiene **6.176** and formaldehyde, and between 1-substituted butadienes **6.177** and nitrosobenzenes.

The most notable Diels-Alder reaction in which the heteroatom is part of the diene system is the dimerisation of acrolein **6.178** giving the adduct **6.179**. As in the reaction of butadienecarboxylic acid **6.154** with acrylic acid **6.155** [see (Section 6.5.2.2) page 229], the two 'electrophilic' carbon atoms are the ones which have become bonded.

The first calculations of the frontier orbitals for acrolein gave the HOMO coefficients on the C=C double bond of acrolein, with the $\alpha$ carbon having the larger coefficient. This failed to explain the regiochemistry, but only because the simple Hückel theory that was used is notoriously weak in dealing with electron distribution in heteroatom-containing systems. Later calculations gave a better set of coefficients, as shown in Fig. 6.29.

**Fig. 6.29**  Frontier orbital energies and coefficients for acrolein

Both LUMO$_{(diene)}$/HOMO$_{(dienophile)}$ and HOMO$_{(diene)}$/LUMO$_{(dienophile)}$ interactions have to be looked at, because the two energy separations are the same for dimerisations. The former interaction is directly appropriate for the formation of the observed product, as shown on the left of Fig. 6.29, but the latter interaction, as shown on the right, has no obvious polarisation in the diene— the C and O atoms have accidentally identical coefficients. However, the resonance integral, $\beta$, for the formation of a C—O bond is smaller than the resonance integral for the formation of a C—C bond when the atoms are more than 1.75 Å apart. Thus the $(c\beta)^2$ term of Equation 3.4 is smaller at oxygen than at carbon in this orbital, and consequently this interaction also explains the regioselectivity. The regioselectivity in this reaction is delicately balanced, but it also matches several cycloadditions between $\alpha,\beta$-unsaturated aldehydes, ketones and imines with C-, Z- and X-substituents on the dienophile, although some of these reactions may well be stepwise, and not pericyclic.

#### 6.5.2.4  The Stereoselectivity of Diels-Alder Reactions.

One of the most challenging stereochemical findings is Alder's endo rule for Diels-Alder reactions. The favoured transition structure **6.180** has the electron-withdrawing substituents in the more hindered environment, under the diene unit, giving the kinetically more favourable but thermodynamically less favourable adduct **6.181**. Heating eventually equilibrates the adducts in favour of the exo adduct **6.182**, by a retro-cycloaddition re-addition pathway.

Any reaction in which a kinetic effect overrides the usual thermodynamic effect on reaction rates is immediately interesting, and demands an explanation, and the most cited is based on the signs of the coefficients in the frontier orbitals—the HOMO of the diene and the LUMO of the dienophile. As a model, we use $\psi_2$ of butadiene for the former, and $\psi_3^*$ of acrolein as the latter (Fig. 6.30a). If we place these orbitals in the appropriate places for the endo reaction, we see that there is the usual primary interaction (solid bold lines), with overlap of orbitals with matching signs, consistent with the rules, but there is an additional bonding interaction (dashed line) between the lower lobe of the p orbital on C-2 of the diene and the upper lobe on the carbonyl carbon of the dienophile, since they have the same sign. This interaction, known as a *secondary orbital interaction*, does not lead to a bond, but it might make a contribution to lowering the energy of this transition structure relative to that for the exo reaction, where it must be absent.

Secondary orbital interactions have also been invoked to explain regiochemistry as well as stereochemistry. Whereas 1-substituted dienes sometimes have only a small difference between the coefficients on C-1 and C-4 in the HOMO, they can have a relatively large difference between the coefficients on C-2 and C-3. Noticing this pattern, Alston suggested that the regioselectivity in Diels-Alder reactions may be better attributed, not to the primary interactions of the frontier orbitals on C-1 and C-4 that we have been using so far, but to a

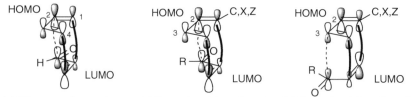

(a) Primary and secondary frontier orbital interactions

(b) Secondary interactions for formation of an 'ortho' adduct

(c) Secondary interactions for formation of a 'meta' adduct

**Fig. 6.30**    Secondary interactions and the endo rule for the Diels-Alder reaction

secondary interaction with the orbital on C-2 (Fig. 6.30b) being stronger than the secondary interaction with the smaller lobe on C-3 (Fig. 6.30c), even though it is not forming a bond.

Explanations for stereochemistry and regiochemistry based on secondary orbital interactions, like frontier orbital theory itself, have not stood the test of higher levels of theoretical investigation. Calculations, low level and high level, produce transition structures which have an endo preference, and a substantial degree of asynchronicity, with the bond from C-4 of the diene to the $\beta$ carbon of the acrolein 0.3 Å shorter than the bond from C-1 to the $\alpha$ carbon. One consequence of this is that the atoms which would be involved in any secondary overlap are too far apart, at about 2.8 Å, for their orbitals to have any significant overlap. A few experimental observations also call secondary orbital interactions into question. Whereas acrolein unexceptionally gives mainly (74:26) the adduct **6.183** with the formyl group endo, methacrolein gives mainly (83:17) the adduct **6.184**, in which the methyl group is endo rather than the formyl group. Furthermore, 1-methoxycyclohexadiene **6.185** and fumaroyl chloride give more of the adduct **6.187** which, if secondary overlap were important, would benefit from secondary overlap with the orbital with the smaller coefficient on C-3 than of the adduct **6.186** which would benefit from secondary overlap with the p orbital on C-2. Also, there is a substantial solvent effect on endo:exo ratios, with the preference for the endo adduct significantly increased in polar solvents, implying that electrostatic interactions are playing some part.

**6.183**

**6.184**

**6.185**    **6.186**    28:72    **6.187**

One possibility for an electrostatic effect is a repulsion in the exo transition structure. Since H—C bonds are polarised towards the carbon, hydrogen atoms carry a partial positive charge, and therefore the in-hydrogens and the carbonyl carbon, which also carries a partial positive charge, will repel each other **6.188**. There will be even greater repulsion from the methylene group of cyclopentadiene **6.189** than from the in-hydrogens of butadienes, which agrees with the greater propensity for forming the endo isomer with this diene. There will be a similar Coulombic repulsion in the methacrolein reaction between the methyl group and the methylene group of cyclopentadiene, and an attraction, a weak hydrogen bond, between the methyl group and C-2. Working in the opposite direction—a Coulombic attraction favouring exo attack—the oxygen atom of a furan, carrying a partial negative charge, will be attracted to the carbonyl group of cyclopropenone **6.190**, and an exo adduct is both kinetically and thermodynamically favoured.

Another approach to explaining the endo rule draws attention to some of the details in a transition structure calculated for the dimerisation of cyclopentadiene (Fig. 6.31).[23] As usual this reaction is kinetically in favour of the formation of the endo adduct **6.192**. The transition structure **6.191** calculated for the dimerisation shows a high degree of asymmetry in the formation of the two σ bonds. The leading bond between C-1 and C-1' is 1.96 Å long, whereas the C—C bond that is still to form between C-4 and C-2' is 2.90 Å. The interesting feature of the transition structure is that the C—C distance between C-2 and C-4' is also 2.90 Å, and the reaction can continue in two equally probable directions, one closing the bond from C-4 to C-2' and the other closing the bond from C-2 to C-4'. The products **6.192** and **6.193**, whichever of the two bonds develops, have the

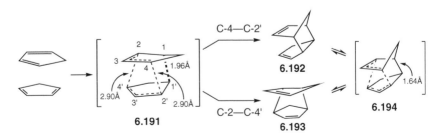

**Fig. 6.31**  The transition structure for the dimerisation of cyclopentadiene

same structure. These products are also connected by a Cope rearrangement, but the transition structure for that reaction **6.194** is different, having a full bond between C-1 and C-1′, and is slightly lower in energy than the transition structure **6.191** for the Diels-Alder reaction.

Another way of looking at the bonds in this transition structure is to see that they develop from the best frontier orbital overlap: the leading bond comes from overlap between the large lobes on C-1 and C-1′ in both the HOMO/LUMO interaction marked in bold in **6.195** and the equally effective LUMO/HOMO interaction marked in bold in **6.196**. The two partly formed bonds, marked with dashed lines, come from overlap between a large lobe on C-4 and a small lobe on C-2′ and between a large lobe on C-4′ and a small lobe on C-2, either of which can develop into the full bond of the product. The traditional secondary orbital interaction in Fig. 6.30, redrawn with the orientation and numbering we are using here as **6.197**, is effectively between C-3 and C-3′. Both of these atoms have small lobes in the frontier orbitals, whereas now, at least in this case with two identical partners, the secondary orbital interaction between C-2 and C-4′, as in the drawing **6.198**, uses one large and one small lobe.

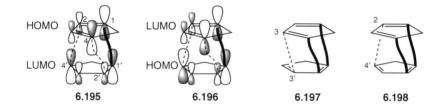

In conclusion, the standard secondary orbital interaction depicted in Fig. 6.30 is not fully accepted, yet it remains a simple and much cited explanation for the endo rule.

### 6.5.2.5 *The Effect of Lewis Acids on Diels-Alder Reactions.*

Diels-Alder reactions are little influenced by polar factors, such as changing the solvent polarity, yet Lewis acids exert a strong catalysing effect. Furthermore, Lewis acid-catalysed Diels-Alder reactions are more stereoselective and more regioselective than the uncatalysed reactions. Thus, cupric ion was used to catalyse the Diels-Alder reaction of the 5-substituted cyclopentadiene **6.201** giving the 7-substituted intermediate **6.202** in Corey's prostaglandin synthesis. The rate constants $k_1$ for the 1,5-hydrogen transfers, which interconvert the cyclopentadienes **6.199–6.201**, are unaffected by the Lewis acid, but the rate constant $k_2$ for the Diels-Alder reaction is increased considerably. In the absence of the Lewis acid, the other dienes, which have an extra X-substituent, are more reactive than the isomer **6.201**.

The effects of Lewis acids on regioselectivity are shown by the reaction of piperylene with methyl acrylate giving mainly the 'ortho' adducts **6.203**, as usual, but this preference is increased with Lewis acid catalysis. Similarly, as an example of enhanced stereoselectivity, the reaction of cyclopentadiene with methyl acrylate gives the endo adduct **6.205** as the major product, but the proportion of this isomer is higher when aluminium chloride is present.

**6.199**   **6.200**   **6.201**   **6.202**

|  | 6.203:6.204 |
| --- | --- |
| without AlCl$_3$ | 90 : 10 |
| with AlCl$_3$ | 98 : 2 |

**6.203**   **6.204**

|  | 6.205:6.206 |
| --- | --- |
| without AlCl$_3$ at 0 °C | 88 : 12 |
| with AlCl$_3$ at 0 °C | 96 : 4 |
| with AlCl$_3$ at −80 °C | 99 : 1 |

**6.205**   **6.206**

All the features of Lewis acid catalysis can be explained by the effect of the Lewis acid on the LUMO of the dienophile. The Lewis acid forms a salt, which is the more active and selective dienophile. For simplicity, protonated acrolein is used as a model dienophile for calculations instead of the Lewis salt. When we were trying to estimate the energies and polarities of the frontier orbitals of acrolein itself [see (Section 2.1.2.2) page 62], we added to the orbitals of a simple diene a contribution from the allyl cation-like character of acrolein. The effect of adding a proton to acrolein is to enhance its allyl cation-like character, and so we add a larger contribution from the allyl cation. The results, reported already on page 141, are: (i) both HOMO and LUMO are even lower in energy; (ii) the HOMO will have the opposite polarity at the C=C double bond, the contribution from the allyl cation now outweighing the contribution from butadiene-like character; and (iii) the LUMO will have even greater polarisation, the $\beta$ carbon carrying an orbital with an even larger coefficient, and the $\alpha$ carbon carrying an orbital with an even smaller coefficient. The lowering in energy of the LUMO makes the $E_{\text{LUMO(dienophile)}} - E_{\text{HOMO(diene)}}$ a smaller number, increasing the rate. The increased polarisation of the LUMO of the C=C double bond increases the regioselectivity. Finally, the increased LUMO coefficient on the carbonyl carbon makes the secondary orbital interaction (Fig. 6.30a) and the electrostatic repulsion **6.188** greater, accounting for the greater endo selectivity whichever explanation we use for endo selectivity.

We can conclude this section on Lewis acid catalysis with a striking example of its use in solving a problem in steroid synthesis. We saw earlier [see (Section 6.5.2.2) page 232] how 2,6-xyloquinone **6.171** added to the diene **6.170** with inappropriate regioselectivity for steroid synthesis. When boron trifluoride was added to the reaction mixture, it formed a salt **6.207** at the more basic carbonyl group, the one conjugated to the two methyl groups. The result was that the polarisation of the LUMO of the C=C double bond was reversed, the major adduct was the appropriate one **6.208**, and a steroid synthesis could be completed.

**6.5.2.6   The Site Selectivity of Diels-Alder Reactions.**   Site selectivity is another kind of regioselectivity, in which a reagent reacts at one site (or more) of a polyfunctional molecule when several sites are, in principle, available. Thus butadiene reacts faster with the quinone **6.209** at C-2 and C-3 than at C-5 and C-6. The cyano groups will lower the coefficients at C-2 and C-3 more than those at C-5 and C-6. The dimer of hexatriene is **6.210** and not **6.211**, which we can similarly explain by looking at the coefficients of the frontier orbitals, essentially narrowing the problem down to assessing the $\Sigma c^2$ term in Equation 3.4.

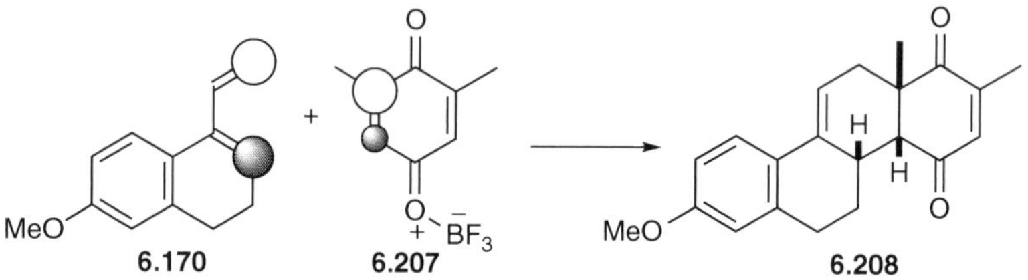

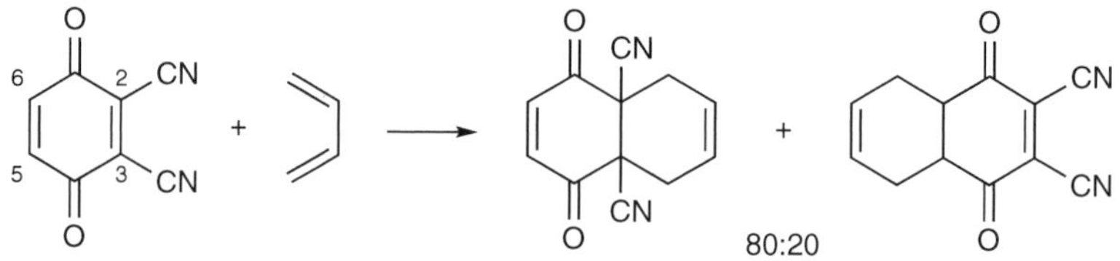

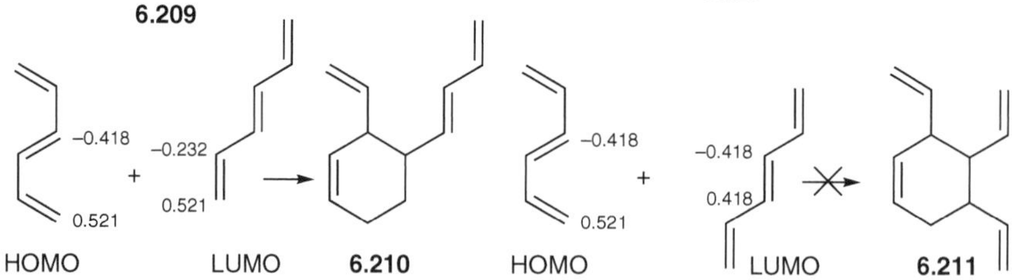

Thus we have the coefficients for the HOMO and LUMO from Fig. 1.35, and the $\Sigma c^2$ term for the observed reaction is given by:

$$\Sigma c^2 = (-0.418 \times -0.232)^2 + (0.521 \times 0.521)^2 = 0.083$$

and for the reaction which is not observed by:

$$\Sigma c^2 = (-0.418 \times -0.418)^2 + (0.521 \times 0.418)^2 = 0.078$$

Since the dimer **6.210** retains a conjugated diene system, it is likely to be lower in energy than **6.211**, making it the thermodynamically favoured product. However, the dimerisation of 2-phenylbutadiene gives mainly the less stable dimer **6.212**. Furthermore, its formation involves attack at the more crowded double bond. The frontier orbitals have the coefficients shown, and the major pathway has the leading bond formed between the two largest coefficients (0.625) and a much higher $\Sigma c^2$-value (0.178) than the minor pathway (0.103) giving the dimer **6.213**.

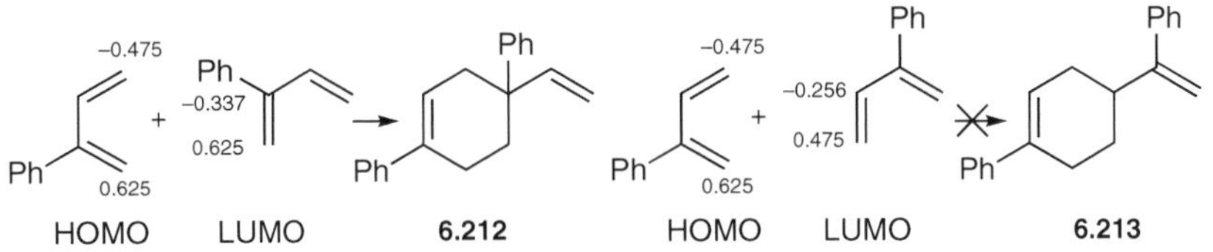

The formation of Thiele's ester **6.217** is a remarkable example of several of the kinds of selectivity that we have been seeing in the last few sections, all of which can be explained by frontier orbital theory. The particular pair of cyclopentadienes which do actually react together **6.215** and **6.216** are not the only ones present. As a result of the rapid 1,5-sigmatropic hydrogen shifts [see (Section 6.3.1.3) page 197], all three isomeric cyclopentadiene carboxylic esters are present, and any combination of these is in principle possible. As each pair can combine in several different ways there are, in fact, 72 possible Diels-Alder adducts.

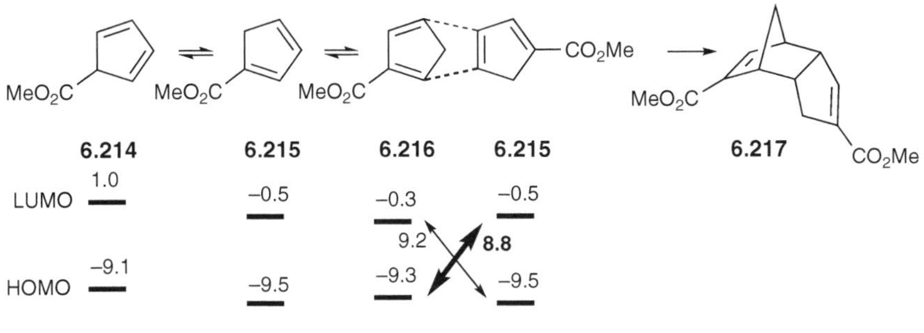

The regioselectivity is a vinylogous version of the combination of a 2-Z-substituted diene with a Z-substituted dienophile, for which the 'para' isomer is expected. The HOMO and LUMO energies of the three isomeric dienes will be close to the representative values used in Fig. 6.22, and these are repeated under the structures. The isomer **6.214** ought to be the most reactive diene, because it has the highest-energy HOMO, but it is known to be present only to a very small extent—evidently too low a concentration to be a noticeable source of products. Leaving this isomer out of account, the smallest energy separation is between the HOMO of **6.216** and the LUMO of **6.215**. These isomers, therefore, will be the ones to combine, and they can react in either of two main ways: **6.216** as the diene and **6.215** as the dienophile or the other way round. The second of these suffers from steric hindrance, because two fully bonded carbon atoms would have to be joined, always a difficult feat. The endo selectivity is normal. Now, knowing about the site selectivity of the reaction, we finally see that it will be the terminal double bond of **6.215** that reacts.

### 6.5.3   1,3-Dipolar Cycloadditions

The range of 1,3-dipolar cycloadditions is large, as we saw with the generic dipole **6.12** and the generic dipolarophile **6.13**. Fig. 6.32 shows the parent members of the most important 1,3-dipoles, and gives their names. Their reaction partners in dipolar cycloadditions are called dipolarophiles, by analogy with the dienophiles in Diels-Alder reactions.

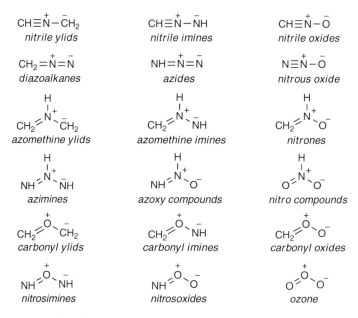

**Fig. 6.32**   The parent members of the most important 1,3-dipoles

***6.5.3.1   The Rates of 1,3-Dipolar Cycloadditions.***   Some 1,3-dipoles, like diazomethane, have high-energy HOMOs, and react faster with alkenes carrying electron-withdrawing substituents. These dipolarophiles have low-energy LUMOs, just as they had when we called them dienophiles. 1,3-Dipoles like diazomethane can be described as nucleophilic in character, and the frontier orbital interactions resemble those of a normal Diels-Alder reaction. Other 1,3-dipoles, like azomethine imines, have low-energy LUMOs, and react faster with dipolarophiles carrying electron-donating substituents, which have high-energy HOMOs. They can be described as electrophilic in character, and their frontier orbital interactions resemble those of a Diels-Alder reaction with inverse electron demand. Attaching electron-donating or electron-withdrawing substituents to a dipole changes this pattern. For example, a diazoketone is an electrophilic dipole, reacting faster with X-substituted alkenes instead of with Z-substituted alkenes. Putting an acyl group, a Z-substituent, onto the diazoalkane lowers the energy of the LUMO, changing the balance of frontier orbital interactions.

A few dipoles, like phenyl azide **6.218**, are neither particularly nucleophilic nor particularly electrophilic. Like tetracyclone **6.147** in its Diels-Alder reactions with styrenes, phenyl azide reacts more slowly with simple alkenes than with alkenes having either an electron-withdrawing group or an electron-donating group. Sustmann obtained a U-shaped curve (Fig. 6.33) when he plotted the rate constants for this reaction against the energy of the HOMO of a wide variety of alkenes. An upwardly curved plot is characteristic of a change of mechanism (as distinct from a change of rate-determining step). The change of mechanism in this case is the change from a dominant HOMO$_{(dipole)}$-LUMO$_{(dipolarophile)}$ interaction on the left to a dominant LUMO$_{(dipole)}$-HOMO$_{(dipolarophile)}$ interaction on the right.

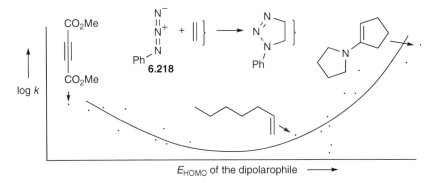

**Fig. 6.33** Correlation between the HOMO energy of dipolarophiles and reaction rate with PhN$_3$

C-, Z- and X-Substituents affect the energies of the molecular orbitals of both dipoles and dipolarophiles in the same way that they affect the orbitals of dienes and dienophiles, but it is not possible to estimate the energies of the frontier orbitals of the parent dipoles in the crude but easy way that it was possible for dienes. Even if we restrict ourselves to the elements carbon, nitrogen and oxygen, we still have many possible types of unsymmetrical dipole (Fig. 6.32), and with very few of them can we find simple arguments from which to deduce the relative energies of the frontier orbitals. Fortunately, Houk has calculated representative energies for the HOMOs and LUMOs of a wide range of dipoles, with and without C-, Z- and X-substituents.[24] They are presented in Table. 6.1, together with Houk's calculations of 'coefficients' needed for the discussion of regioselectivity in the next section. For now, we can see that diazoalkanes have a high-energy

HOMO, and that azomethine imines have a low-energy LUMO, matching the assertions above. Also, phenyl azide has neither a particularly high-energy HOMO nor a particularly low-energy LUMO, explaining the pattern of its reactivity with alkenes seen in Fig. 6.33. The data in the two columns labelled HOMO refer to the 4-electron $\pi$ orbital taking part in the reaction, and not to any nonbonding lone pairs that might in some cases be higher in energy. Likewise the important unoccupied orbital for cycloadditions is not always the lowest in energy of the unoccupied orbitals, which in several cases is an orbital at right angles to the reaction plane. The data in the two columns labelled LUMO therefore refer to the lowest of the unoccupied orbitals which is capable of participating in a 1,3-dipolar cycloaddition.

### 6.5.3.2 The Regioselectivity of 1,3-Dipolar Cycloadditions.

Regiochemistry in a 1,3-dipolar cycloaddition can be illustrated by the reaction of diazomethane with methyl acrylate, in which the first-formed intermediate **6.221** undergoes tautomerism to give the conjugated system **6.222**. There is no guessing the regiochemistry by just looking at the reagents. If we draw the resonance structures **6.219** and **6.220**, they show that both the C and the N termini are simultaneously nucleophilic and electrophilic, as the arrows show. It is an ambident nucleophile. The total charge distribution is likely to have more negative charge at the electronegative end of the dipole, but diazoalkanes are known to react with electrophiles like carbonyl groups at the carbon end. Both sets of arrows are reasonable, both describe what happens, but we cannot use them to decide which atom, C or N, will bond to the electrophilic $\beta$ carbon of the dipolarophile. The same problem arises with all the other unsymmetrical dipoles. Before the frontier orbital arguments appeared, explaining the regioselectivity of 1,3-dipolar cycloadditions had been an outstanding challenge.

Because the bonds being made are not always C—C bonds, as they are in the common Diels-Alder reactions, we must allow for the change in the resonance integral, $\beta$, when making N—C and O—C bonds, as well as estimating the coefficients of the atomic orbitals, $c$. Fortunately, all this work has been done for us by Houk, and we shall take his figures in Table 6.1 on trust. Instead of straightforward coefficients for the atomic orbitals, as we had for dienes and dienophiles, he has calculated $(c\beta)^2$ values in their place, and divided them by 15 to bring the numbers close to 1. They are calculated assuming that the new bonds are being made to *carbon* atoms in the dipolarophile from the carbon,

**Table 6.1** Energies and 'coefficients' of 1,3-dipoles

| Dipole | HOMO | | LUMO | |
|---|---|---|---|---|
| | Energy (eV) | $(c\beta)^2/15$ | Energy (eV) | $(c\beta)^2/15$ |
| Nitrile ylids | −7.7 | $CH\equiv\overset{+}{N}-\overset{-}{CH_2}$  1.07  1.50 | 0.9 | $CH\equiv\overset{+}{N}-\overset{-}{CH_2}$  0.69  0.64 |
| $PhC\equiv\overset{+}{N}-\overset{-}{CH_2}$ | −6.4 | | 0.6 | |
| Nitrile imines | −9.2 | $CH\equiv\overset{+}{N}-\overset{-}{NH}$  0.90  1.45 | 0.1 | $CH\equiv\overset{+}{N}-\overset{-}{NH}$  0.92  0.36 |
| $PhC\equiv\overset{+}{N}-\overset{-}{NH}$ | −7.5 | | −0.5 | |
| Nitrile oxides | −11.0 | $CH\equiv\overset{+}{N}-\overset{-}{O}$  0.81  1.24 | −0.5 | $CH\equiv\overset{+}{N}-\overset{-}{O}$  1.18  0.17 |
| $PhC\equiv\overset{+}{N}-\overset{-}{O}$ | −10.0 | | −1.0 | |
| Diazoalkanes | −9.0 | $CH_2=\overset{+}{N}=\overset{-}{N}$  1.57  0.85 | 1.8 | $CH_2=\overset{+}{N}=\overset{-}{N}$  0.66  0.56 |
| Azides | −11.5 | $HN=\overset{+}{N}=\overset{-}{N}$  1.55  0.72 | 0.1 | $HN=\overset{+}{N}=\overset{-}{N}$  0.37  0.76 |
| $PhN=\overset{+}{N}=\overset{-}{N}$ | −9.5 | | −0.2 | |
| Nitrous oxide | −12.9 | $N\equiv\overset{+}{N}-\overset{-}{O}$  0.67  1.33 | −1.1 | $N\equiv\overset{+}{N}-\overset{-}{O}$  0.96  0.19 |
| Azomethine ylids | −6.9 | $CH_2=\overset{H}{\underset{}{\overset{\cdot}{N}^+}}\overset{-}{CH_2}$  1.28  1.28 | 1.4 | $CH_2=\overset{H}{\underset{}{\overset{\cdot}{N}^+}}\overset{-}{CH_2}$  0.73  0.73 |
| $RO_2CCH=\overset{Ar}{\underset{}{\overset{+}{N}}}\overset{-}{CHCO_2R}$ | −7.7 | | −0.6 | |
| Azomethine imines | −8.6 | $CH_2=\overset{H}{\underset{}{\overset{\cdot}{N}^+}}\overset{-}{NH}$  1.15  1.24 | −0.3 | $CH_2=\overset{H}{\underset{}{\overset{\cdot}{N}^+}}\overset{-}{NH}$  0.87  0.49 |
| $PhCH=\overset{+}{N}\overset{-}{NPh}$ | −5.6 | | −1.4 | |
| $CH_2=\overset{+}{N}\overset{-}{NCOR}$ | −9.0 | | −0.4 | |
| Nitrones | −9.7 | $CH_2=\overset{H}{\underset{}{\overset{\cdot}{N}^+}}\overset{-}{O}$  1.11  1.06 | −0.5 | $CH_2=\overset{H}{\underset{}{\overset{\cdot}{N}^+}}\overset{-}{O}$  0.98  0.32 |
| $CH_2=\overset{R}{\underset{}{\overset{+}{N}}}\overset{-}{O}$ | −8.7 | | 0.3 | |
| $PhCH=\overset{H}{\underset{}{\overset{+}{N}}}\overset{-}{O}$ | −8.0 | | −0.4 | |
| Carbonyl ylids | −7.1 | $CH_2=\overset{+}{O}\overset{-}{CH_2}$  1.29  1.29 | 0.4 | $CH_2=\overset{+}{O}\overset{-}{CH_2}$  0.82  0.82 |
| $Ar(NC)C=\overset{+}{O}\overset{-}{C(CN)Ar}$ | −6.5 | | −0.6 | |
| $(NC)_2C=\overset{+}{O}\overset{-}{C(CN)_2}$ | −9.0 | | −1.1 | |
| Carbonyl imines | −8.6 | $CH_2=\overset{+}{O}\overset{-}{NH}$  1.04  1.34 | −0.2 | $CH_2=\overset{+}{O}\overset{-}{NH}$  1.06  0.49 |
| Carbonyl oxides | −10.3 | $CH_2=\overset{+}{O}\overset{-}{O}$  0.82  1.25 | −0.9 | $CH_2=\overset{+}{O}\overset{-}{O}$  1.30  0.24 |
| Ozone | −13.5 | | −2.2 | |

nitrogen or oxygen atoms of the dipole. They would need to be changed if the bond forming is between *two* heteroatoms. Because $\beta$ contains $S$, the overlap integral, it is a distance-dependent function, which is also dependent upon whether the element is C, N or O [see (Section 1.7.2) page 45]. The values chosen by Houk involve a reasonable guess about the distance apart of the atoms in the transition structure.

To account for the regioselectivity of 1,3-dipolar cycloadditions, we must first assess whether the reaction that we are looking at has a smaller separation between the HOMO of the dipole and the LUMO of the dipolarophile or between the LUMO of the dipole and the HOMO of the dipolarophile. The former is called *dipole-HO controlled* and the latter *dipole-LU controlled*. We can do this simply by taking the energies of the dipoles in Table 6.1, and the energies of the representative dipolarophiles in Fig. 6.22, and transferring them to a diagram like that in Fig. 6.34. Here we see that diazomethane has frontier orbital energies with the smallest separation in energy of all of the possible frontier orbital interactions (the double-headed arrows) for the reaction between diazomethane and a Z-substituted alkene, for which $E_{\text{LUMO(dipolarophile)}} - E_{\text{HOMO(dipole)}}$ is 9 eV. Reactions of diazomethane with electron-deficient alkenes are the fastest and most often encountered of the cycloaddition reactions of diazoalkanes, and we can now see that the strong frontier orbital term for this particular combination accounts for the chemoselectivity.

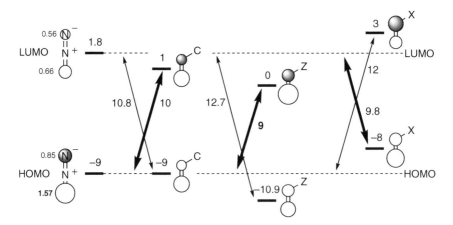

**Fig. 6.34** Frontier orbitals for diazomethane and representative dipolarophiles

This reaction is therefore dipole-HO-controlled, and we can turn to the 'coefficients', the $(c\beta)^2$ values, for the HOMO of the dipole from Table 6.1 and the coefficients of the LUMO of the dipolarophile from Fig. 6.22. Regioselectivity follows in the usual way from the large-large/small-small interaction **6.223**, which has the carbon end of the dipole bonding to the $\beta$ carbon of the Z-substituted alkene, as observed.

**6.223**                          **6.221**                          **6.222**

The reactions of diazomethane with C- and X-substituted alkenes are much slower, and consequently there are fewer known examples. The slower rate of reaction is explained easily by the larger energy separation in the frontier orbitals (10 and 9.8 eV, respectively, in Fig. 6.34). The regioselectivity, however, is the same: $\Delta^1$-pyrazolines like **6.225** and **6.227** with the substituent at C-3 are obtained with both C- and X-substituted dipolarophiles. This at first sight surprising observation can be explained by the change from dipole-HO-control in the cases of the Z- and C-substituted alkenes **6.223** and **6.224** to dipole-LU-control **6.226** in the case of the X-substituted alkene ethyl vinyl ether.

**6.224**        **6.225**        **6.226**        **6.227**

Substituents present on a dipole modify the energies and coefficients shown for the unsubstituted cases in Table 6.1, but it can be easy to predict how these substituents will affect the energies and $(c\beta)^2/15$ values, by taking advantage of our understanding of the effects of C-, Z-, and X-substituents on alkenes. Table 6.1 shows the effect of a few of the commonly found substituents on the energy of the frontier orbitals—phenyl groups do indeed raise the HOMO and lower the LUMO energy, and ester and cyano groups lower both the HOMO and the LUMO energy.

The effect of substituents on the diazoalkane can be accounted for using simple reasoning. X-Substituents raise both the HOMO and the LUMO energies, and speed up reactions with Z-substituted alkenes, making alkyl diazomethanes more reactive than diazomethane in cycloadditions. Z-Substituents, which lower both the HOMO and the LUMO energies, speed up the normally slow reactions with X-substituted alkenes. Furthermore, a Z-substituent on the carbon atom of diazomethane reduces the coefficient on the carbon atom in the LUMO, just as it does in the LUMO of an alkene. Since the $(c\beta)^2/15$ terms for the LUMO of diazomethane are rather similar **6.228**, the Z-substituent is enough to polarise them decisively in the opposite sense **6.229**. This reaction is now dipole-LU-controlled, and the regioselectivity changes to that shown by the pyrazole **6.230**.

The following discussion is limited to a couple of other dipoles, in order to illustrate some of the less straightforward cases. The balance of factors leading to a particular chemo- and regioselectivity is often close—the choice of which pair of frontier orbitals to take is sometimes difficult, the fact that some frontier orbitals are not strongly polarised forces us to judge each case carefully on its merits, and the outcome is not quite always in agreement with the experimental results. There is more discussion of these and all the other cases in Houk's papers.[24]

Let us look at azides. Phenyl azide is the usual dipole in this class, with the phenyl group a C-substituent, which will raise the energy of the HOMO and lower that of the LUMO. We can see from Table 6.1 that in the HOMO of hydrazoic acid the nitrogen carrying the substituent (H in hydrazoic acid, Ph in phenyl azide) has the larger coefficient, and that in the LUMO of hydrazoic acid it has a smaller value. In consequence, the phenyl group is more effective in raising the energy of the HOMO than in lowering the energy of the LUMO. The result, taking the values from Table 6.1, is shown in Fig. 6.35. The smallest energy separation (7.8 eV) is with X-substituted dipolarophiles, which implies that azide cycloadditions will usually be dipole-LU-controlled, and fast with C- and X-substituted dipolarophiles. The orientation is that shown by styrene **6.231** as a C-substituted alkene giving the triazoline **6.232** and by dihydrofuran as an X-substituted alkene **6.233** giving the triazoline

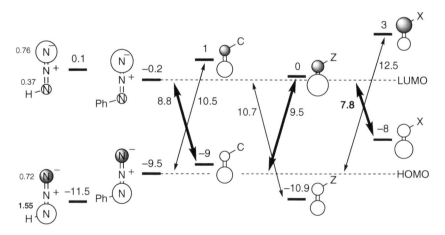

**Fig. 6.35**  Frontier orbitals for phenyl azide and representative dipolarophiles

**6.234**. However, when the dipolarophile is phenylacetylene instead of styrene, the regioselectivity is sharply reduced, and nearly equal amounts of the 1,5-diphenyltriazole and the 1,4-diphenyltriazole were obtained.

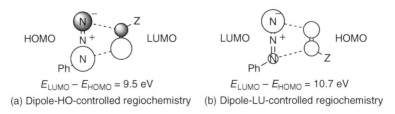

This puzzling observation can be explained, by the frontier orbitals. The second $\pi$ bond of an acetylene is stronger than the first, because it is made between two atoms held close together by the first $\pi$ bond. The overlap of the p orbitals on carbon is therefore stronger, and an acetylene has a lower-energy HOMO than ethylene. This argument is supported by photoelectron spectroscopy, where the HO level is generally found to be 0.4–0.9 eV lower than that of the corresponding alkene. We can also relate this observation to the familiar notion that alkynes are less reactive towards electrophiles like bromine than are the corresponding alkenes. Curiously, the LUMO is not raised for alkynes relative to alkenes. This is shown by UV spectroscopy, where phenylacetylene ($\lambda_{max}$ 245 nm) and styrene ($\lambda_{max}$ 248 nm) would appear to have rather similar separations of their HOMOs and LUMOs. The effect in going from styrene to phenylacetylene is therefore to lower both the HOMO and the LUMO by about, say, 0.5 eV. This changes a dipole-LU-controlled reaction for styrene into one which is affected by both interactions for phenylacetylene, and dipole-HO-control leads to the opposite regioselectivity.

With Z-substituted dipolarophiles and phenyl azide, the situation is again delicately balanced and only just dipole-HO-controlled (9.5 eV against 10.7 eV). For the dipole-HO-controlled reaction, we should expect to get adducts oriented as in Fig. 6.36a. However, a phenyl group reduces the coefficient at the neighbouring atom both for the HOMO and for the LUMO, and this will reduce the polarisation of the HOMO. Conversely, it will increase the polarisation for the LUMO and hence increase the effectiveness of the interaction of the LUMO of the dipole with the HOMO of the dipolarophile, as in Fig. 6.36b. The difference in

**Fig. 6.36** Regioselectivity for phenyl azide reacting with a Z-substituted alkene

energy for the two cases is small enough that firm prediction is not really possible. In practice, dipole-HO-control appears to be dominant, as shown by the formation of the triazoline with the methoxycarbonyl group on C-4 in the reaction with methyl acrylate, but it only needs a small change in the structure of the dipolarophile, such as the addition of an α-methyl group, for some dipole-LU-control to become evident in the formation of some of the adduct with the methoxycarbonyl group on C-5 from methyl methacrylate. Perhaps the methyl group has raised the energy of both the HOMO and the LUMO of the dipolarophile, making the HOMO/LUMO separations still more nearly equal.

Two dipoles especially important in organic synthesis are nitrile oxides and nitrones. The frontier orbital picture for a simple nitrile oxide is shown in Fig. 6.37, where we can see that the easy reactions ought to be decisively dipole-LU-controlled, and fast with C- and X-substituted alkenes. This matches well with the reactions of benzonitrile oxide with styrene, terminal alkenes, enol ethers and enamines which all give only the 5-substituted isoxazolines **6.235**.

In the reactions with Z-substituted alkenes, however, the frontier orbitals are not strongly in favour of either of the pairings, and the polarisation gives opposite predictions, with dipole-LU-control marginally in favour of the isoxazoline **6.236** and dipole-HO-control strongly in favour of the isoxazoline **6.237**. In practice, the regiochemistry with Z-substituted alkenes is delicately balanced, and it may be that the decisive factor is simply steric, since the two ends of the nitrile oxide are very different in their steric demands.

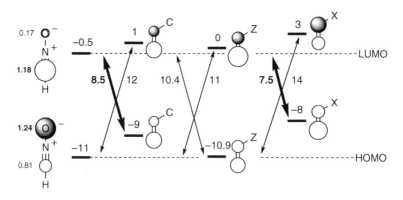

**Fig. 6.37**   Frontier orbitals for a nitrile oxide and representative dipolarophiles

The frontier orbital picture for a simple nitrone is shown in Fig. 6.38, where we can see that the easy reactions will be dipole-LU-controlled with X-substituted alkenes and dipole-HO-controlled with Z-substituted alkenes. In practice, phenyl, alkoxy, and methoxycarbonyl substituents speed up the cycloadditions. Any substituent on the carbon atom of the dipole introduces a steric element in favour of the formation of the 5-substituted isoxazolidines **6.238**. The selectivity with monosubstituted alkenes is in favour of this regioisomer, decisively so with C- and X-substituents, but delicately balanced with Z-substituents, since the HOMO of the dipole is not strongly polarised. With methyl crotonate, both adducts have the methyl group on the 5-position and the ester group on the 4-position.

Dipolarophiles with electronegative heteroatoms such as carbonyl groups, imines and cyano groups also show an orientation in agreement with frontier orbital theory. Because heterodienophiles all have low-energy LUMOs, their reactions will usually be dipole-HO-controlled. The reaction of diazomethane with an imine giving the triazoline **6.240** and the final step in the formation of an ozonide

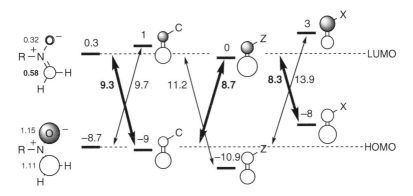

**Fig. 6.38** Frontier orbitals for a nitrone and representative dipolarophiles

**6.241** fit this pattern, and are representative of many others that might have been chosen.

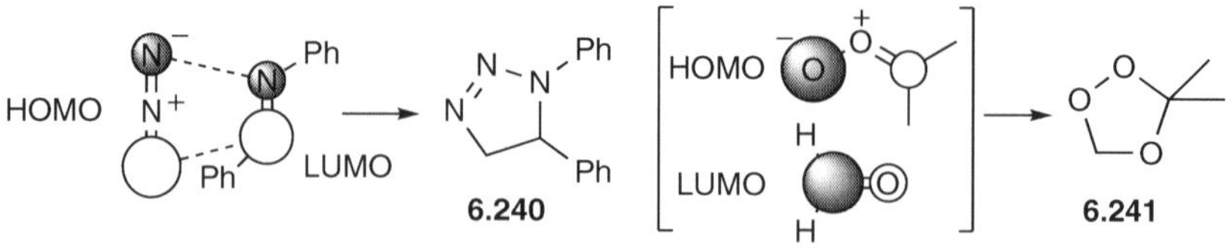

**6.240**     **6.241**

We are far from exhausting the subject of regioselectivity in dipolar cycloadditions with these few examples. Frontier orbital theory, for all its success in accounting for most of the otherwise bewildering trends in regioselectivity, is still fundamentally defective. We should keep in mind that the frontier orbitals used here must reflect some deeper forces than those that we are taking into account in this essentially superficial approach. Nevertheless, no other easily assimilated theory, whether based on polar or steric factors, or on the possibility of diradical intermediates, has had anything like such success.

***6.5.3.3 The Stereoselectivity of 1,3-Dipolar Cycloadditions.*** There is no endo rule for 1,3-dipolar cycloadditions like that for Diels-Alder reactions. Stereoselectivity, more often than not, is low, as shown by the reactions of *C,N*-diphenylnitrone—both regioisomers **6.238** and **6.239** (R=CO$_2$Et) from the reaction with ethyl acrylate are mixtures of exo and endo isomers, only a little in favour of the exo product. Similarly, the reactions of methyl crotonate with nitrones favour the exo product **6.242** over the endo **6.243**. In contrast, other reactions are endo selective, as in the cycloaddition **6.244** of an azomethine ylid to dimethyl maleate giving largely the endo adduct **6.245**.

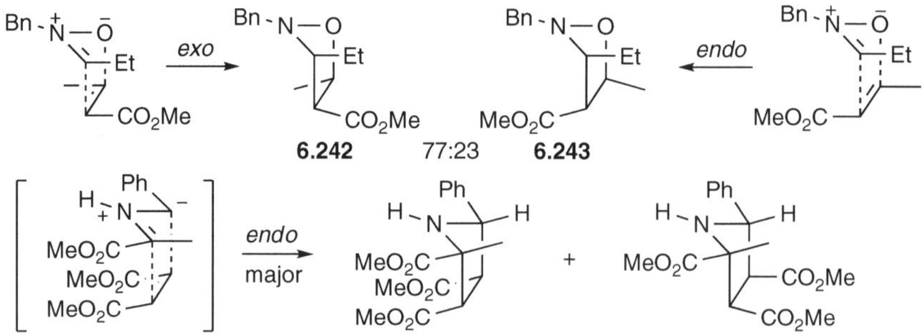

**6.242**   77:23   **6.243**

**6.244**     **6.245**     80:20   **6.246**

### 6.5.4 Other Cycloadditions

***6.5.4.1 [4 + 6] Cycloadditions.*** Secondary orbital interactions have been cited as an explanation for the stereochemistry of the [4 + 6] cycloaddition between cyclopentadiene and tropone **6.20** → **6.21**, which favours the exo transition structure **6.247**. The frontier orbitals have a repulsive interaction (wavy lines) between C-3 and C-4 on the tropone and C-2 on the diene (and between C-5 and C-6 on the tropone and C-3 on the diene) in the endo transition structure **6.248**. However, in this reaction the exo adduct is thermodynamically favoured, the normal repulsion between filled orbitals in the endo transition structure is an adequate explanation. There is no real need to invoke secondary orbital interactions.

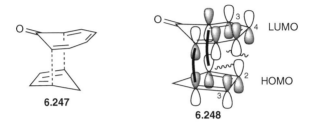

**6.247**

**6.248**

*6.5.4.2 Ketene Cycloadditions.* As we saw earlier [see (Section 6.3.2.8) pages 211 and 212], ketenes undergo cycloadditions to double bonds **6.118** (repeated below) to give cyclobutanones. In practice, the reaction is faster and cleaner when the ketene has electron-withdrawing groups on it, as in dichloroketene, and when the alkene is relatively electron-rich, as in cyclopentadiene. The product from this pair of reagents is the cyclobutanone **6.249**.

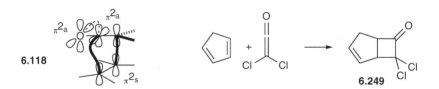

**6.118**

**6.249**

Already we can see that the effects of substituents follow the same pattern as in the Diels-Alder reaction, and we can explain them using the interaction between the LUMO of the ketene and the HOMO of the ketenophile. The conjugation of the C—Cl bonds with the carbonyl group of the ketene will lower, by negative hyperconjugation, the energy of the LUMO, which is more or less $\pi^*_{CO}$. A C- or X-substituent on the alkene will raise the energy of its HOMO, and the energy separation between the frontier orbitals is reduced. This interaction contributes to the bonding represented by the solid line on the left in the transition structure **6.118**. In addition, the HOMO of the ketene is more or less the $\pi$-bonding orbital of the C=C double bond conjugated to the lone pair on the oxygen atom, which is an X-substituent raising its energy, and the LUMO of the ketenophile will be lowered by the C-substituent. This frontier orbital interaction may contribute to the bonding represented by the solid line on the right in the transition structure **6.118**.

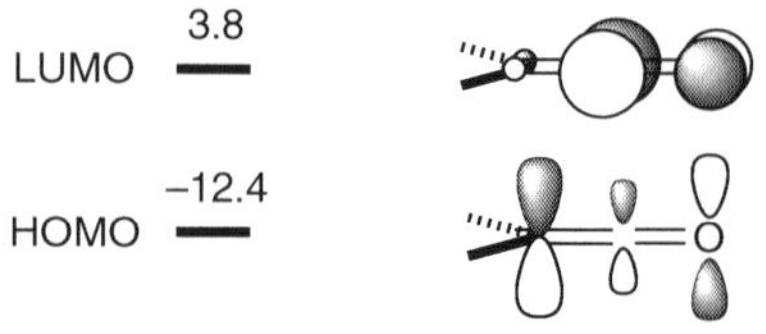

**Fig. 6.39**   Energies and coefficients of the frontier orbitals of ketene

The energies and coefficients of the frontier orbitals of ketene are shown in Fig. 6.39. The regioselectivity in the reaction between cyclopentadiene and dichloroketene giving the cyclobutanone **6.249** is explained by the overlap from the large LUMO coefficient on the central atom of the ketene and the larger coefficient at C-1 in the HOMO of the diene.

An intriguing case is presented by the reaction of ketenes with enamines, such as that between the enamine **6.250** and dimethylketene. This reaction between a strongly nucleophilic alkene and the inherently electrophilic ketene might likely to be stepwise, with the regioselectivity largely determined by the formation of the well stabilised zwitterionic intermediate **6.251**. However, the interesting aspect of this reaction is that the ring closure to give the cyclobutanone **6.252** ought not to be easy, because it would be 4-*endo-trig* at the enolate end (**6.251** arrows) (see pages 163–165). It may well be that this pathway, even though the zwitterion **6.251** is formed, is sterile. There is in fact evidence that this reaction at least in part takes a concerted pathway directly to the cyclobutanone **6.252**, in which case it is another obedient case of the regioselectivity shown by an X-substituted alkene with a ketene.

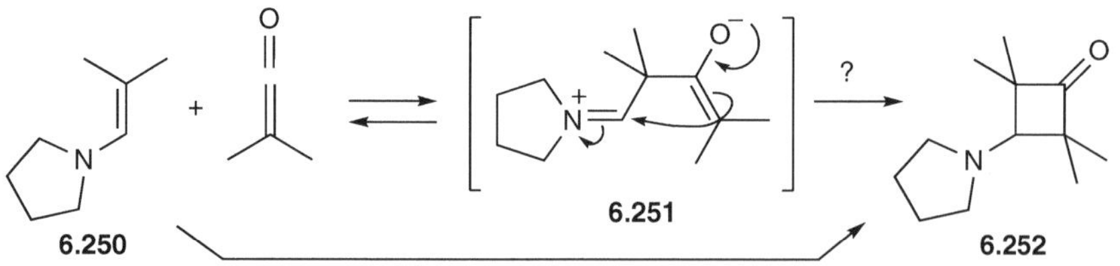

When the two substituents on the ketene are different, as in methylketene **6.253**, the stereoselectivity is usually in favour of the product **6.254** with the larger substituent in the more hindered endo position. This follows from the approach **6.255**, in which the ketene is tilted so that both bonds can develop simultaneously (solid lines), and tilted so that the smaller substituent, the hydrogen atom, is closer to the C=C bond. As the bonds develop further, the methyl group moves down into the more hindered environment, but this must only become perceptible after the transition structure has been passed.

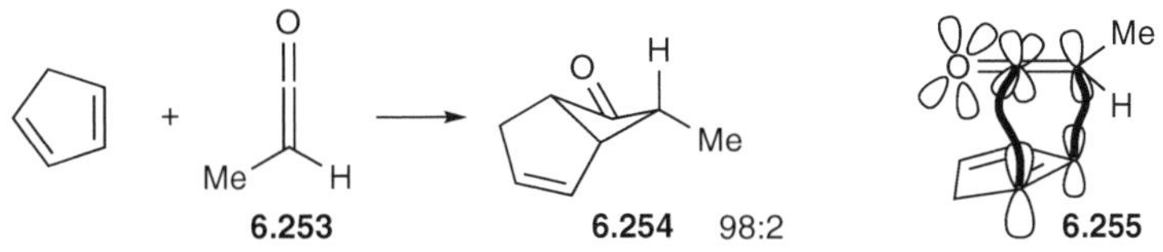

**6.253**          **6.254**   98:2          **6.255**

The cycloaddition of ketenes to carbonyl compounds also shows the expected regioselectivity. Both HOMO$_{(ketone)}$/LUMO$_{(ketene)}$ and LUMO$_{(ketone)}$/HOMO$_{(ketene)}$ interactions may be important, but they lead to the same conclusions about regioselectivity. Lewis acid catalysis is commonly employed in this reaction; presumably the Lewis acid lowers the energy of the LUMO of the ketene (or that of the ketone) in the same way that it does with dienophiles. Ketenes also dimerise with ease, since they are carbonyl compounds. The regiochemistry, whether it is forming a $\beta$-lactone **6.256**, **6.257** or a 1,3-cyclobutanedione **6.258**, is that expected from the frontier orbitals of Fig. 6.39.

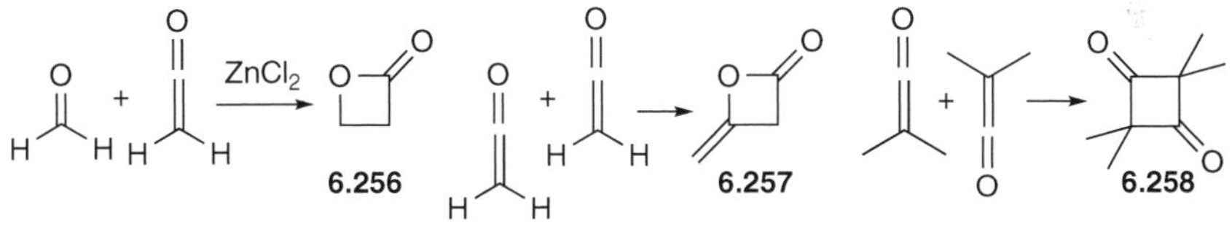

**6.256**          **6.257**          **6.258**

The reaction of the imine **6.259** with the ketene **6.260**, one of many Staudinger reactions, is more plausibly stepwise. The imine is nucleophilic enough to attack a ketene carbonyl group directly from the lone pair on the nitrogen atom. However, the ring closure of the intermediate **6.261** to give the $\beta$-lactam **6.262**, even though it is 4-*endo-trig* at both ends, will not be difficult, because it is an electrocyclic reaction. Electrocyclic reactions do not seem to suffer unduly from the strictures of Baldwin's rules.

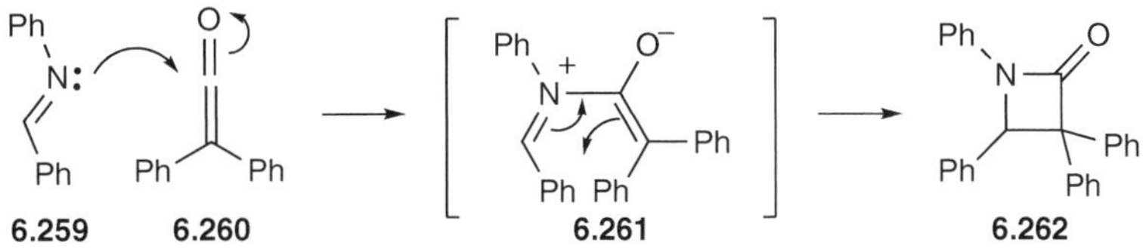

**6.259**     **6.260**          **6.261**          **6.262**

*6.5.4.3 Allene Cycloadditions.* As we saw earlier [see (Section 6.3.2.7) page 211], allenes undergo cycloadditions similar to those of ketenes, except that allenes react faster if they have X-substituents and the alkene has Z-substituents. The HOMO and LUMO of allenes are higher in energy than those of ketene, and they are polarised with larger coefficients on the terminal atoms in the degenerate HOMOs (Fig. 6.40). The degenerate LUMOs are essentially unpolarised.

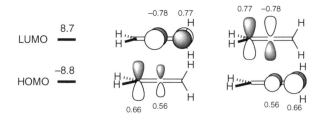

**Fig. 6.40** Energies and coefficients of the frontier orbitals of allene

The regiochemistry of the cycloadditions of allenes is not easily explained by these frontier orbitals. Penta-2,3-diene and acrylonitrile give the adducts **6.263** and **6.264** in which the central carbon of the allene, with the smaller HOMO coefficient, has bonded to the $\beta$ carbon of the Z-substituted alkene. Equally unexplained is the regiochemistry of the reaction between allene itself and diazomethane, which gives more of the adduct **6.265** than of its regioisomer **6.266**. Energetically this reaction can be expected to be dipole-HO-controlled ($E_{\mathrm{LUMO}} - E_{\mathrm{HOMO}} = 15.5$ eV), but this should impart little regiocontrol since the LUMO of the allene is so little polarised. The $\mathrm{LUMO_{(dipole)}/HOMO_{(allene)}}$ interaction ($E_{\mathrm{LUMO}} - E_{\mathrm{HOMO}} = 16.75$ eV) might come into play, since the allene HOMO is polarised, but the orbitals, with the LUMO of diazomethane having the larger 'coefficient' on C, do not match up to explain the high level of regioselectivity. It is possible that steric effects play a major part. Steric effects are manifestly important when enantiomerically enriched penta-2,3-diene is used in the reaction with acrylonitrile—the adducts **6.263** and **6.264** have the absolute configuration corresponding to attack on the allene from the lower surface, as drawn, approaching C-2 from the opposite side from the methyl group on C-4.

This raises the question of the direction of twist in the methylene group not undergoing attack, and for that we need to return to **6.119**. A revealing reaction is the dimerisation of buta-1,2-diene **6.267**. The stereochemistry in the major product **6.269** is comfortably accounted for using the $[_{\pi}2_s + _{\pi}2_s + _{\pi}2_s]$ picture, with the $[_{\pi}2_s]$ component at the bottom in **6.268** approaching the upper allene from the side opposite to the methyl group on C-3 (bold lines), but with its own methyl group orthogonal and offering little steric hindrance. The direction of rotation at C-3, determined by the dashed curve, brings the methyl group towards the viewer, and into the inside position of the diene in the product making it a Z-alkene. The corresponding product with the E-double bond will be lower in energy, yet it is a relatively minor product (18%). The next most abundant product is that from the reaction **6.270** in which both methyl groups are orthogonal to the bonds forming at C-3. The C-3 atoms will approach each other to minimise steric interactions by having the methyl groups mutually in the sectors between the hydrogen and methyl groups **6.271**, which leads to the *trans* arrangement of the methyl groups in the product **6.272**.

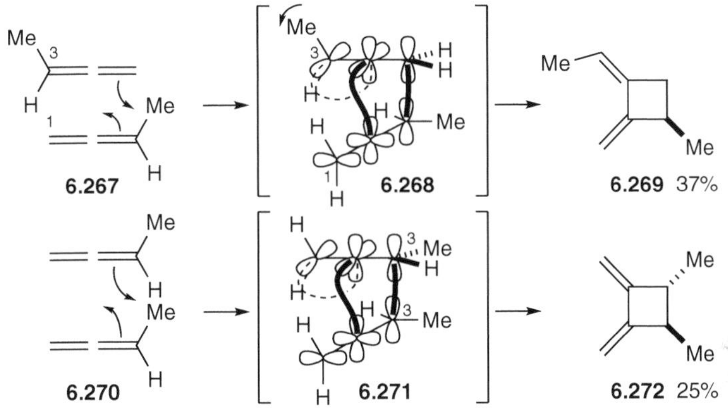

### 6.5.4.4 Carbene Cycloadditions.

Carbene and carbenoid cycloadditions show substantial and orderly regioselectivity. The reaction of the Simmons-Smith carbenoid **6.273** with isoprene **6.274** takes place on the double bond with the highest coefficient in the HOMO. Dichlorocarbene, an electrophilic carbene, reacts at the terminal double bond of cycloheptatriene **6.275**, and at the central double bond of heptafulvalene **6.276**. In both cases the site of attack is the double bond having the largest coefficient in the HOMO. In the former, the $\Sigma c^2$ values would lead one to predict attack at the central double bond [$(0.418^2 + 0.418^2)$ is a little larger than $(0.521^2 + 0.232^2)$], but it is likely that the asymmetry of carbene cycloadditions (see page 214) makes the single largest coefficient disproportionately important.

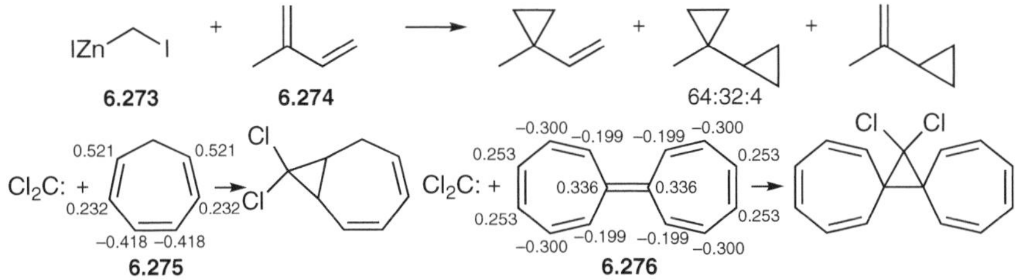

Nucleophilic carbenes, which might show a different site selectivity, rarely undergo cycloadditions, but methoxychlorocarbene, an ambiphilic carbene, adds to the exocyclic double bond of 6,6-dimethylfulvene **6.278**, to give the cyclopropane **6.277**, whereas dichlorocarbene adds to one of the ring double bonds to give the cyclopropane **6.279**. These match the sites with the largest coefficients in the LUMO and HOMO, respectively.

**6.277**  MeOClC:    LUMO **6.278** HOMO    Cl₂C:    **6.279**

### 6.5.4.5 Epoxidations and Related Cycloadditions.

The so-called K-region of polycyclic hydrocarbons like benzanthracene **6.280** is implicated in the carcinogenicity of these compounds. It is believed that the hydrocarbon is epoxidised at the K-region; the product **6.281** is then an electrophilic species capable of alkylating the pyrimidines and purines of nucleic acids. On the whole, but only very approximately, the more nucleophilic the K-region (i.e. the larger the coefficients and the higher the energy of the HOMO), the more carcinogenic the hydrocarbon proves to be, presumably at least partly because it is epoxidised more readily.

HOMO coefficients    biological oxidation    K-region    **6.280**    **6.281**

Epoxidation is like a cycloaddition, in that the new bonds to the oxygen atom are formed simultaneously from each of the carbon atoms. In cycloaddition reactions in general, in which both bonds are being made to more or less the same extent in the transition structure, the highest single coefficient is not the most important, as it was in the reactions of carbenes. Instead, it is the *highest adjacent pair* of coefficients, which is often found in the K-region. Thus if benzanthracene **6.280** is to take part in a cycloaddition reaction in which it provides the 2-electron component, a high value of $\Sigma c^2$ is found in the K-region. ($0.298^2 + 0.288^2$ is larger than $0.194^2 + 0.324^2$ or any other sum of adjacent coefficients not involving the angular carbons; reaction at the angular carbon atoms presumably involves the loss of too much conjugation.)

The same idea accounts for the site selectivity in the reactions of the carcinogenic hydrocarbons **6.282** and **6.283**, both of which react with osmium tetroxide in the K-region. The contrast is with the behaviour of these hydrocarbons with other oxidising agents, like lead tetraacetate, chromic acid and sulfuryl chloride, which react only at one site at a time: none of the hydrocarbons **6.280**, **6.282** or **6.283** reacts in the K-region with these reagents. Instead, reaction takes place at the site with the highest single coefficient in the HOMO, just as we would expect for an electrophilic substitution.

6.282 6.283

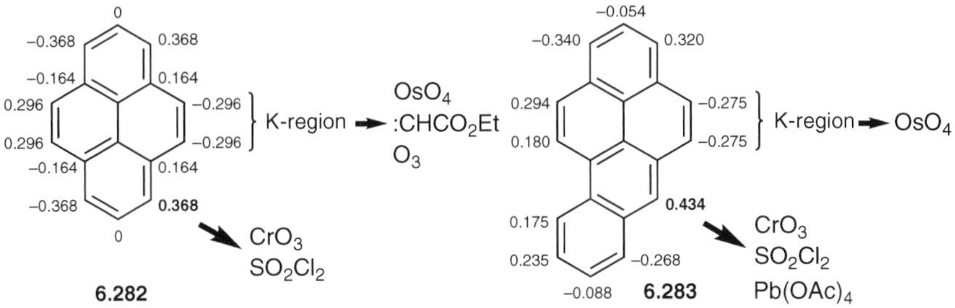

## 6.5.5 *Other Pericyclic Reactions*

***6.5.5.1 Sigmatropic Rearrangements.*** It is more difficult to explain the effect of substituents on the rates, and on the regio- and stereoselectivities of unimolecular reactions. For example, Cope and Claisen rearrangements take place with chair-like transition structures, other things being equal, but we cannot strictly define the secondary orbital interactions, because the molecule does not have a HOMO of one component and a LUMO of the other, as we had with bimolecular reactions. Similarly, we cannot properly use frontier orbitals to explain the effects of peripheral electron-donating and -withdrawing substituents on the rates of sigmatropic and electrocyclic reactions, which are inherently unimolecular, yet the effects can be profound.

The most striking examples are those of an anion or cation substituent on the periphery of [3,3]-sigmatropic rearrangements. The rates of the Cope rearrangements of the dienes **6.284** and **6.286** are much greater than the rates without the oxyanion or carbocation substituent at C-3 of the Cope system. The bold lines in the starting ions **6.284** and **6.286** and in the first-formed products **6.285** and **6.287** emphasise the 1,5-hexadiene systems characteristic of Cope rearrangements. The oxyanion —OM in **6.284** is an X-substituent, and the carbocation in **6.286** is a Z-substituent (but without having any double bond character in the way that a carbonyl group does). Any explanation of the substituent effect must be able to encompass this change from a strong electron donor to a strong electron withdrawer.

6.284 6.285 6.286 6.287

| M | $k_{rel}$ |
|---|---|
| H | 1 |
| K | $10^{12}$ |
| K + 18-C-6 | $10^{17}$ |

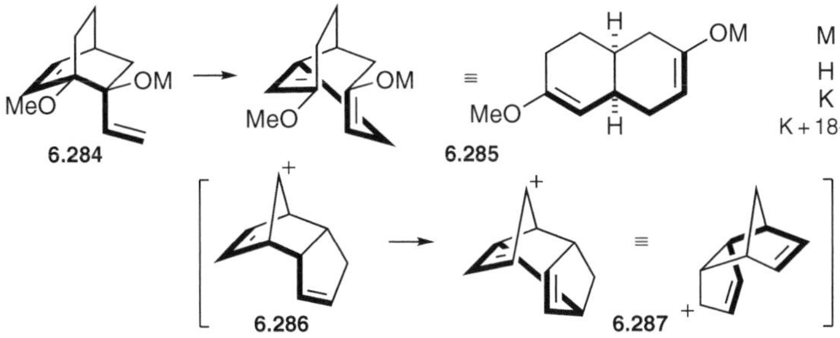

One approach to explaining the substituent effect has been suggested by Carpenter.[25] It uses simple Hückel molecular orbitals, and avoids the need to rely on frontier orbitals. The basis of the idea is to return to the picture of a pericyclic transition structure as having aromatic character [see (Section 6.4.1) page 214].

To provide a baseline from which to compare the substituted system, we first compute the $\pi$ energy of the unsubstituted starting material on the left in Fig. 6.41a and the fully aromatic version of the corresponding transition structure on the right. The starting material will have the $\pi$ energy of two independent $\pi$ bonds, each doubly occupied, $4\beta$ below the $\alpha$ level (Fig. 1.26), and the transition structure will have the $\pi$ energy of benzene, $8\beta$ below the $\alpha$ level (Fig. 1.37). If there is an anionic substituent on C-3, the starting material in Fig. 6.41b, will still have the $\pi$ energy of two independent $\pi$ bonds, $4\beta$ below the $\alpha$ level, but the aromatic version of the transition structure will have the $\pi$ energy of a benzyl anion, which is $8.72\beta$ below the $\alpha$ level. The presence of the substituent lowers the energy barrier between the starting material and the transition structure by $0.72\beta$ relative to the unsubstituted case. The $\pi$ stabilisation of a benzyl cation is the same as the benzyl anion, because the highest of the orbitals in the anion is on the $\alpha$ level, and makes no contribution to the $\pi$ energy. The calculation for having a cationic substituent is therefore exactly the same as for the anionic substituent.

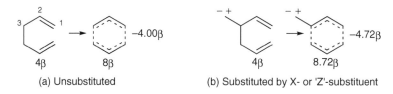

(a) Unsubstituted          (b) Substituted by X- or 'Z'-substituent

**Fig. 6.41**    $\pi$-Energy changes for a Cope rearrangement with a substituent on C-3

A substituent on C-2 has much less effect, and the empty or filled p orbital is again expected to be the same. The unsubstituted system is the same as in Fig. 6.41a. The starting material with a C-2 substituent will have $\pi$ energy of one isolated $\pi$ bond, $2\beta$ below the $\alpha$ level and one allyl system, $2.83\beta$ below the $\alpha$ level, making a total of $4.83\beta$. The transition structure will be modelled by a benzyl system, $8.72\beta$ below the $\alpha$ level, and the overall stabilisation is $3.89\beta$, which is less than the change of $4\beta$ seen in the unsubstituted case. A donor or withdrawing substituent on C-2 might therefore be expected to slow the rearrangement down. The corresponding calculation for a C-substituent makes it $0.04\beta$ less than the unsubstituted case, and so it ought to be rather less rate-retarding. In practice a phenyl substituent at C-2 is mildly rate-accelerating, but this has been explained as a substituent stabilising a transition structure with radical character on C-2. The main point to be seen here is the dramatically greater effectiveness of a substituent on C-3 than on C-2.

The experimentally observed effects of donor and withdrawing substituents on the rates of Claisen rearrangements are summarised in **6.288** and **6.289**. Predictions based on the simple theory above, whether allowing for the effect of the oxygen lone pair or not, match most of the substituent effects, and more elaborate treatments and calculations account for the anomalous accelerating effect of a donor substituent at C-6.

**6.288**                                      **6.289**

The same is true for Cope rearrangements in general. Most substituents at C-1 and C-3 accelerate the reaction, and with more than one of them, their effects are cooperative. Substituents at C-2, however, shift the balance of the transition structure towards a biradical-like intermediate in which the new $\sigma$ bond is formed ahead of the old one breaking. Substituent effects at this site are not cooperative with substituent effects at C-1 and C-3, because they change the nature of the transition structure rather than contribute to it in the same way.

[1,5]-Sigmatropic shifts have another kind of selectivity—the migratory aptitude of the groups that are migrating. It is clear that hydrogen migrates exceptionally easily, and carbon groups are more reluctant. Hydrogen bonding (see pages 90–92) is a paradigm for the exceptional ease with which hydrogen migrates, and another factor is the relative ease with which bonding can develop *in any direction* towards an s orbital. Carbon atoms with bonding made from s and p orbitals have much stricter limits on the direction from which a bond can be made. In [1,5]-sigmatropic rearrangements, hydrogen does not have to obey the constraints of retention or inversion, but can lean over towards the carbon atom to which it is starting to form a bond. Likewise with antarafacial [1,7]-sigmatropic rearrangements, there is no difficulty for the hydrogen atom, moving from the top surface at one end to the bottom surface at the other (Fig. 6.11), to develop overlap so that it sits halfway between, making a nearly linear connection between the two ends.

In the most simple of sigmatropic carbon shifts, the [1,2]-shift in a carbocation, or the related Beckmann, Curtius and Baeyer-Villiger reactions, major factors are the capacity of the migrating carbon to carry partial positive charge, together with the capacity of the carbon atom from which it is migrating to take up the developing positive charge—the [1,2]-shift is accelerated when either or both are capable of supporting an electron deficiency because both are electron deficient in the transition structure **6.290** with two half-formed bonds. However, other things being equal, phenyl and vinyl groups migrate more easily than alkyl, even though phenyl and vinyl are less stabilised cations than alkyl. They do so by participation in a stepwise event—the new bond forms to the p orbital of the $\pi$ bond to give a cyclopropane **6.291** with three full bonds, which then breaks the old bond to give the product. When the overlap creating the $\sigma$ bond in the intermediate **6.291** is geometrically impossible, phenyl and vinyl groups are much slower to migrate than alkyl.

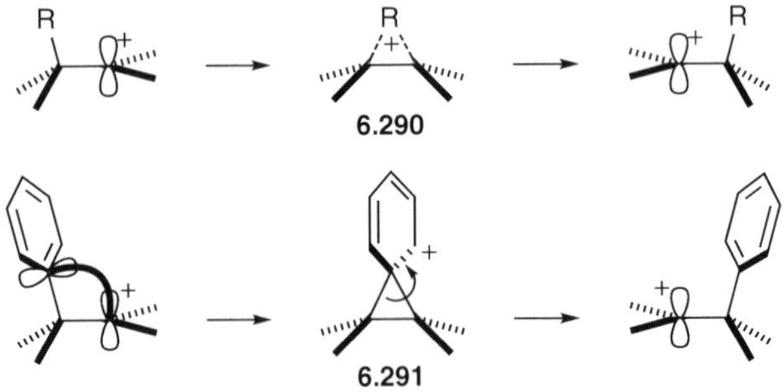

6.290

6.291

### 6.5.5.2 Electrocyclic Reactions.

An anionic substituent on the periphery also speeds up electrocyclic reactions. For example, the benzocyclobutyl oxide **6.292**, an intermediate obtained by trapping benzyne with the lithium enolate of acetaldehyde, undergoes a rapid electrocyclic opening, even at 0 °C, to give the diene **6.293**, which is trapped *in situ* by another molecule of benzyne to give the anthracene hydrate **6.294**. In contrast, a benzcyclobutene without the benefit of the oxyanion substituent has to be heated in refluxing decane (bp 174 °C) for many hours in order to achieve the ring opening that is rate-determining in the Diels-Alder reactions of this type of masked diene.

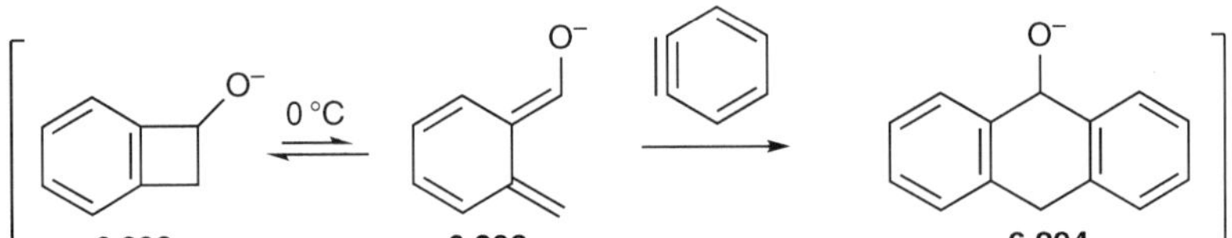

6.292            6.293                    6.294

To explain the increase in the rate of the cyclobutene opening **6.292** → **6.293**, we need to remember that the conrotatory pathway will have a Möbius-like aromatic transition structure, not the anti-aromatic Hückel cyclobutadiene that we saw in Fig. 1.38. We have not seen the energies for this system expressed in $\beta$ terms, nor can we do it easily here, but the numbers are in Fig. 6.42, where we can see that a

(a) Unsubstituted            (b) Substituted by X- or 'Z'-substituent

**Fig. 6.42** $\pi$-Energy changes for a conrotatory cyclobutene opening with a substituent on C-3

donor and a withdrawing group on C-3 can each accelerate the reaction—the number on the right ($-4.29$) is more negative than for the unsubstituted system ($-3.66$).

***6.5.5.3   Alder 'Ene' Reactions.***   Like the Diels-Alder reaction, Alder 'ene' reactions usually take place only when the enophile has a Z-substituent, the regiochemistry is that expected from the interaction of the HOMO of the 'ene' and the LUMO of the enophile, and Lewis acids increase the rate. All these points can be seen in the reaction of $\beta$-pinene with the moderately activated enophile methyl acrylate, which takes place at room temperature in the presence of aluminium chloride, but which would probably not have taken place easily without the Lewis acid.

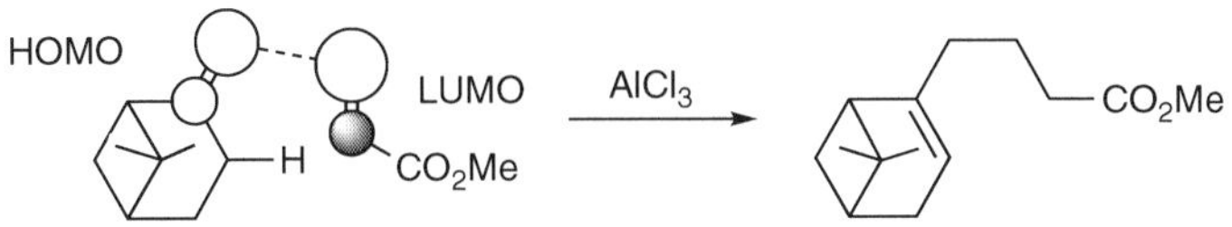

### 6.5.6   Periselectivity

*Periselectivity* is a special kind of site selectivity. When a conjugated system enters into a pericyclic reaction, a cycloaddition for example, the whole of the conjugated array of electrons may be mobilised, or a smaller part of it. The Woodward-Hoffmann rules limit the total number of electrons (to 6, 10, 14, etc., in all-suprafacial reactions, for example), but they do not tell us which of 6 or 10 electrons would be preferred. Thus in the [4 + 6] reaction of cyclopentadiene with tropone giving the tricyclic ketone **6.21**, there is an alternative Diels-Alder reaction, leading to the [2 + 4] adduct **6.295**. The products **6.21** and **6.295** are probably not thermodynamically much different in energy, so that will not be a compelling argument to account for this example of periselectivity, although it may be a factor.

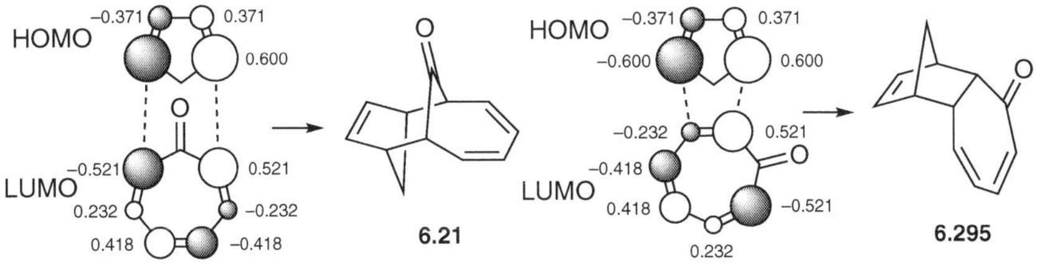

The frontier orbitals, however, are clearly set up to make the longer conjugated system of the tropone more reactive than the shorter. The largest coefficients of the LUMO of tropone (values taken from Fig. 6.28) are at C-2 and C-7 with the result that bonding to these sites lowers the energy more than bonding to C-2 and C-3. In general, the ends of conjugated systems carry the largest coefficients in the frontier orbitals, and we can therefore expect pericyclic reactions to use the longest part of a conjugated system compatible with the Woodward- Hoffmann rules. This proves to be true, with the qualification that the reactions must be geometrically reasonable. Examples of this pattern can be seen in the reactions **6.70 → 6.71**, **6.86 → 6.87** and **6.88 → 6.89** illustrated earlier. Here are further examples, with one each of a cycloaddition **6.296**, an electrocyclic reaction **6.297** and a sigmatropic rearrangement **6.298**, all of which show the largest possible number of electrons being mobilised, when smaller, but equally allowed numbers might have been used instead.

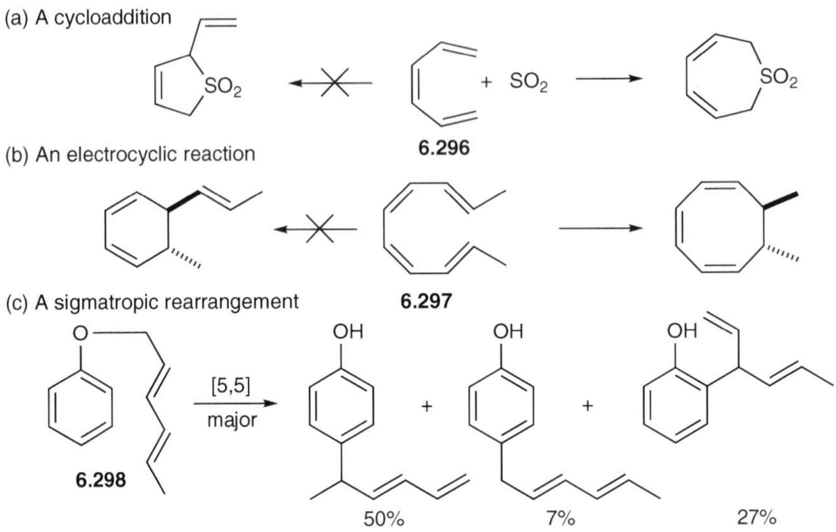

(a) A cycloaddition

**6.296**

(b) An electrocyclic reaction

**6.297**

(c) A sigmatropic rearrangement

**6.298**

50%　　　　7%　　　　27%

However, carbenes react with dienes to give vinylcyclopropanes **6.302**, avoiding the [2 + 4] cycloaddition with a linear approach giving cyclopentenes **6.299**. We have seen that the cyclopropane-forming reaction is allowed when it uses a nonlinear approach **6.130**, but we need to consider why the nonlinear approach is preferred when the linear approach giving a cyclopentene could profit from overlap to the atomic orbitals with the two large coefficients at the ends of the diene.

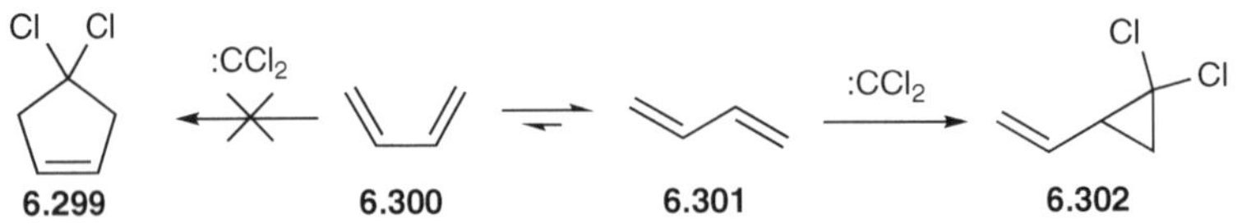

**6.299**　　　　**6.300**　　　　**6.301**　　　　**6.302**

One factor which must be quite important is the low probability that the diene is in the s-*cis* conformation **6.300** necessary for overlap to develop simultaneously at both ends. Since the cyclopropane-forming reaction can take place in any conformation, it goes ahead without waiting for the diene to change from the s-*trans* **6.301** to the s-*cis* conformation. Cyclic dienes like cyclopentadiene, fixed in an s-*cis* conformation, also react to give cyclopropanes, probably because the alternative would create the strained bicyclo[2.1.1]hexene ring system. Cyclic dienes in larger rings also form the cyclopropanes, but they have the two ends of the diene held so far apart that they cannot easily be reached from the one carbon atom of the carbene.

Ketenes also seem to be avoiding the higher coefficients in their reactions with dienes. We have already seen [see (Section 6.3.2.8) page 211 and (Section 6.5.4.2) page 253] that they can undergo [2 + 2] cycloadditions in an allowed manner giving adducts like **6.305**, but we also have to account for why they do so, even when a [2 + 4] reaction is available. Either of the [2 + 4] reactions giving the adducts **6.303** or **6.304** would involve the higher coefficients in the HOMO of the diene, which seemingly ought to make these reactions faster.

The reason why the C=C double bond of a ketene does not react as the $_\pi 2_s$ component of a $[_\pi 2_s + _\pi 4_s]$ reaction, giving the adduct **6.304**, is that the orbital localised on the C=C double bond is at right angles to the p orbitals of the C=O double bond. Consequently, the C=C π bond does not have a low-lying LUMO. Both its HOMO and LUMO are probably raised in energy by conjugation with the lone pair on the oxygen atom, and it is not, therefore, a good dienophile. In the [2 + 2] reaction, however, it is the LUMO of the C=O π bond that is involved in forming the leading C—C bond, and this is low-lying in energy. The [4 + 2] isomer in which the carbonyl group is the dienophile, giving the ether **6.303**, is not so obviously unfavourable, and a [4 + 2] cycloaddition of this type is known for other ketenes. A [3,3]-Claisen rearrangement connects this product to the [2 + 2] adduct **6.305**, and this pathway may be involved in some ketene cycloadditions. 1-Methoxybutadiene and diphenylketene work the other way round, giving initially the [2 + 2] adduct **6.306**, which undergoes a [3,3]-Claisen rearrangement to give what looks like the [4 + 2] adduct **6.307**.

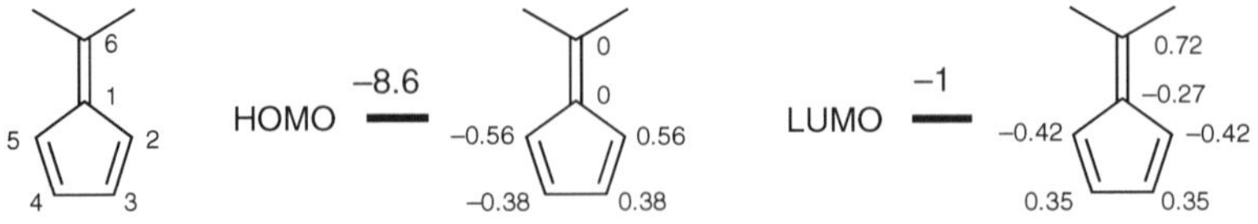

**Fig. 6.43** Frontier orbitals of 6,6-dimethylfulvene

The reactions of dimethylfulvene **6.308** also provide examples where the longest conjugated system is not always involved in the cycloaddition, but this time frontier orbital theory is rather successful in accounting for the experimental observations. The frontier orbital energies and coefficients are illustrated in Fig. 6.43, which shows that there is a node through C-1 and C-6 in the HOMO. The result is that when a relatively unsubstituted fulvene might react either as a $_\pi 6$ or $_\pi 2$ component with an electron deficient (low-energy LUMO) diene, it should react as a $_\pi 2$ component because of the zero coefficient on C-6 in the HOMO. This is the usual reaction observed with electrophilic dienes and dichloroketene. Similarly, when it reacts with tropone **6.20** as a $_\pi 4$ component, it does so at C-2 and C-5, where the coefficients in the HOMO are largest, giving a tricyclic ketone similar to **6.21**. In contrast, if the important frontier orbital in the cycloaddition is the LUMO of the fulvene, and if the fulvene is to react as a $_\pi 2$ or $_\pi 6$ component, it will now react as a $_\pi 6$ component, because the largest coefficients are on C-2 and C-6. For this to be feasible, its partner must have a high-energy HOMO. This is found with diazomethane—the $LU_{(fulvene)}/HO_{(diazomethane)}$ interaction is probably closer in energy than the $HO_{(fulvene)}/LU_{(diazomethane)}$, and the adduct actually obtained **6.309** is the one expected from these considerations.

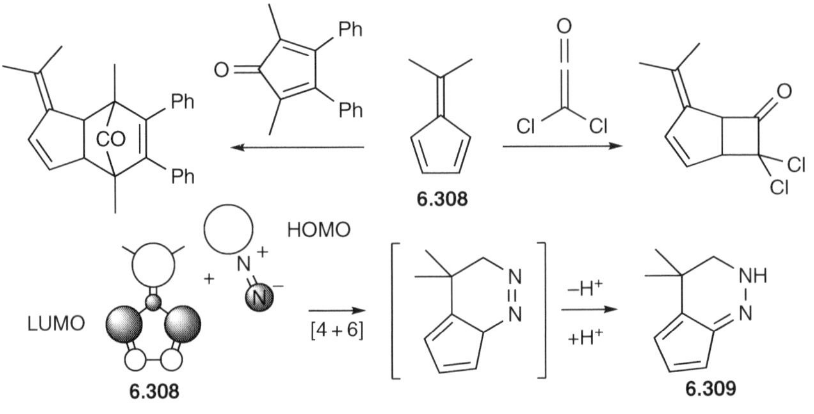

## 6.5.7 Torquoselectivity

Torquoselectivity applies only to electrocyclic reactions, and refers to which is the faster of the two possible senses in which an electrocyclic ring opening can take place. In a conrotatory ring opening, the two substituents R, *cis* to each other in **6.46** in Fig. 6.3, can follow each other to the right in a clockwise rotation, as illustrated, or they can follow each other to the left in an anticlockwise rotation. If the two groups R are the same, there is no difference in the energies, since the transition structures are enantiomeric. If they are different, there is a remarkable level of torquoselectivity determining which substituent shall turn up on the *cis* double bond and which on the *trans*. The observation is that X-substituents selectively move outwards to give a *trans* double bond. This is not simply a steric effect, for the more powerfully electron-donating the X-substituent is in the cyclobutenes **6.310**, the stronger the preference for the formation of *trans*-**6.311**.

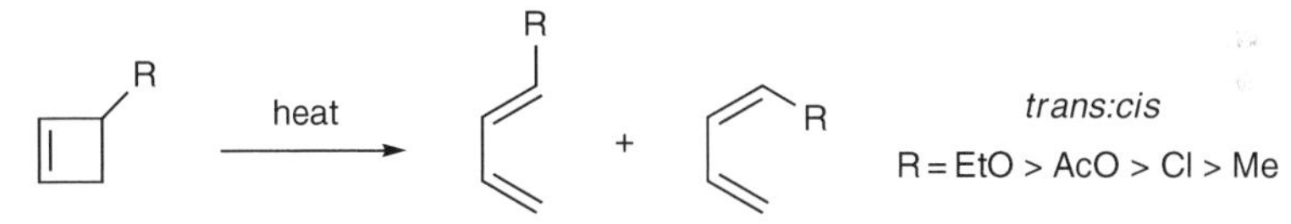

| | | |
|---|---|---|
| **6.310** | *trans*-**6.311**  *cis*-**6.311** | *trans:cis*  R = EtO > AcO > Cl > Me |

More remarkably, Z-substituents move inwards to be on a *cis* double bond, as in the formation of the *cis,cis*-diene **6.313**, in which both trifluoromethyl groups have moved inwards, and the formation of the *cis*-pentadienal **6.315**, in which the aldehyde group has moved inwards. Steric effects are not absent, since the corresponding methyl ketone, with a larger Z-substituent, gives the *trans*-butadienyl ketone, but in the presence of Lewis acids, when coordination to the carbonyl group makes it into a more powerfully electron-withdrawing substituent, the ring opening gives the *cis*-butadienyl ketone.

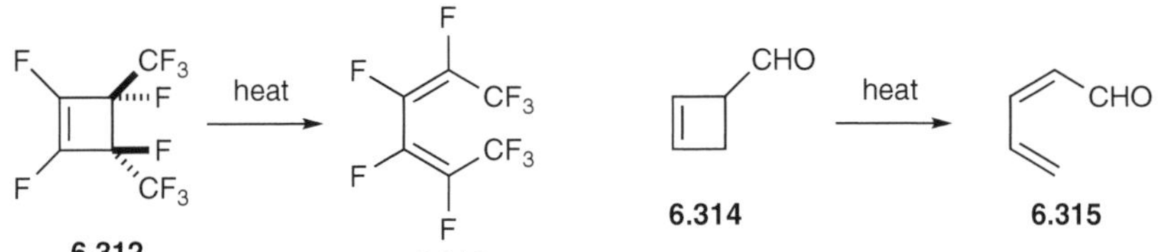

The simplest explanation[26] is to note that the transition structure for conrotatory opening with a filled p orbital inside **6.316** has a three-atom, four-electron conjugated system, which will be anti-aromatic, whereas an empty orbital inside **6.317** has a three-atom, two-electron conjugated system, which will be aromatic. Houk's calculations indicate that there is very little involvement of the p orbitals of the $\pi$ bond in the transition structure, but even if they are included, the conjugated system is then of the Möbius kind and the systems are still anti-aromatic and aromatic, respectively.

**6.316**     **6.317**

With two more electrons, the disrotatory ring opening of a hexatriene, with a total of $(4n+2)$ electrons, has the two upper substituents R in **6.44** able to move outwards, as illustrated for the reaction going from right to left in Fig. 6.3, or able to move inwards. In general, steric effects seem to dominate, and the larger substituents move outwards. More usually, the reaction seen is in the other direction, and the question is then: which reacts faster, a hexatriene with one substituent on a *cis* double bond and the other on a *trans*, or to have them both on a *trans* double bond? The former leads to a cyclohexadiene with the two substituents *trans* to each other, which is usually the lower in energy. Nevertheless the ring closure **6.318** → **6.319** is slower than the ring closure **6.320** → **6.321**.

Ph Ph 20× Ph Ph
6.318 → 6.319 slower 6.320 → 6.321
Ph ''Ph than Ph Ph

**6.318**     **6.319**     **6.320**     **6.321**

Houk has explained this by pointing out that the same considerations apply as in cyclobutene openings—an X-substituent on the inside will contribute to an anti-aromatic transition structure, and a Z-substituent on the inside will contribute to an aromatic transition structure. In both cases the effect will be less than it was for the conrotatory opening and closing, electronically because the orbitals in a disrotatory process **6.322** and **6.323** will be less well aligned for overlap, whether energy-raising as in **6.322** or energy-lowering as in **6.323**. Furthermore, a transition structure looking like **6.324** must have a substantial steric clash between two of the substituents, making the steric component deterring any substituent from occupying the inside position more severe than it was for a conrotatory process. The prediction is that the electronic nature of the substituents will have only a small contribution, and that steric effects are likely to be more important than they were in cyclobutene openings.

**6.322**     **6.323**     **6.324**

Torquoselectivity of a different but powerful kind is found in the ring opening of cyclopropyl halides. These reactions are formally related to the disrotatory opening of a cyclopropyl cation to give an allyl cation, but the opening is concerted with the loss of the leaving group. Cyclopropyl cations themselves are high energy species and are not themselves intermediates, as can be seen in the reactions of the stereoisomeric halides **6.325–6.327**, which give the stereoisomeric cations **6.328–6.330**, respectively. These cations are configurationally stable at the low temperatures used. Had the free cyclopropyl cation been involved the cyclopropyl halides **6.325** and **6.326** would have given the same allyl cations instead of one giving the W-shaped cation **6.328** and the other giving the U-shaped cation **6.329**. In all three cases, although only the first and second prove it, the torquoselectivity is such that the chloride ion leaves from the same side as that in which the substituents move towards each other.

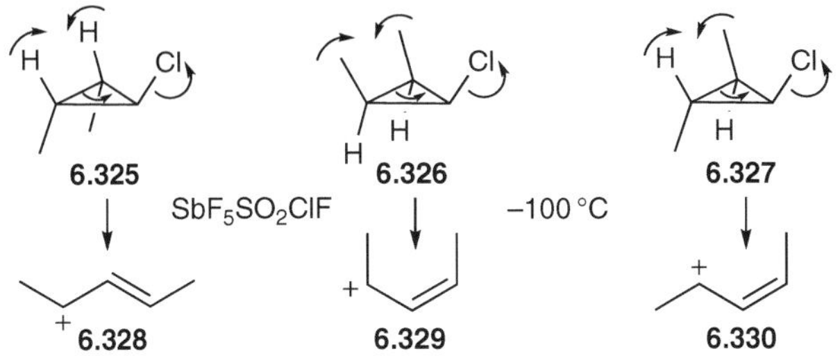

The simple explanation is that, if the substituents on the same side as the leaving group are moving towards each other, and the substituents on the opposite side are moving apart, then the bulk of the electron population from the breaking $\sigma$ bond is moving downwards **6.331** (arrow) through a transition structure **6.332** to the allyl cation **6.333**, effectively providing a push from the backside of the C—Cl bond.

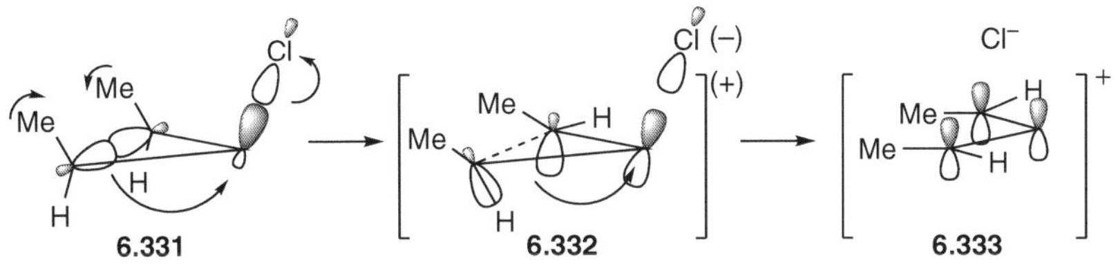

The reverse reaction of this general class—an allyl cation giving a cyclopropyl cation—is found in Favorskii rearrangements. The diastereoisomeric $\alpha$-chloro enolates **6.334** and **6.337** give the cyclopropanones **6.335** and **6.338**, respectively. Thus the reaction is stereospecific, at least in a nonpolar solvent. Evidently the allyl cation is not formed, otherwise the two chlorides would give the same product, or mixture of products. The cyclisation step is presumably disrotatory with the torquoselectivity determined by which side of the allyl system the chloride leaves from. The cyclopropanone is not isolated, because the alkoxide attacks the carbonyl group with subsequent cleavage of the bond towards the methyl group giving the esters **6.336** and **6.339**.

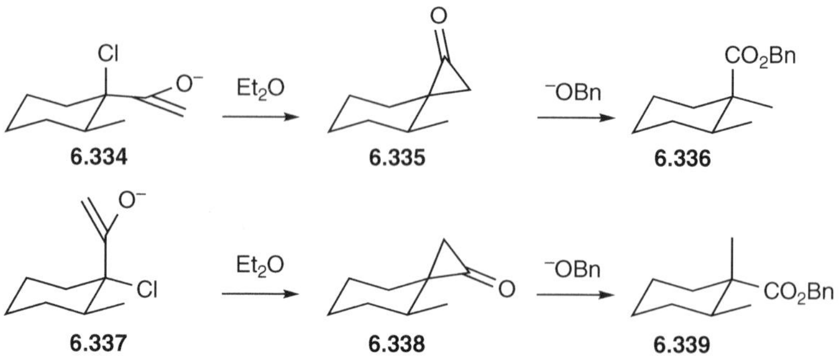

In a more polar solvent, Favorskii reactions cease to be stereospecific, and presumably take place by ionisation of the chloride to give the same cation from each diastereoisomer. Whether the reaction takes place by way of the cation or with concerted loss of the chloride ion, this reaction presented a serious puzzle before its pericyclic nature was recognised. The $\sigma$ overlap of the p orbital on C-2 of the enolate with the p orbital at the other end of the allyl cation **6.340** or with the orbital of the C—Cl bond **6.341** looked forbiddingly unlikely—it is 3-*endo-trig* at C-2. It is made possible by its pericyclic nature, where the tilt of the orbitals can begin to sense the development of overlap. The torquoselectivity in the development of overlap **6.341**, however improbable it looks, corresponds to inversion of configuration at the carbon atom from which the chloride departs.

## 6.6   Exercises

1. Account for the change of product ratio in the following reaction as a result of the change of solvent from polar to nonpolar (note that the solvents are inert in the conditions used for these reactions).

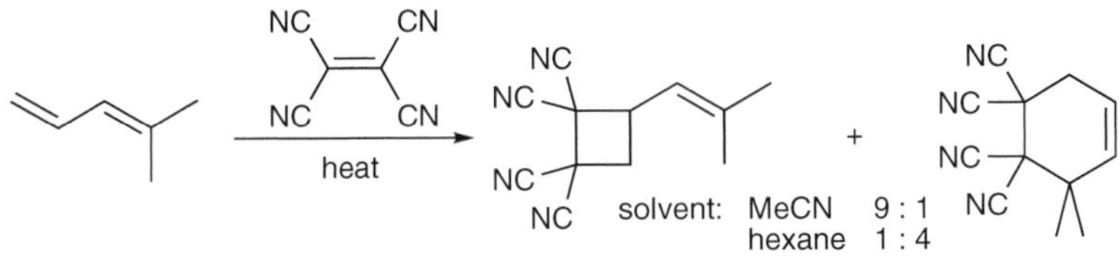

2. Assign the following thermal pericyclic reactions to their class, and predict the relative stereochemistry of the products.[27]

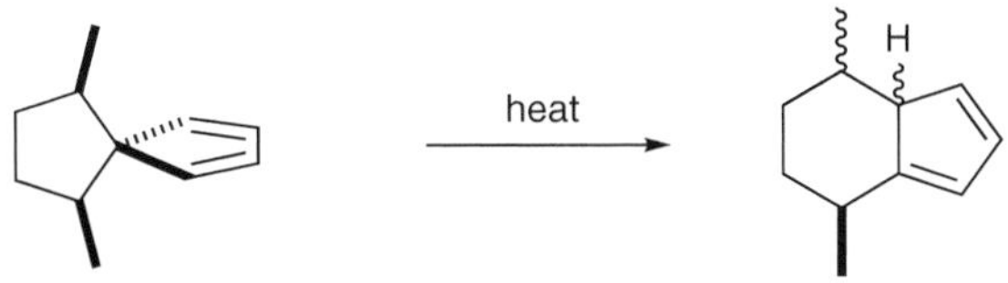

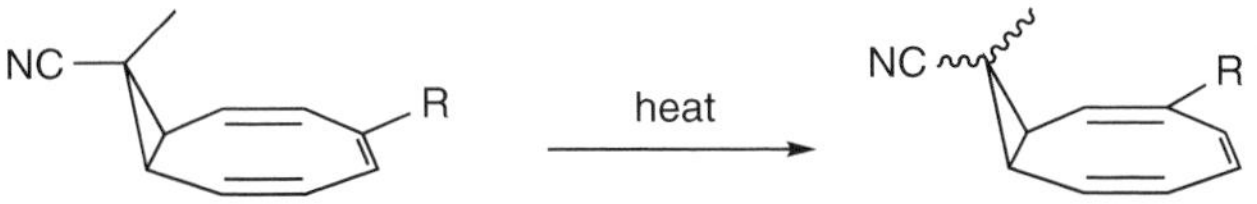

(note: the R group does not move)

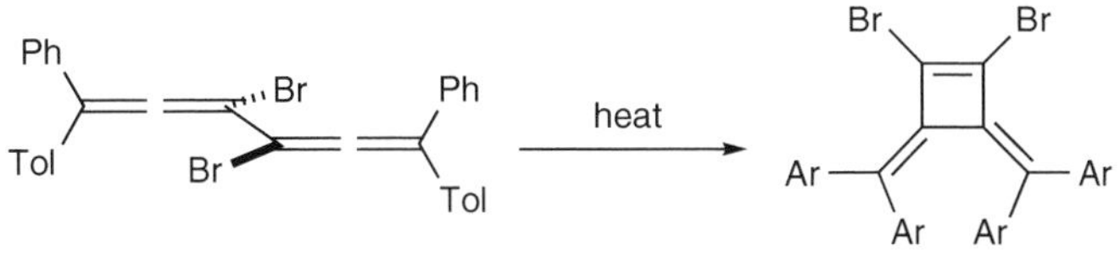

3. The following thermal reactions take place in two pericyclic steps, in which the intermediates could not be detected. Suggest what pathway is followed. Identify the intermediates **A–C**, identify the classes of the pericyclic reactions, and account for the stereochemistry of the first two products.[28]

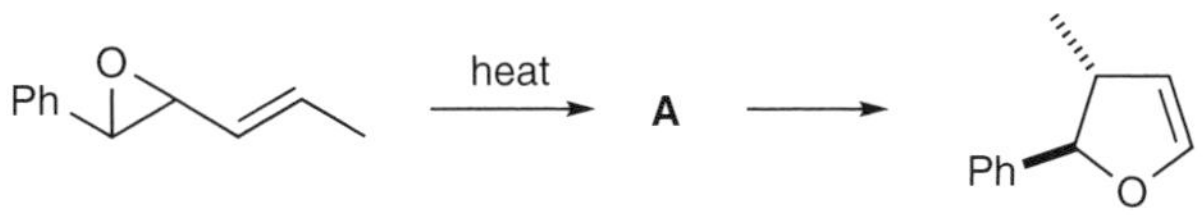

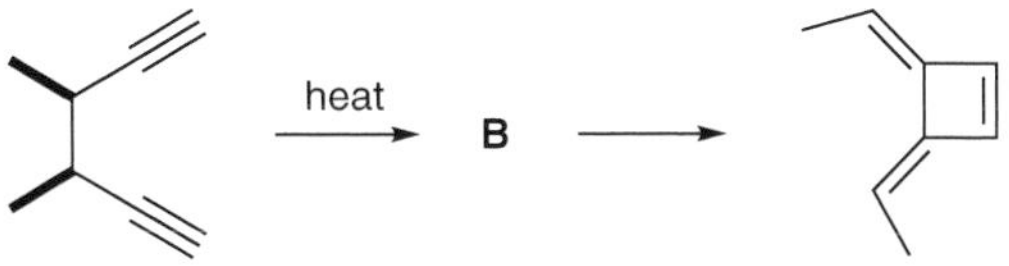

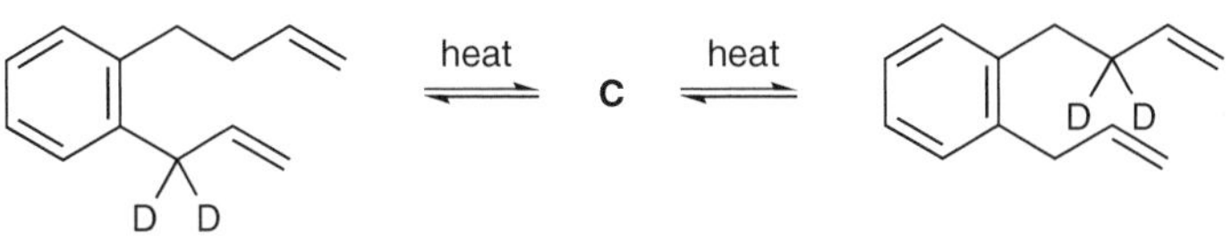

4. The following reactions take place in three or four pericyclic steps. Suggest what pathway is followed, identify the intermediates **A–G**, and identify the classes of the pericyclic reactions.[29]

(a)

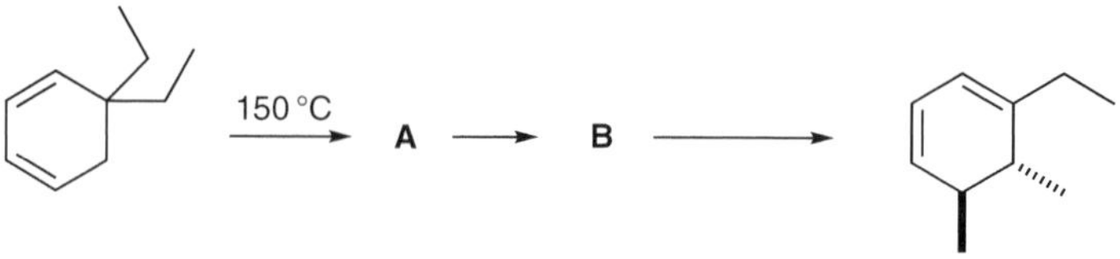

150 °C

**A** ⟶ **B** ⟶

(b)

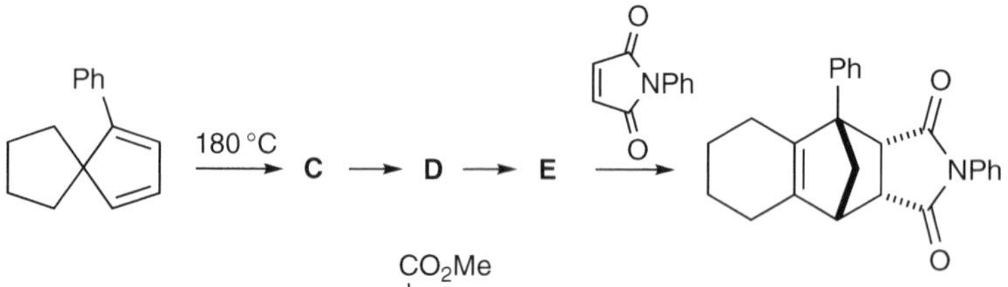

180 °C

**C** ⟶ **D** ⟶ **E** ⟶

(c)

140 °C

**F** ⟶ **G** ⟶

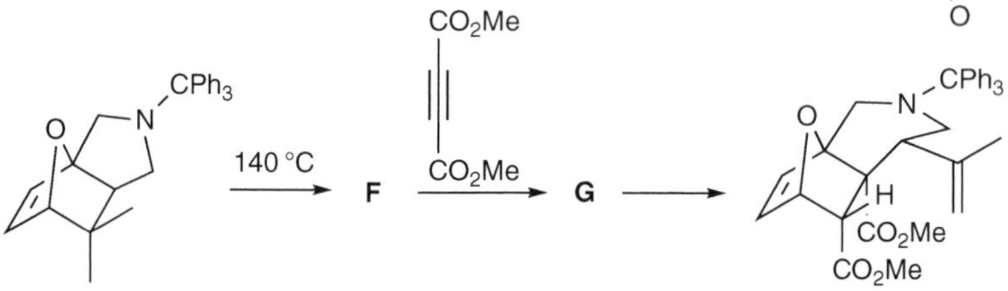

5. The following reactions all involve one or more ionic reactions, as well as one or more thermal pericyclic reactions.[30] In each case, draw out the whole reaction sequence, identify the pericyclic steps, assign them to their classes, and predict the relative stereochemistry at the bonds marked with a wavy line.

(a)

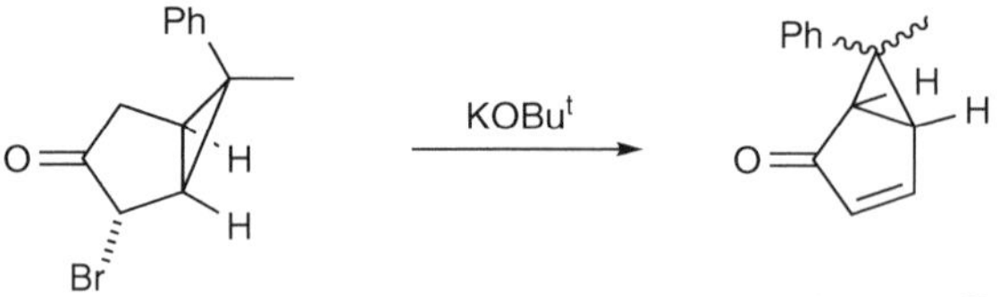

KOBu$^t$

(b)

1. BuLi
2. heat
3. H$_2$O

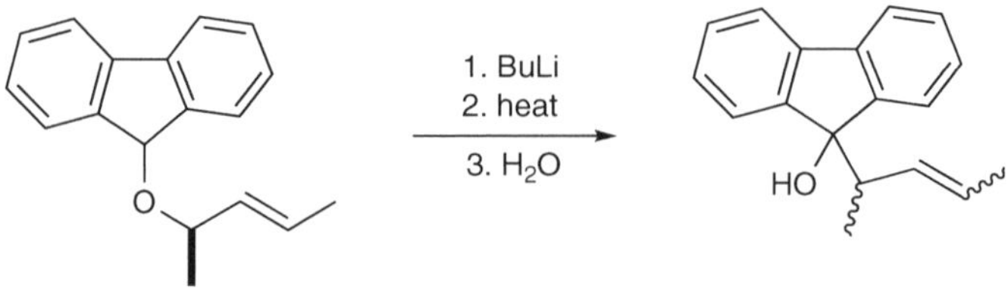

(c)

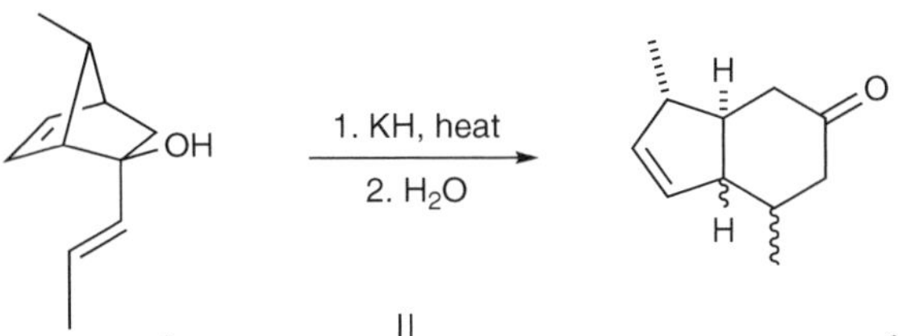

(d)

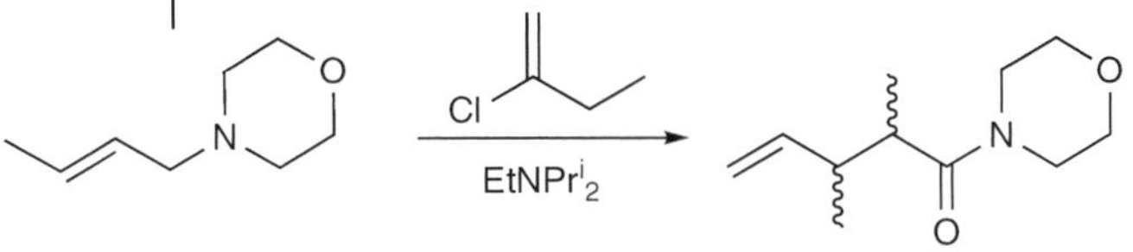

(e)

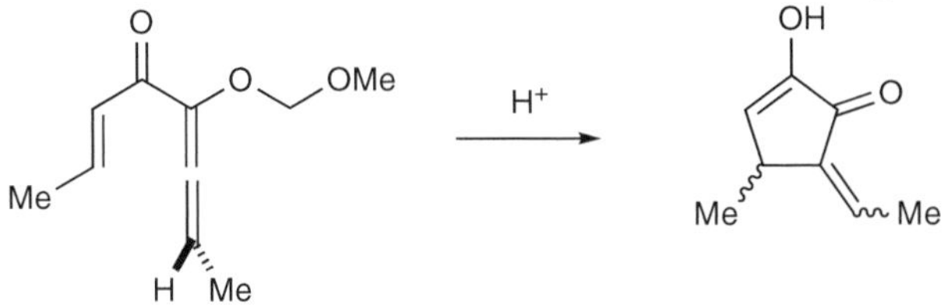

6. The thermal rearrangement of the alcohol **6.342** into the alcohol **6.343** might reasonably take place by either of two pathways. One involves two successive pericyclic reactions and the other a single pericyclic reaction. Suggest what the two pathways are, identifying the nature of the pericyclic steps. The diastereoisomeric alcohol **6.344** gives the diastereoisomeric product **6.345**. Show how this identifies which of the two pathways is actually followed.

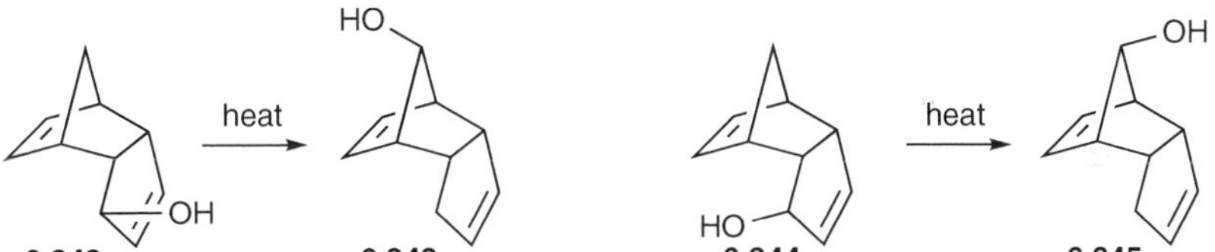

**6.342**     heat     **6.343**     **6.344**     heat     **6.345**

7. Identify the intermediates **A** and **B** and hence explain why the *tert*-butyl groups are *cis*, in the thermodynamically less favourable arrangement, both in the episulfide **6.346** and in the cyclopropanone **6.347**.

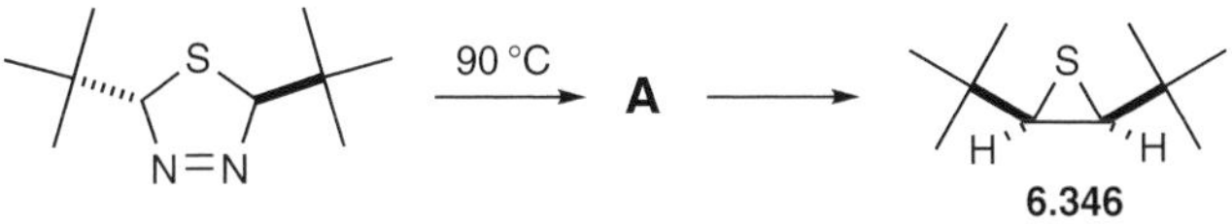

90 °C    **A**    **6.346**

−78 °C    **B**    **6.347**

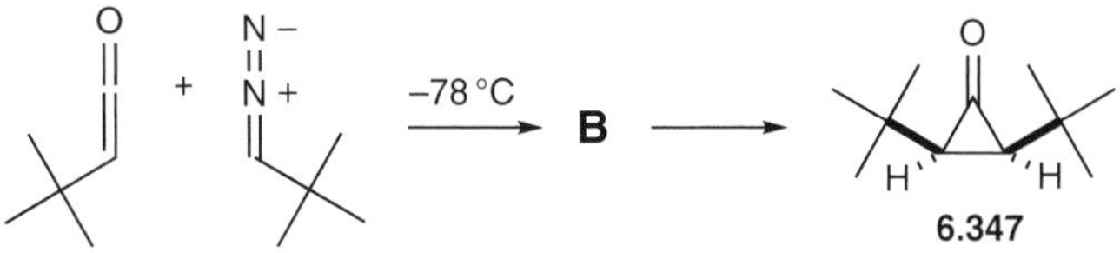

8. Account for the regioselectivity in this contrasting pair of reactions:

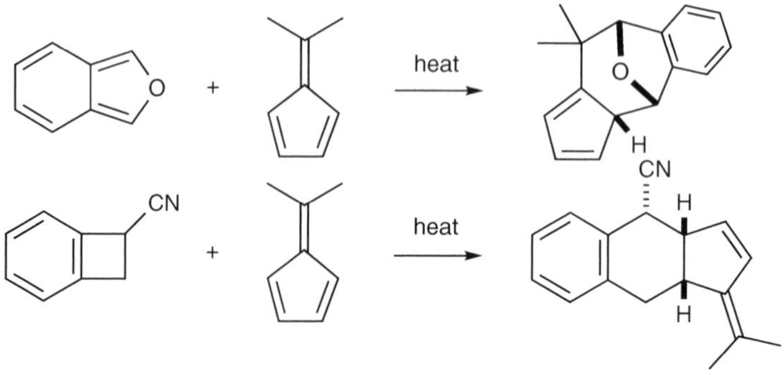

9. Account for the stereochemistry of the following reaction.

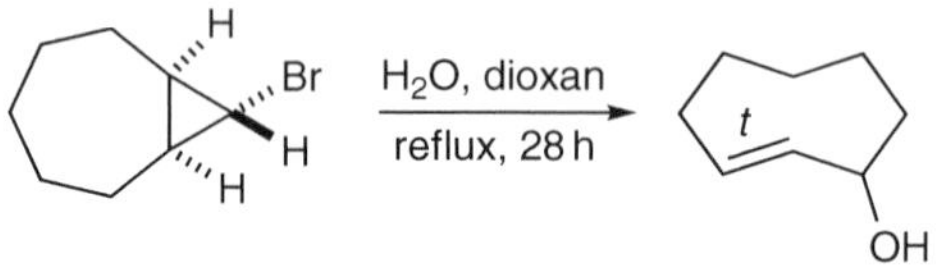

# 7 Radical Reactions

Much of the selectivity seen in radical reactions may be explained by frontier orbital theory, in contrast to ionic reactions, where it makes a relatively small contribution. Frontier orbital theory may not be well founded as a fundamental treatment, but it is appropriate that it might come to the fore with radicals, where Coulombic forces are usually small, orbital interactions likely to be strong, and the key steps usually exothermic. Most of the discussion in this chapter will use frontier orbital theory, and will seem to do so uncritically.[31] It is important to remember that it is not as sound as its success in this area will make it seem.

## 7.1 Nucleophilic and Electrophilic Radicals

We saw in Chapter 2 (see pages 67–68), that all substituents, C-, Z- or X-, stabilise radicals, that carbon-based radicals are usually pyramidal, with a low barrier to inversion of configuration, and that the energy of the singly occupied molecular orbital (SOMO) was inherently close to the nonbonding $\alpha$ level, unchanged by C-substitution, lowered by Z-substitution and raised by X-substitution. In contrast to the frontier orbitals in ionic and pericyclic reactions, the SOMO can interact with both the HOMO and the LUMO of the reaction partner to lower the energy of the transition structure (Fig. 7.1). Plainly the interaction with the LUMO will lead to a drop in energy ($E_3$ in Fig. 7.1b); but so does the interaction with the HOMO, and, for that matter, with each of the filled orbitals. Because there are two electrons in the lower orbital and only one in the upper, there will usually be overall a drop in energy ($2E_1 - E_2$) from this interaction. We can combine these effects in the frontier orbital picture in Fig. 7.1c. Radical reactions are consequently fast, and, in favourable cases, are even diffusion controlled, having little or no activation enthalpy.

Radicals are soft: most of them do not have a charge, and in most chemical reactions they react with uncharged molecules. Thus the Coulombic forces are usually small while the orbital interactions remain large. The clean polymerisation of methyl methacrylate demonstrates this typically soft pattern of behaviour. Radicals attack at the conjugate position of $\alpha,\beta$-unsaturated carbonyl compounds such as methyl methacrylate **7.1**, rather than at the carbonyl group, and the attack

*Molecular Orbitals and Organic Chemical Reactions: Student Edition*   Ian Fleming
© 2009 John Wiley & Sons, Ltd

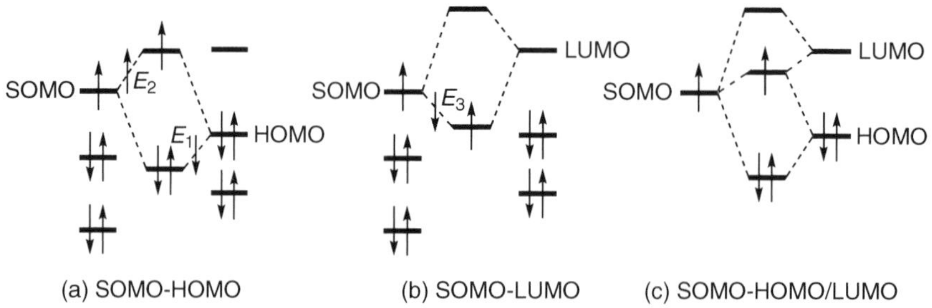

(a) SOMO-HOMO  (b) SOMO-LUMO  (c) SOMO-HOMO/LUMO

**Fig. 7.1** The interaction of the SOMO with the HOMO and the LUMO of a molecule

by the ambident $\alpha$-carbonylmethyl radical **7.2** takes place from the carbon atom, not from the oxygen atom.

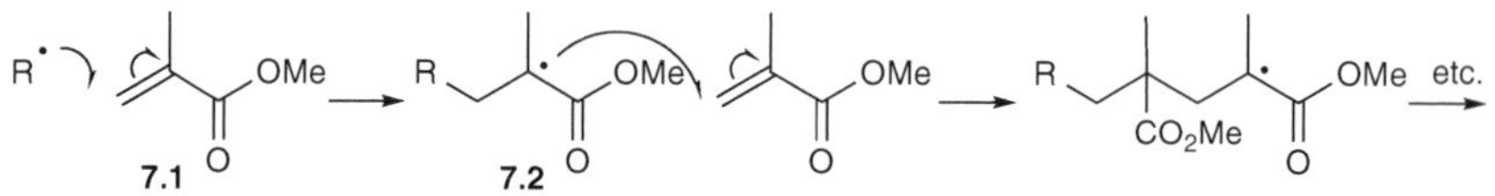

Highly reactive species like radicals are not usually expected to show high levels of selectivity (the reactivity-selectivity principle), and so it has always been something of a puzzle why they did, nevertheless, have nucleophilic and electrophilic character—some radicals showing higher reactivity towards reagents normally thought of as electrophilic, and others higher reactivity towards reagents normally thought of as nucleophilic. These observations are easily explained by frontier orbital theory. Radicals with a high-energy SOMO (Fig. 7.2a) will react fast with molecules having a low-energy LUMO, characteristic of electrophiles, and radicals with a low-energy SOMO (Fig. 7.2b) will react fast with molecules having a high-energy HOMO, characteristic of nucleophiles. The former are therefore the nucleophilic radicals and the latter are the electrophilic radicals.

(a) High-energy SOMO—a nucleophilic radical  (b) Low-energy SOMO—an electrophilic radical

**Fig. 7.2** Frontier orbital interactions for a nucleophilic and an electrophilic radical

This insight is strikingly illustrated by the observation of alternating copolymerisation. The radical-initiated polymerisation of a 1:1 mixture of dimethyl fumarate **7.3** and vinyl acetate **7.5** takes place largely to give a polymer in which the fragments derived from the two monomers alternate along the chain. In this case it is evident that a growing radical such as **7.4** attacks vinyl acetate rather than fumarate; but the new radical **7.6**, so produced, attacks fumarate rather than vinyl acetate. The radical **7.4**, because it is flanked by a carbonyl group, in other words by a Z-substituent, will have a low-energy SOMO, and will be an electrophilic radical. It therefore reacts faster with the molecule having the higher energy HOMO, namely the X-substituted alkene **7.5**. The new radical **7.6** is next to an oxygen atom, in other words an X-substituent, and will have a high-energy SOMO. It will be a nucleophilic radical, closer in energy to a low-lying LUMO. Of the two alkenes **7.3** and **7.5**, the fumarate, because it is a Z-substituted alkene, has the lower energy LUMO [see (Section 2.1.2.2) page 61], and it is therefore this molecule which reacts with the radical **7.6**—and so on, as the polymerisation proceeds.

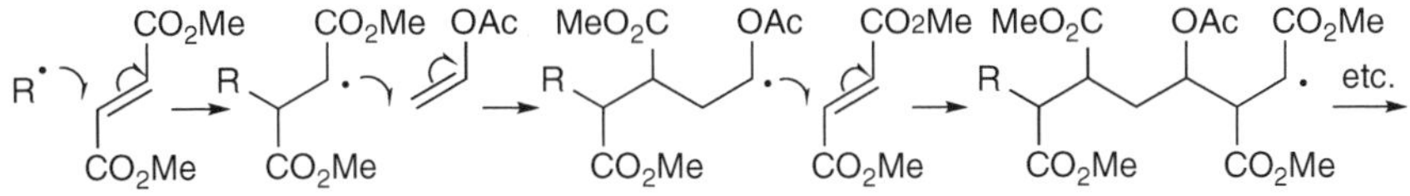

> *In general: radicals with a high-energy SOMO show nucleophilic proper-ties, and radicals with a low-energy SOMO show electrophilic properties.*

Radicals show three types of reaction: substitution **7.7**, addition to double bonds **7.8**, and radical-with-radical combination **7.9**, and the reverse of each of these reactions. We shall now look at these in turn to see how the various kinds of selectivity in each of them can be explained.

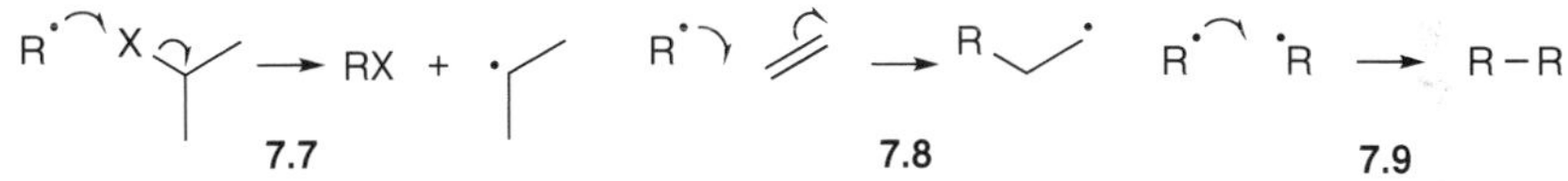

## 7.2 The Abstraction of Hydrogen and Halogen Atoms

### 7.2.1 The Effect of the Structure of the Radical

Nucleophilic and electrophilic radicals abstract different hydrogen atoms from butyrolactone **7.10**: the *tert*-butoxy radical selectively abstracts a hydrogen atom from the methylene group adjacent to the oxygen atom, whereas a boryl radical abstracts a hydrogen atom from the methylene group α to the carbonyl group.

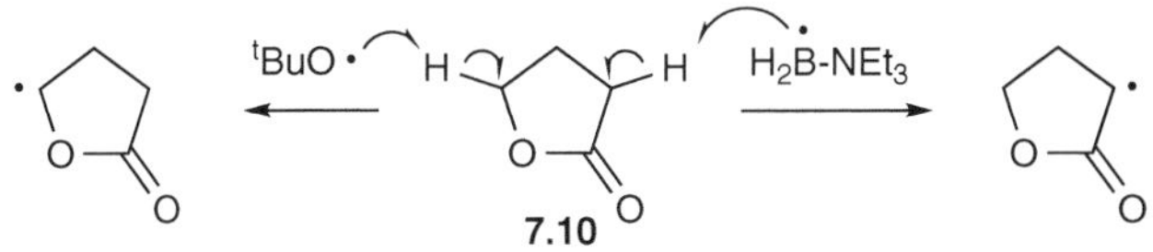

7.10

The frontier orbitals are the SOMO of the radical and the local $\sigma$ or $\sigma^*$ orbitals of the C—H bonds. The *tert*-butoxy radical, based on an electronegative element, will have a low-energy SOMO, and will have a stronger interaction with the $\sigma$ orbital which is high in energy for having an adjacent X-substituent. In contrast, the boryl radical, based on an electropositive element, will have a high-energy SOMO, and it will have a stronger interaction with the low-energy $\sigma^*$ orbital of the C—H bond adjacent to the Z-substituent.

Except for extreme cases like this, radical reactions are generally not subject to strong polar effects. From the picture of C—H bonding in Chapter 1, we can deduce that the SOMO of a methyl radical is close to halfway between the local $\sigma$ and $\sigma^*$ orbitals of a C—H bond, so that the interactions should be more or less equally the SOMO with the HOMO and with the LUMO. In agreement, methyl radicals abstracting hydrogen atoms are found to be only marginally electrophilic.

### 7.2.2   The Effect of the Structure of the Hydrogen or Halogen Source

***7.2.2.1   Selectivity Affected by the Nature of the Radical.***   When a tributyltin radical has a choice of a C—S, a C—Se, or a C—halogen bond, they are selected, other things being equal, in the order $I > Br > SeAr > Cl > SAr > SMe$. This is roughly in the order of the strengths of the Sn—X bond being made, and is explained simply as a consequence of the most exothermic reaction being the most rapid [see (Section 3.3) page 103]. Furthermore, these reactions are faster than attack on a C—H bond, since the halogens and Se and S are soft sites, and can accept bonding to a radical ahead of the bond breaking—the interaction of the SOMO with the nonbonding, lone-pair orbitals is likely to be stronger than with $\sigma$ or $\sigma^*$ of a C—H bond, as well as forming a stronger bond. The relatively less nucleophilic methyl radical, however, abstracts a hydrogen atom from benzyl chloride rather than the chlorine atom. Even more subtle examples of selectivity come when it is a question of which kind of C—H bond is attacked.

Most radicals attack hydrogen atoms in the order: allylic > tertiary > secondary > primary. The most important factor here is again that the faster reactions are producing the product with the lower energy. In addition, the more neighbouring groups a C—H bond has, the more overlap (hyperconjugation) can be present. Since such overlap is between filled orbitals and filled orbitals, the effect is to raise the energy of the HOMO. This effect therefore puts the energy of the HOMOs of the C—H bonds in the same order as their ease of abstraction.

Selectivity between hydrogen atom abstraction and addition to an alkene [see (Section 7.3) page 281] is dependent upon the structures of the radical and of the substrate. Simple alkyl radicals attack H—Sn bonds competitively with their conjugate addition to Z-substituted alkenes, showing that there is a fairly delicate balance, even though the H—Sn bond is notably weak. *tert*-Butoxy radicals remove allylic hydrogens faster than they add to the terminus of simple alkenes, but quite small changes, to perfluoroalkoxy radicals for example, reverse this selectivity.

***7.2.2.2   Selectivity Affected by Stereoelectronic Effects.***   Molecules with a more or less rigid relationship between a lone pair and a C—H bond can be used to probe the effect of conjugation between the two. Ethers, acetals and orthoesters show a range of reactivity towards hydrogen atom abstraction by *tert*-butoxy radicals, with some telling stereochemical features. The acetal **7.11** shows a selectivity between the three different kinds of hydrogen atom that matches the energy of the radicals produced. The most stable is the tertiary radical **7.12** flanked by two oxygen atoms, which is produced nearly seven times faster than the secondary **7.13**, which is flanked by only one. It is normal to correct for the statistical factor that there are four times as many hydrogens that can produce the secondary radical as the tertiary, and so the selectivity for tertiary is actually 27 times the secondary. The third possibility would be the primary radical, with no lone pair stabilisation, produced by abstraction from the methyl group, which is not observed at all. However, the rigid acetal **7.14** loses a hydrogen atom only from the secondary position to give the radical **7.15**, which is stabilised by syn overlap with one of the lone pairs, whereas the tertiary radical that would be created at the bridgehead **7.16** would not be stabilised, because the singly occupied orbital would be gauche to all the lone-pair orbitals.

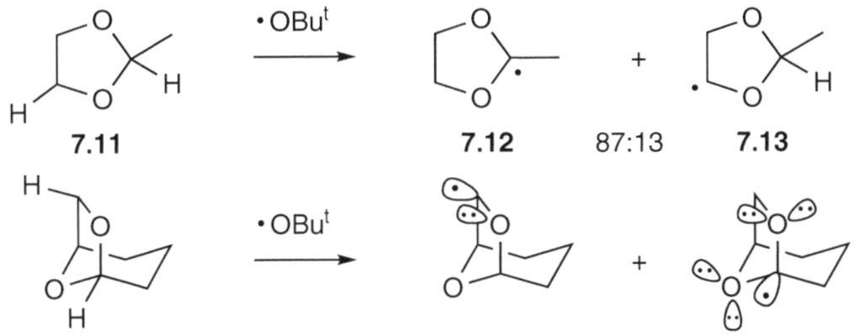

## 7.3 The Addition of Radicals to π Bonds

### 7.3.1 Attack on Substituted Alkenes

There is a great deal of information available about the addition of radicals to π bonds, since it is such an important step in radical polymerisation,[32] as we have already seen. A lot of these reactions are easily explained: the more stable 'products' **7.2**, **7.6** and **7.17**, with the radical centre adjacent to the substituent are almost always obtained, and the site of attack usually has the higher coefficient in the appropriate frontier orbital.

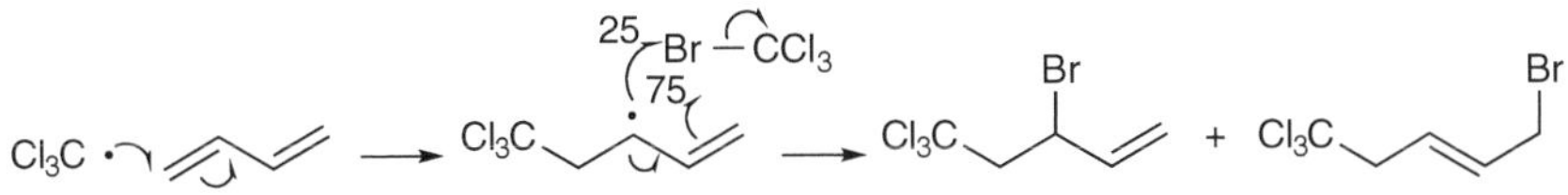

With C- and Z-substituted alkenes, the site of attack will be the same regardless of which frontier orbital is the more important—both have the higher coefficient on the carbon atom remote from the substituent (Fig. 2.4). With X-substituted alkenes, however, the HOMO and the LUMO are polarised in opposite directions (Fig. 2.6). For most X-substituted alkenes, the HOMO will be closer in energy to the SOMO of the radical, because X-substituted alkenes generally have high-energy HOMOs and high-energy LUMOs [see (Section 2.1.2.3) pages 63–65]. Together with the usual pattern for forming the more stabilised radical, this explains the direction of addition, as we saw with the electrophilic radical **7.4** adding to the unsubstituted terminus of vinyl acetate **7.5**. Similarly, oxygen atoms, which will also be electrophilic, attack but-l-ene to give more butanal than 2-butanone.

Although the regioselectivity is usually high in all these reactions, the relative rates reveal that orbital interactions are important, in addition to the thermodynamic factors favouring the formation of the more stable radical. Thus, a plot of the Hammett $\rho$ values for addition of a range of radicals to substituted styrenes correlates with the ionisation potential of the radicals. Similarly, the radical **7.19** produced by addition of an alkyl radical to the vinylphosphonate **7.18** will be very similarly stabilised no matter what the alkyl group R is, yet the relative rates for the different radicals are in the order: $^t$Bu > $^i$Pr > Et > Me. This is opposite to the usual expectation that the more stable the radical the less reactive it is. The simplest explanation is that the more-substituted radical has the higher-energy SOMO, which is therefore closer to the low-energy LUMO of a Z-substituted alkene.

| | | Me | Et | Pr$^i$ | Bu$^t$ |
|---|---|---|---|---|---|
| $k_{rel}$ | | 1 | 1.04 | 4.8 | 23.6 |

In contrast, for electrophilic radicals attacking an X-substituted alkene, adding an X-substituent like methyl to the Z-substituted radicals **7.20** lowers the rate of attack on 1-decene **7.21**. Thus the radical **7.20** ($R^1 = R^2 = H$) reacts 4 times faster than **7.20** ($R^1 = Me, R^2 = H$), and the radical **7.20** ($R^1 = R^2 = Cl$) reacts 2.5 times faster than **7.20** ($R^1 = Me, R^2 = Cl$). The other numbers here are not so easy to interpret, since the chlorine atoms, although $\pi$-donating, are also $\sigma$-withdrawing, and it is more than likely that steric effects are also contributing to these results.

| $R^1, R^2$ | H,H | H,Me | H,Cl | Cl,Cl | Cl,Me |
|---|---|---|---|---|---|
| $k_{rel}$ | 45 | 11.2 | 4.5 | 2.5 | 1 |

Varying the alkene instead of the radical leads to the same pattern. The lower the energy of the LUMO of the alkene **7.23**, the faster a nucleophilic radical like the cyclohexyl radical **7.22** will add to it, but an electrophilic radical like the malonate radical **7.24** adds more rapidly the higher the energy of the HOMO.

| R | CN | CO$_2$Et | Ph | Me |
|---|---|---|---|---|
| $k_{rel}$ | 440 | 42 | 3.6 | 1 |

| R | CO$_2$Et | Ph | Me | MeO | Me$_2$N |
|---|---|---|---|---|---|
| $k_{rel}$ | 1 | 3.5 | 3.7 | 2.7 | 23 |

The general rule, therefore, is that radicals add to the less substituted end of C-, Z- or X-substituted alkenes to give the more stable radical; this usually matches the coefficients in the appropriate frontier orbitals, and the relative rates are usually in line with the appropriate frontier orbital separations.

There are exceptions to this pattern, most of them in cyclisation reactions. Hex-5-enyl radicals **7.26** (R$^1$ = R$^2$ = H) cyclise to give the less stable, primary radical **7.25** (R$^1$ = R$^2$ = H) with a selectivity of 98:2.

Radicals attack alkenes with an obtuse Bürgi-Dunitz-like angle, for the same reasons that anions and cations do (see pages 159–162), and for this reason Baldwin's rules can be expected to apply. They are called the Baldwin-Beckwith rules when applied to radicals. Although the 6-*endo-trig* pathway is not explicitly disfavoured, it is more strained than the observed 5-*exo-trig* pathway giving the radical **7.25**, countering the electronic control of regioselectivity seen in open-chain systems.

Three situations where the 6-*endo-trig* pathway is observed are understandable. Cyclisation to the secondary radical **7.27** takes place when the groups R$^1$ and R$^2$ are electron-withdrawing. The well-stabilised radical **7.26** (R$^1$ = CN, R$^2$ = CO$_2$Et) forms the five-membered ring **7.25** more rapidly, but the latter is able to open again under the reaction conditions to give the thermodynamically preferred product **7.27** with a secondary radical and a six-membered ring. The selectivity is 84:16, but dependent, of course, on how rapidly the two radicals are being quenched in competition with the ring-opening and -closing. It may also be that the ring closure in the 6-*endo-trig* sense (**7.26** → **7.27**) is assisted when the substituents are electron-withdrawing: the electrophilic radical **7.26** (R$^1$ = CN, R$^2$ = CO$_2$Et) will have a relatively low-energy SOMO, and it will therefore be more sensitive to the polarisation of the HOMO of the alkene group. A methyl radical attacks propylene with a regioselectivity for attack at the terminus of 5:1, whereas the relatively electrophilic trifluoromethyl radical is more selective (10:1) for attack in this sense. In the radical without electron-withdrawing groups **7.26** ($R^1 = R^2 = H$), the interaction with the $\pi$ bond will be more affected by the LUMO of the alkene, and this might be part of the reason for contrathermodynamic cyclisation. When $R^1$ is a hydrogen atom and $R^2$ an aryl ring, the substituents on the aromatic ring affect the ratio, with Z-substituents increasing the proportion of endo closure and X-substituents decreasing it, showing that there is an electronic component to the selectivity, and not just the extent of reversibility.

The second category where the 6-*endo-trig* pathway is observed is when there is a second substituent on the inside carbon. The radical **7.29** adds at the terminus faster than at the inside carbon by a factor of 1.6, because the steric effect slows down attack at the more substituted carbon, and the tertiary radical produced **7.30** will be more stabilised than it was without the substituent. When the substituent is at the terminus, it naturally speeds up exo attack, especially when the substituent is electron-withdrawing.

The third category where the 6-*endo-trig* pathway is observed is exemplified by the silicon-containing radicals **7.31** and **7.32**. Neither of these reactions is thermodynamically controlled—the radical addition step is not reversible in these reactions, so they must both be under kinetic control.

The probable explanation is that the long Si—C bonds ease the strain in the 6-*endo-trig* transition structures. It is also true in the second case **7.32** that the silyl group is effectively a Z-substituent, which will further encourage attack by the radical on the terminus which has the higher coefficient in the HOMO.

### 7.3.2 Attack on Substituted Aromatic Rings

The rates of attack of radicals on aromatic rings correlate with ionisation potential, with localisation energy and with superdelocalisability (see page 130), a picture reminiscent of the situation in aromatic electrophilic substitution. As in that field, there are evidently a number of related factors affecting reactivity. Frontier orbitals provide useful explanations for a number of observations in the field.

The partial rate factors of Table 7.1 show that a phenyl radical reacts with nitrobenzene and anisole faster than it does with benzene. This can readily be explained if the energy levels come out, as they plausibly might, in the order shown in Fig. 7.3.

**Table 7.1**  Partial rate factors for radical attack on benzene rings; $f$ is the rate of attack at the site designated relative to the rate of attack at one of the carbon atoms of benzene itself

| Attacking radical | Ring attacked | $f_o$ | $f_m$ | $f_p$ |
|---|---|---|---|---|
| $p\text{-}O_2NC_6H_4\cdot$ | $PhNO_2$ | 0.93 | 0.35 | 1.53 |
| $Ph\cdot$ | $PhNO_2$ | 9.38 | 1.16 | 9.05 |
| $p\text{-}O_2NC_6H_4\cdot$ | PhOMe | 5.17 | 0.84 | 2.30 |
| $Ph\cdot$ | PhOMe | 3.56 | 0.93 | 1.29 |

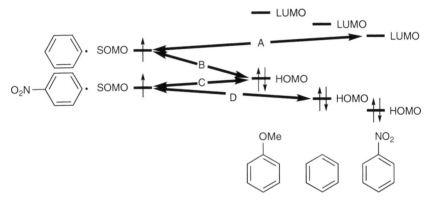

**Fig. 7.3**  Interactions for the attack of an aryl radical on substituted benzene rings

With anisole, the SOMO/HOMO interaction (B) is strong, and with nitrobenzene the SOMO/LUMO interaction (A) is strong, but with benzene neither is stronger than the other. Product development control can also explain this, since the radicals produced by attack on nitrobenzene and anisole will be more stabilised than that produced by attack on benzene. However, this cannot be the explanation for another trend which can be seen in Table 7.1, namely that a $p$-nitrophenyl radical reacts faster with anisole and benzene than it does with nitrobenzene. This is readily explained if the SOMO of the $p$-nitrophenyl radical is lower in energy than that of the phenyl radical, making the SOMO/HOMO interactions (C and D) strong with the former pair.

In hydrogen atom abstractions, alkyl radicals change, as the degree of substitution increases, from being mildly electrophilic (the methyl radical) to being mildly nucleophilic (the *tert*-butyl radical). In addition reactions to pyridinium cations, the Minisci reaction, they are all relatively nucleophilic, as shown by their adding exclusively to the 2-position **7.33**. This change is reasonable, because the LUMO of an aromatic ring like this will be substantially lower in energy than that of a C—H bond and the SOMO of the radical can interact more favourably with it. The addition is the rate-determining, as well as the site-determining, step. The second step, the removal of the hydrogen atom, is usually easy, and resembles the loss of the proton in aromatic electrophilic substitution.

The more substituted radicals continue to be measurably the more nucleophilic. The relative rates with which the various alkyl radicals react with the 4-cyan-opyridinium cation (**7.33**, $Y = CN$) and the 4-methoxypyridinium cation (**7.33**, $Y = OMe$) are given in Table 7.2. The LUMO of the former will obviously be lower than that of the latter. The most selective radical is the *tert*-butyl, which reacts 350 000 times more rapidly with the cyano compound than with the methoxy. This is because the *tert*-butyl radical has the highest-energy SOMO, which interacts (B in Fig. 7.4) very well with the LUMO of the 4-cyanopyridi-nium ion, and not nearly so well (A) with the LUMO of the 4-methoxypyridinium ion. At the other end of the scale, the methyl radical has the lowest-energy SOMO, and hence the difference between the interactions C and D in Fig. 7.4 is not so great as for the corresponding interactions (A and B) of the *tert*-butyl radical. Therefore, it is the least selective radical, reacting only 50 times more rapidly with the cyano compound than with the methoxy.

**Table 7.2**  Relative rates of reaction of alkyl radicals with the pyridinium cations **7.33**

| Attacking radical | $k_{Y = CN}/k_{Y = OMe}$ | SOMO energy ($-IP$ in eV) |
|---|---|---|
| Me$^•$ | 46 | $-9.8$ |
| *n*-Bu$^•$ | 203 | $-8.0$ |
| *sec*-Bu$^•$ | 1300 | $-7.4$ |
| *tert*-Bu$^•$ | 350 000 | $-6.9$ |

The most vexed subject in this field is the *site* of radical attack on substituted aromatic rings. Some react cleanly where we should expect them to. Phenyl radicals add to naphthalene **7.36**, to anthracene **7.37** and to thiophene **7.38**, with the regioselectivity shown on the diagrams. In all three cases, the frontier orbitals are clearly in favour of this order of reactivity (because of the symmetry in these systems, both HOMO and LUMO have the same absolute values for the coefficients).

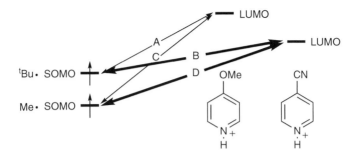

**Fig. 7.4**    Interactions of frontier orbitals for the reaction of alkyl radicals with pyridinium cations

However, there is a lot of evidence that radicals are much less selective than cations and anions. Thus, dimethylamino radicals attack toluene to give 10% *ortho*-, 47% *meta*- and 43% *para*-dimethylaminotoluenes; phenyl radicals attack pyridine with little selectivity, and chlorine atoms attack naphthalene unselectively. Since all substituents stabilise radicals, substituted benzenes usually (but not invariably, see Table 7.1) react faster than benzene itself, and most of them, whether C-, Z- or X-substituted, show some preference for *ortho/para* attack, no doubt because attack at these sites gives the more stable intermediates. In assessing the contribution of the frontier orbitals, we are back with the problem (see page 131) of how to describe the orbitals of substituted benzene rings—in other words, how to estimate the relative importance of the two high-lying occupied orbitals ($\psi_2$ and $\psi_3$) and the two low-lying unoccupied orbitals ($\psi_4^*$ and $\psi_5^*$). Thus the HOMO and the LUMO shown in Fig. 7.3, for example, are best thought of, not as single orbitals, but as composites of the kind discussed in Chapter 4.

One trend seems clear, and it is a trend readily explained by the frontier orbitals. In an X-substituted benzene, the proportion of *meta* attack falls as the energy of the SOMO of the attacking radical rises (Table 7.3). Because the HOMO and LUMO energies of X-substituted benzenes will be raised, we can expect that the HOMO of the aromatic ring is the more important frontier orbital. Thus, the lower the energy of the SOMO of the radical, the better the interaction with the weighted combination of orbitals making up the HOMO, and hence the more *ortho* and *para* attack there is. The *meta*-selectivity of the dimethylamino radical comes at least in part because its SOMO energy is low.

**Table 7.3**   Regioselectivity in the attack of a range of radicals on anisole

| Attacking radical | %o | %m | %p | (%o+2%p)/%m | SOMO energy ($-$IP in eV) |
|---|---|---|---|---|---|
| $Me_3Si\cdot$ | 62 | 31 | 7 | 2.5 | $-7$ |
| $cyc\text{-}C_6H_{11}\cdot$ | 67 | 28 | 5 | 2.8 | $-7.8$ |
| $Ph\cdot$ | 69 | 18 | 13 | 5.3 | $-9.2$ |
| $Me\cdot$ | 74 | 15 | 11 | 6.4 | $-9.8$ |
| $HO_2CCH_2\cdot$ | 78 | 5 | 17 | 22.4 | $-10.9$ |

## 7.4   Synthetic Applications of the Chemoselectivity of Radicals

To be useful in organic synthesis, radical reactions must be designed to take advantage of what is known about radical reactivity and selectivity—combining nucleophilic radicals with electrophilic partners, and electrophilic radicals with nucleophilic partners. Here we shall look at three telling short synthetic sequences showing how selectivity at each step allows a chain of events to take place, to give largely one product, with a number of different radical intermediates, each of which does what is wanted in competition with other pathways.

One commonly used sequence is the conjugate addition of a radical derived from an alkyl halide to an electrophilic alkene. The first step is the thermal cleavage of AIBN to give the initiator radical **7.39**, which faces three possible reaction partners: the tin hydride **7.40**, the alkyl halide **7.41** and methacrylonitrile **7.43**. It is neither particularly nucleophilic nor electrophilic, having two weak X-substituents and one Z-substituent, which roughly cancel each other out. However, the highest rate constant for reaction with any of the substrates present is for attack on the tin hydride, because it has a weak H—Sn bond (bond dissociation energy 308 kJ mol$^{-1}$, compared with a range from 385 to 435 kJ mol$^{-1}$ for the H—C bonds in alkanes) with a high-energy HOMO. The tin radical, with a high-energy SOMO, is powerfully nucleophilic, and tin forms a strong bond to halogens. Accordingly it selectively attacks the alkyl halide **7.41** to displace the alkyl radical **7.42**. This mildly nucleophilic radical selects the electrophilic alkene **7.43** with its low-energy LUMO, and gives the new radical **7.44**, with the usual regioselectivity. This radical is constitutionally similar to the initiator radical **7.39**, and so it continues the chain by abstracting a hydrogen atom from the tin hydride to give the product **7.45** together with the tin radical, which can recirculate by attacking another molecule of alkyl halide. AIBN is only needed in catalytic amounts to initiate this chain of reactions. It is a key feature that the radical **7.44** is not nucleophilic enough to add rapidly to another molecule of the electrophilic alkene **7.43**, propagating its polymerisation. The rate constant for the hydrogen atom abstraction is approximately 300 times larger than that for the attack on the alkene. Thus each of the three key steps has well matched components with high selectivity, and the overall yield is correspondingly high.

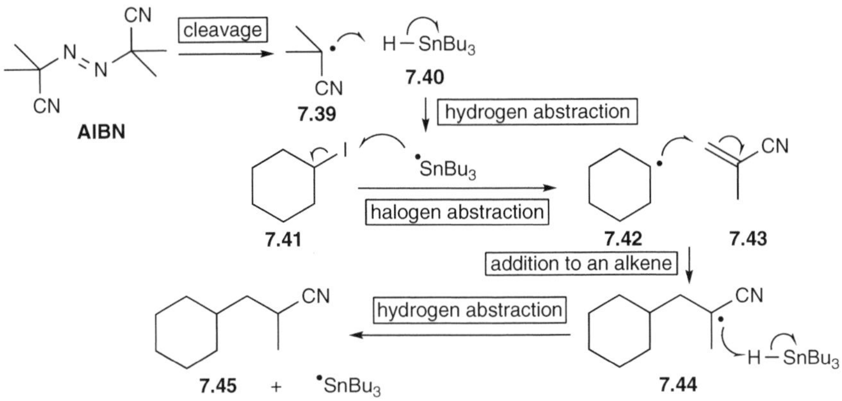

A complementary sequence uses an alkyl halide **7.46** with a Z-substituent to create an electrophilic radical **7.47** in the presence of a nucleophilic alkene **7.48**. In this case, the radical **7.49** expels the low-energy tributyltin radical to regenerate the tin radical achieving overall the allylation of the ester, catalytic in both the AIBN and the tin hydride. The Z-substituent in the radical **7.47** is necessary for an efficient reaction—in its absence the allylstannane has to be used in large excess.

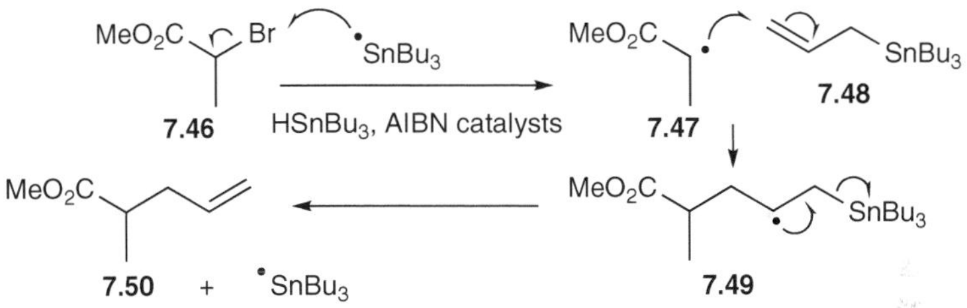

A similar sequence can add another step without loss of control, by using an intramolecular step. Thus, the tin radical selectively removes the halogen from the iodide **7.51** to produce the primary radical **7.52**, which undergoes cyclisation to give the secondary radical **7.53**, not because the reaction of a simple alkyl radical with an unconjugated double bond is inherently fast, but because it is intramolecular. Only then does the radical **7.53** continue on its normal course of reacting with the electrophilic alkene **7.54**, which has been equipped with a stannyl leaving group to give the prostaglandin precursor **7.55** and complete the cycle that makes the whole sequence catalytic in tin hydride.

## 7.5  Stereochemistry in some Radical Reactions[33]

Unlike alkyl halides in $S_N2$ reactions or alkyllithiums in $S_E2$ reactions, radicals have no intrinsic stereochemical preference at the reactive centre, except that trigonal radicals, and even cyclopropyl radicals, provided that there is an $\alpha$ electronegative substituent, can retain the configuration of their halide precursors. Otherwise, radicals, being somewhere in between cations and anions, show some of the same stereochemical preferences discussed in Chapter 5, such as attack on the exo face of bicyclic systems, and on the less hindered face of open-chain double bonds, which is useful when that can be identified with confidence. Radical additions to alkenes have transition structures early on the reaction coordinate, with the bond being formed still quite long making steric effects relatively weak. This has the advantage that bonds between heavily substituted centres can often be made more easily by radical reactions than by ionic reactions, but it also means that diastereocontrol in radical reactions is sometimes rather worse.

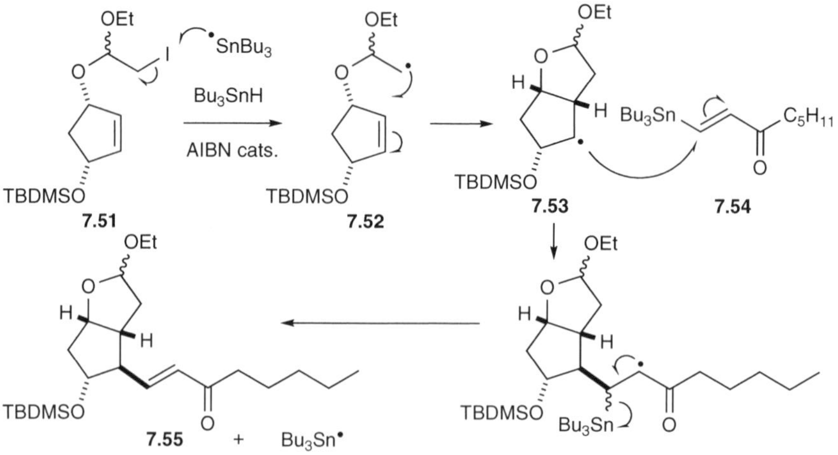

The stereochemistry of exo closure in a case like the radical **7.56**, giving the *cis* product **7.58** (*cis:trans* 72:28), is controlled by the usual preference for the resident substituent to adopt an equatorial orientation **7.57** and for the chain of atoms to adopt a chair-like conformation.

Radical reactions have some stereochemical features that can be compared directly with their ionic counterparts, especially when the radical centre is adjacent to an existing stereogenic centre. The tris(trimethylsilyl)silyl radical adds to chiral ketones like 3-phenyl-2-butanone **7.59** to give a radical **7.60** flanked by a stereogenic centre. The hydrogen atom abstraction from a thiol, determines the relative stereochemistry, and the products **7.61** and **7.62** are analogous to those from the hydride reduction of the ketone. They are formed in the same sense, and the stereochemistry is explained by the Felkin-Anh picture **7.60**.

7.59 7.60 7.61 74:26 7.62

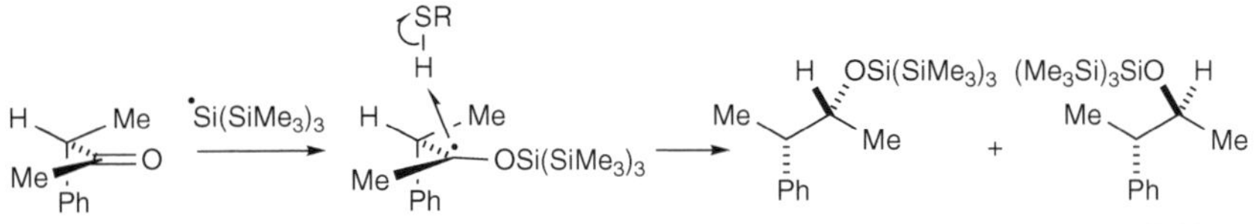

In more rigid systems with nucleophilic radicals, the cyclohexyl radical **7.63** attacks reagents with a small steric demand to give the cyclohexanes **7.64** with the incoming group axial, but, with more sterically demanding reagents like acrylonitrile, the preference changes to equatorial attack, in line with the reactions of cyclohexanones with sterically demanding nucleophiles [see (Section 5.2.1.2) page 173]. However, the anomeric radical **7.65**, attacks most reagents axially with high selectivity, anti to the lone pair, to give the *C*-glycoside **7.66**. In the abstraction of a hydrogen atom from an anomeric carbon, which is the reverse of this type of reaction, we have already seen the highly selective removal of a hydrogen atom anti-periplanar to two lone pairs in the acetal **7.14**. This stereoelectronic effect in radical reactions is called the '$\beta$-oxygen effect.'

7.63 7.64

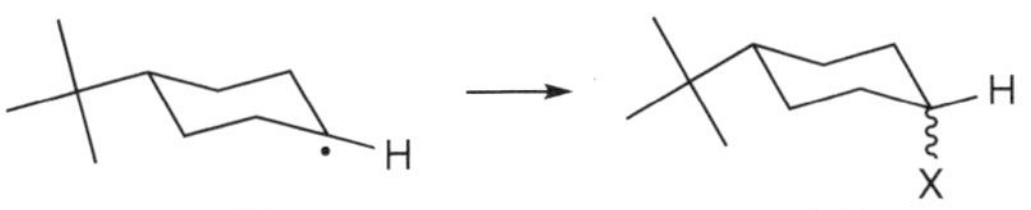

| X | axial:equatorial |
|---|---|
| D | 70:30 |
| OH | 80:20 |
| Cl | 77:23 |
| $CH_2CH_2CN$ | 45:55 |

7.65 7.66 98:2

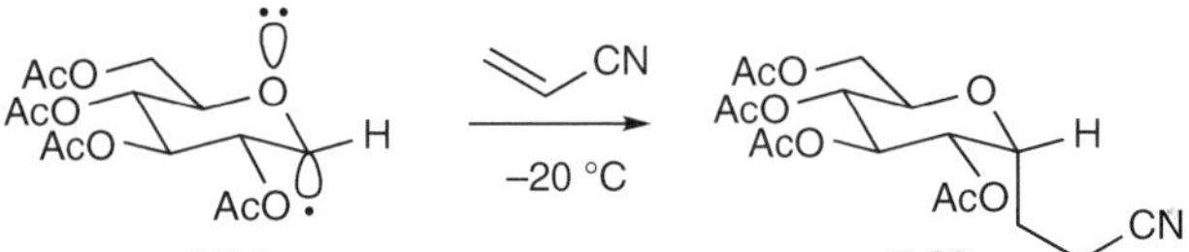

In open-chain systems with electrophilic radicals, the conformation for attack on the enol radicals **7.67** and **7.68** is similar to that for enolate alkylation **5.135** and protonation **5.136**, respectively [see (Section 5.2.3.1) page 179].

7.67 62:38

7.68 96:4

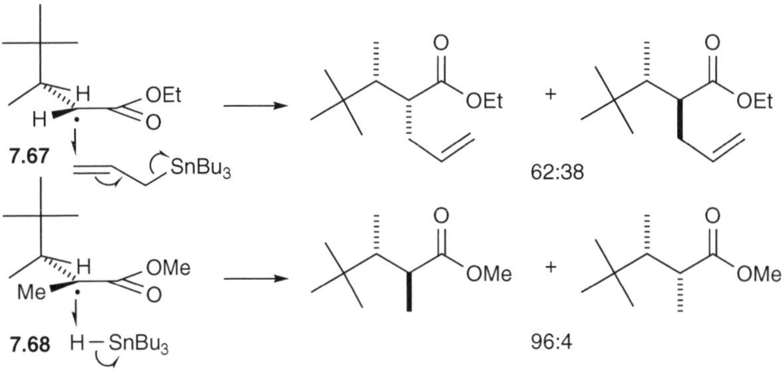

## 7.6   Ambident Radicals

### 7.6.1   Neutral Ambident Radicals

The regioselectivity of attack on ambident radicals should be determined on the product side of the reaction coordinate by the relative stability of the regioisomeric products, and on the starting material side by the coefficients of the SOMO. This is commonly what is observed. A monosubstituted allyl radical, generated by adding a radical to a diene, usually reacts at the unsubstituted end of the allyl radical, as in the bromination of the radical **7.17**, because it gives the more substituted alkene, and the X-substituted allyl radical has the larger coefficient in the SOMO at the terminus [see (Section 4.3.3.1) page 126]. We have also seen that the $\alpha$-carbonylmethyl (enol) radical **7.47** reacts at the carbon atom, which has the higher coefficient in the SOMO, just as it does for the HOMO of the enolate ion.

If orbital and product development arguments are in conflict, we might expect orbital effects to be more important, since the key steps of radical reactions are so often strongly exothermic. The cyclohexadienyl radical **7.69** should have a higher coefficient at C-3 [see (Section 4.3.3.2) page 128] than at C-1, and indeed it seems that this site most readily extracts a hydrogen atom from another molecule to give the thermodynamically less stable unconjugated product **7.70**. Cathodic electrolysis of pyridinium ions **7.71** causes an electron to be added to the ring. This electron is in an orbital resembling the LUMO of pyridine (see page 138), which has the largest coefficient at the 4-position. This is the site of dimerisation giving largely the 4,4-dimer **7.72**.

The best-known neutral ambident radicals are phenoxy radicals. Many substituted phenoxy radicals couple to give the polymers in wood, and dimerise or couple intramolecularly in the biosynthesis of a number of alkaloids, It is believed that many phenoxy radical couplings are actually the coupling of radical with radical and not of radical with neutral molecule, although the attack of a radical on a phenate ion may occasionally be an important pathway. A radical coupling with a radical is inherently a very fast process, since it is so exothermic, but only relatively stable radicals live long enough to encounter another radical; phenoxy radicals may well belong in this class. In any event, we can expect a large number of possible products, but *o-p, p-p,* and *O-O* are rather more common than *O-p* and *O-o,* and these are much more common than *O-O.* This is in line with the aptitude for enol radicals to react at carbon rather than oxygen. We can obtain a measure of the electron distribution in the SOMO of a phenoxy radical from ESR, which shows high coupling constants to the hydrogen atoms at the *ortho* and *para* positions. Using the McConnell equation [see (Section 1.9.4) page 56], the electron distribution in the SOMO of the phenoxy radical has been estimated as 0.25 on oxygen, 0.27 *ortho,* 0.40 *para* and −0.07 *meta.* We should remember that the $\beta$-value for C—O bond formation is less than for C—C bond formation, further enhancing the tendency for bonding to carbon. It is not clear from the experimental evidence whether *p-p* coupling really is preferred, as these numbers would suggest. The problem is that products are often obtained in low yield, and the mass balance is usually poor. In addition there is the statistical effect of there being two *ortho* positions to one *para.* We can guess that there will not be much in it, and that does seem to be the case, judging by the natural products that can be found with all these patterns.

### 7.6.2 Charged Ambident Radicals

**7.6.2.1 Radical Cations.** Radical cations are usually prepared using an oxidising agent to remove one electron from substrates with a high ionisation potential, in other words a high-energy HOMO. The result, if the starting material is uncharged, is a radical cation with a SOMO having a similar energy to the original HOMO. Thus the SOMO can interact strongly with the HOMO of the same or another molecule, and in both the large coefficients of the atomic orbitals are the nucleophilic sites. The result is that bonds are often formed between two nucleophilic sites, achieving an umpolung of reactivity.

For example the radical cation **7.73** is generated by oxidation of 2-methyl-naphthalene. The odd electron is in the HOMO of naphthalene, the highest coefficient of which is at C-1. The methyl group, as an X-substituent, will further enhance the coefficient at this site relative to the other $\alpha$ positions; thus, the *total* electron population at this site will be higher than at the other $\alpha$ positions. Nevertheless, the nucleophile, an acetate ion, attacks at this site to give eventually 1-acetoxy-2-methylnaphthalene **7.74**. That an anion should attack a site of relatively high electron population is easily accounted for by a strong SOMO/HOMO interaction. The anti-Markovnikov addition of methanol to some alkenes under photo-oxidation conditions is similar. 1-Phenylcyclopentene, for example, on irradiation in the presence of a single-electron acceptor like 1-cyanonaphthalene loses an electron to give the radical cation **7.75**. The single electron is in the HOMO of a C-substituted alkene, which has the larger coefficient at C-2. This is where the nucleophile attacks and subsequent steps lead to the anti-Markovnikov product **7.76**.

7.73          7.74

2-cyanonaphthalene

7.75          7.76

Anodic electrolysis of aniline and dimethylaniline also gives radical cations, which are stable enough to dimerise, giving predominantly *N-p* and *p-p* coupling. This is just like the phenoxy radical coupling except that, with nitrogen being less electronegative than oxygen, there will be a larger coefficient on nitrogen than there was on oxygen, and there will also be a higher $\beta$-value for N—C bond formation than there was for O—C bond formation.

Enol ethers, which are X-substituted alkenes, can be oxidised by one-electron transfer to reagents like ceric ammonium nitrate (CAN). This gives a radical cation, the SOMO of which is essentially the same as the HOMO of the starting material. Thus the silyl dienol ether **7.77** has a higher energy HOMO than the silyl enol ether **7.79**, and is therefore the more easily oxidised. The radical cation **7.78** then couples with the silyl enol ether **7.79** with regioselectivity that gives, after further oxidation and hydrolysis, the most stable possible product **7.80**, and joins the carbon atoms which have the highest coefficients in the SOMO and HOMO, respectively.

7.77          7.78          7.79          7.80

### 7.6.2.2 Radical Anions.

Radical anions are complementary to radical cations: they are usually prepared using a reducing agent to add one electron to substrates with a high electron affinity, in other words a low-energy LUMO. The result, if the starting material is uncharged, is a radical with a SOMO having a similar energy to the original LUMO. The radical anion can either couple or its SOMO can interact strongly with the LUMO of the same or another molecule, and in both pathways the large coefficients of the atomic orbitals are the sites that were originally or are still electrophilic. The net result is that bonds are often formed between two electrophilic sites, achieving again an umpolung of reactivity, as shown by the pinacol coupling of acetone **7.81** and the $\beta,\beta$-coupling of methyl vinyl ketone **7.82**. In each case, the odd electron has been fed into the orbital which was the LUMO of the starting material; the site of coupling therefore should, and does, correlate with the site at which nucleophiles attack the neutral compounds.

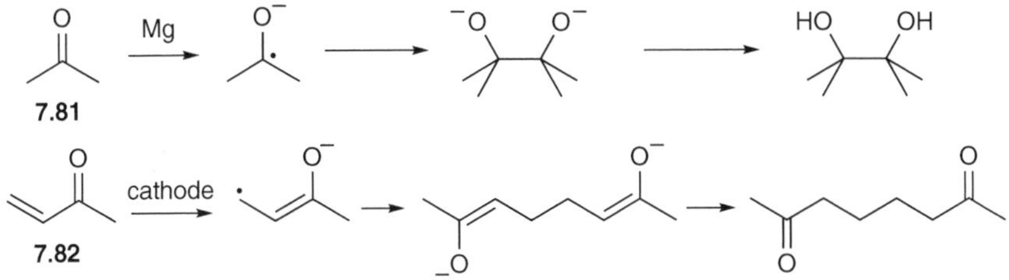

In protic solvents, radical anions derived from ketones can pick up a proton instead of coupling, and the resultant radicals pick up a second electron, in competition with the pinacolisation. When this pathway is observed, the stereochemistry of the alcohol product appears to be determined at the radical anion or anion stage, and not in the final protonation step, which is probably faster than the pyramidal inversion of the anion. The pyramidalisation of the radical anion in a conformation **7.83** like that of Felkin and Anh **5.82** [see (Section 5.2.1.1) page 169], has a steric clash that is greater than it was in the ketone itself. The alternative conformation, with the large group 'inside' **7.84**, and the R group sitting between the small and medium-sized groups, is lower in energy, and leads to the anti-Cram product. The experimental observation is that hydride reduction (Felkin-Anh control) of the ketone **7.85** gives the alcohol **7.86** as the major product, but electron-transfer reduction gives its diastereoisomer **7.87**. In cyclohexanones, the pyramidalisation is similarly greater in the radical anion than it is in the ketone [**5.94**; see (Section 5.2.1.2) page 171], thus explaining the high degree (98:2) of selectivity for the formation of the equatorial alcohol when 4-*tert*-butylcyclohexanone is reduced by lithium in liquid ammonia in the presence of *tert*-butanol.

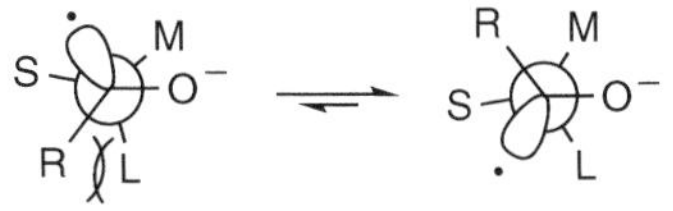

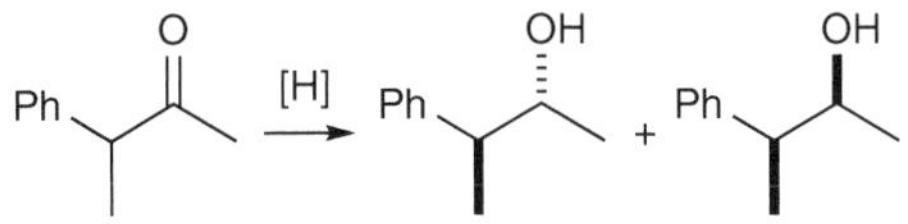

| [H] | 7.86:7.87 |
|---|---|
| LiAlH$_4$ | 74:26 |
| Li, NH$_3$ | 24:76 |

This stereochemistry is similar to that in the reduction of $\alpha,\beta$-unsaturated ketones. Stork found that octalones like **7.88** were reduced exclusively to the *trans*-decalone **7.90**, even in cases when the *trans*- was less stable than the corresponding *cis*-decalin. In this reaction, an electron is fed into the LUMO of the unsaturated ketone, and hence the radical anion **7.89** will have a higher total electron population on C-5 (steroid numbering) than in the starting material, and C-5 can be expected to be more nearly pyramidal than planar trigonal. In other words, it will bend towards a tetrahedral geometry and, in so doing, relieve some of the steric strain in the rest of the molecule. This will be more efficient if the larger lobe of the atomic orbital on C-5 is on the lower surface, because the AB ring system will then be more like that of a *trans*-decalin. Accordingly protonation takes place on the lower surface, and the product is the *trans*-decalone.

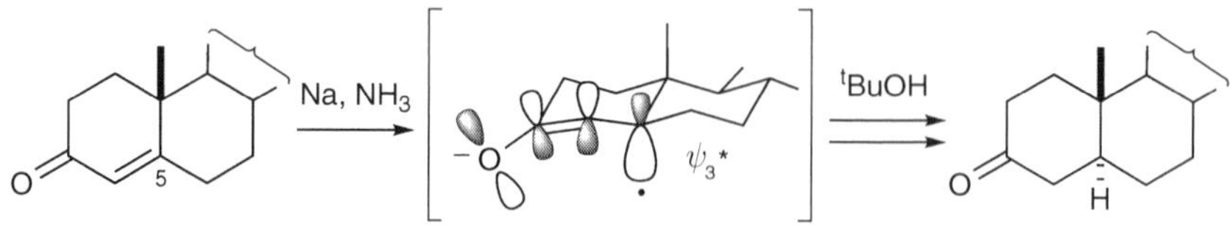

**7.88**                    **7.89**                          **7.90**

Another radical anion is the intermediate in Birch reduction, where aromatic rings are reduced with sodium in liquid ammonia in the presence of an alcohol. The solvated electron adds to the benzene ring of anisole **7.91** to give the radical anion **7.92**. This is protonated by the alcohol present at the *ortho* position. The cyclohexadienyl radical **7.93** from *ortho* protonation is the lowest-energy radical possible, and the *ortho* position has the highest total electron population. The SOMO for **7.92** will mostly be like the orbital $\psi_5^*$ [see (Section 1.5.1) page 33], with large coefficients on both *ortho* and *meta* positions. A computation shows that it has marginally the largest coefficient at one of the *meta* positions, indicating that frontier orbital control is not at work here, and that *ortho*-protonation simply takes place at the site of highest charge, which also leads to the most stable possible intermediate. The radical **7.93** is reduced to the corresponding anion **7.94** by the addition of another electron, and the new anion is protonated at the central carbon atom of the conjugated system, for reasons discussed earlier (see page 127). Thus, we can see why X-substituted benzenes in general are reduced to 1-substituted cyclohexa-1,4-dienes **7.95**.

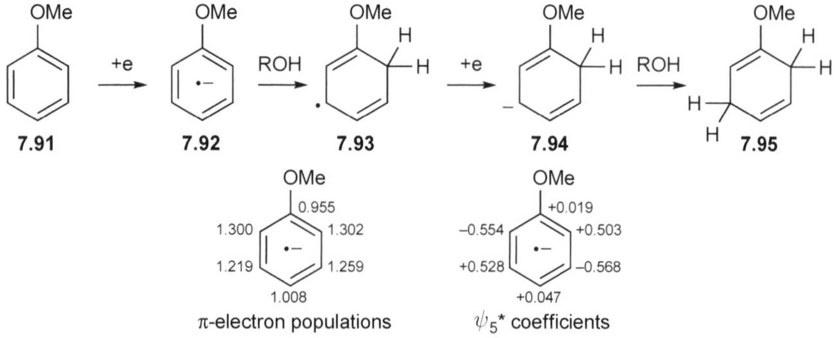

**7.91**          **7.92**          **7.93**          **7.94**          **7.95**

π-electron populations              $\psi_5^*$ coefficients

In contrast, C-substituted benzenes like biphenyl **7.96** are reduced to 3-substituted cyclohexa-1,4-dienes **7.99**, and this too fits the analysis. The Hückel coefficients for the SOMO of the radical anion **7.97** also reflect the total $\pi$-electron distribution, since the other three filled orbitals lead to a more or less even distribution of $\pi$-electron population. So, regardless of whether it is the Coulombic or the frontier orbital term that is more important, both contributions lead to protonation at C-4 to give the radical **7.98**. Reduction and protonation of this intermediate (or possibly a mixture with the 1-protonated isomer) leads to the observed product **7.99**. Further reduction of this molecule then takes place, but now the benzene ring is an X-substituted one. The major final product, accordingly, is the hydrocarbon **7.100**, which has been reduced 1,4 in one ring and 2,5 in the other.

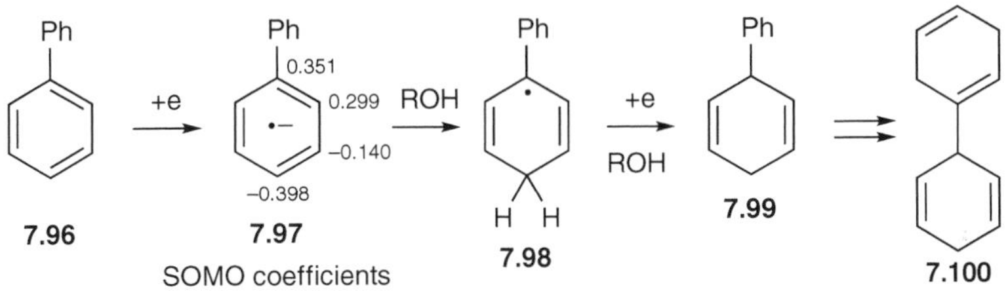

The Birch reduction of benzoic acid is the same type as that of biphenyl, and the product, with protonation successively at C-4 and C-1 is the acid **7.102**. In the reaction medium, it will be benzoate ion **7.101** that is being reduced. As a result of the delocalisation of the negative charge in the benzoate ion, we should probably regard the carboxylate ion more as a C- than as a Z-substituent.

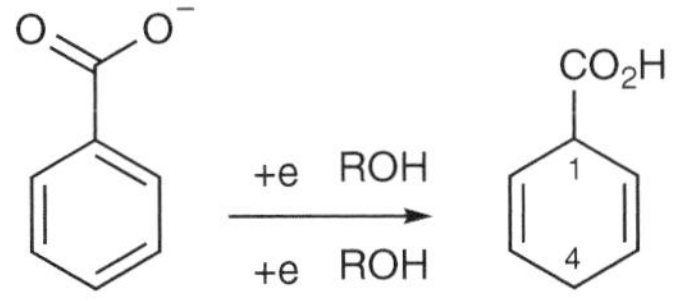

## 7.7   Radical Coupling

Radical coupling is not a common reaction. Most radicals are highly reactive, and it is rare for a high enough concentration of radicals to build up for it to be probable that a radical will collide with another radical before it has collided productively with something else. Benzyl radicals are moderately well stabilised, and their coupling therefore is not uncommon. Benzyl radicals have the highest odd-electron population on the exocyclic atom, and, in contrast to phenoxy radicals, they do give largely dibenzyl.

Radical coupling is more frequently seen when the two radicals are created together, typically within a solvent cage or within the same molecule, so that they are more likely to meet each other before something else can happen to them. If a single electron transfer (SET) takes place from one molecule to another, the result is a pair of radicals which can combine. We saw [see (Section 4.1) pages 111–114] that there are SET reactions like this, and Kornblum's reaction, the coupling of the nitronate anion **4.14** with the nitrobenzene **4.15**, and the reaction of the pinacolate ion **4.16** with bromobenzene **4.17** are certainly among them. We shall see some more examples of radical coupling in the next chapter, where the radicals are created by irradiation with light.

## 7.8   Exercises

1. Suggest an explanation of the change in regiochemistry for the range of radicals attacking methyl crotonate:

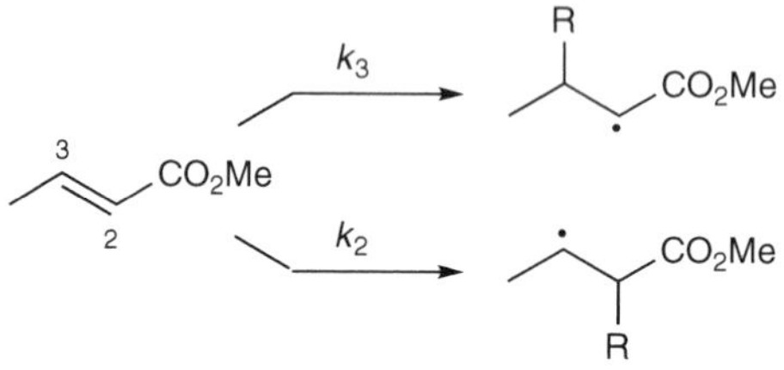

| R | $k_3{:}k_2$ |
|---|---|
| $PhCO_2$ | 45:55 |
| Ph | 63:27 |
| ${}^tBuO$ | 91:9 |
| $c\text{-}C_6H_{11}$ | 92:8 |

2. Explain the selectivity in these reactions:

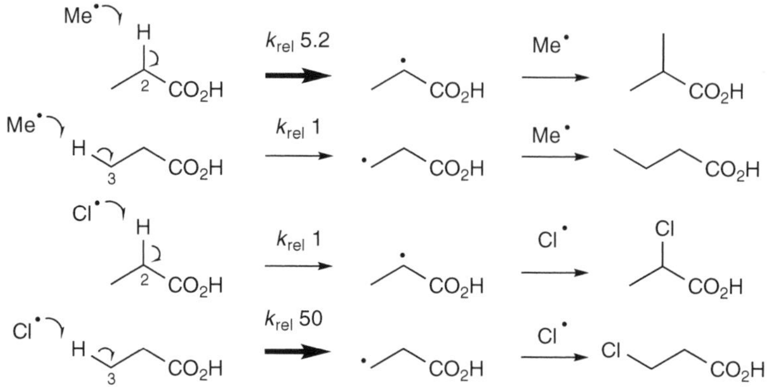

3. Explain why hydrogen atom abstraction from the acetal **7.103** is faster than from the acetal **7.105**, even though they both produce the same radical **7.104**.

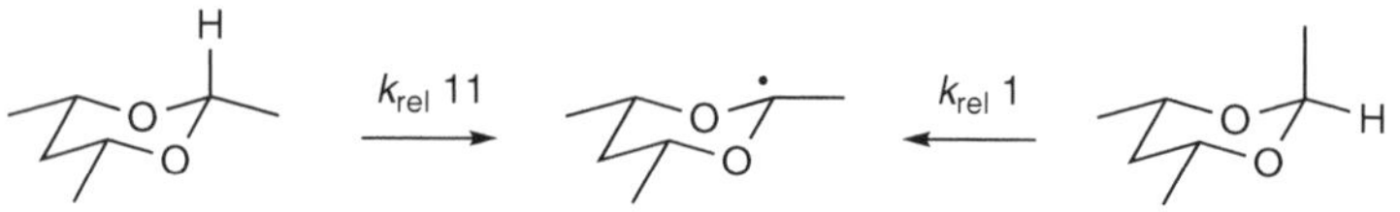

7.103                     7.104                     7.105

4. Explain why the *trans*-chloride **7.106** is produced as the major product in the chlorination of 4-*tert*-butylcyclohexene, in spite of its being less stable than its stereoisomer **7.107**.

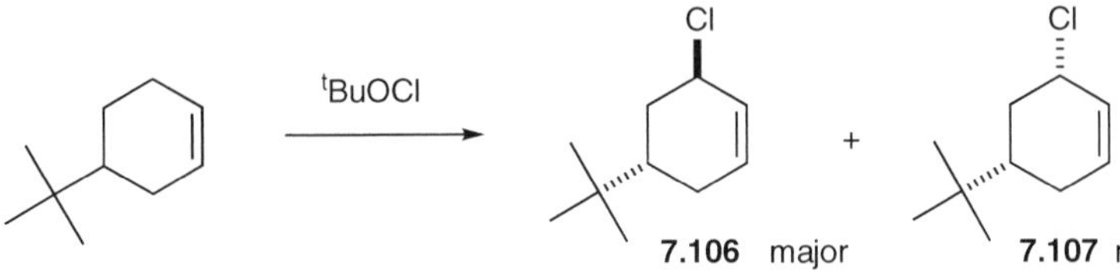

**7.106** major          **7.107** minor

5. Explain why both methyl and trifluoromethyl radicals add to propene to give the more-substituted radical, but the methyl radical attacks ethylene 1.4 times more rapidly than it attacks propene, whereas the trifluoromethyl radical attacks propene 2.3 times more rapidly than it attacks ethylene.

6. Two well known radical-chain reactions are the allylic bromination of alkenes **7.108** → **7.109** using *N*-bromosuccinimide (NBS), and the anti-Markovnikov addition of HBr to alkenes when peroxides are present **7.108** → **7.110**. Radical brominations using NBS are known to take place by the NBS slowly releasing bromine, since the same results can be obtained using bromine in low concentration. In the key step of the allylic bromination using NBS, a bromine atom derived from the bromine molecule abstracts an allylic hydrogen atom **7.111**. In the peroxide-catalysed addition of HBr to propene, the key step is the addition of a bromine atom to the double bond **7.112**. Explain how it *seems* to be possible for the bromine atom to show different selectivity depending upon its source!

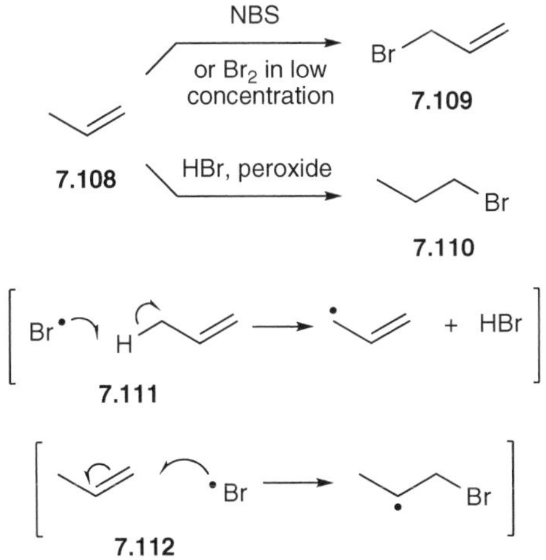

NBS
or Br₂ in low concentration

**7.108**

HBr, peroxide

**7.109**

**7.110**

**7.111**

**7.112**

# 8 Photochemical Reactions

Radicals, as we have just seen in Chapter 7, are highly reactive. Thermodynamically, they are high in energy, since one electron is in a nonbonding orbital instead of being low in energy in a bonding orbital. Furthermore, the single electron is able to interact with both frontier orbitals of the other component, making radicals kinetically unstable too. The same pattern is seen in photochemical reactions, but to a greater degree. The promotion of one electron from a bonding to an antibonding orbital raises the energy even more, making excited states thermodynamically highly unstable. The presence of two singly occupied orbitals in the excited state doubles the number of energetically profitable frontier orbital interactions, and the frontier orbitals are likely to be even closer in energy to the orbitals with which they interact than they were for radical reactions. Photochemistry therefore is replete with remarkable reactions—with the large amount of energy trapped in the excited state, it is hardly surprising that many of the familiar patterns of reactivity in ground-state chemistry are turned on their heads. Nevertheless, we shall see that there is some order in this subject when we take account of the orbitals that are involved.

## 8.1 Photochemical Reactions in General

In most bimolecular photochemical reactions, the first step is the photoexcitation of one component, usually the one with the chromophore which most efficiently absorbs the light. Typically, if a conjugated system of carbon atoms is present in one component, it can absorb a photon of relatively long wavelength, and in doing so an electron leaves the HOMO and arrives in the LUMO. Alternatively, an electron in a nonbonding orbital, like that of the lone pair on the oxygen atom of a ketone, which usually is the HOMO, is promoted from this orbital to the LUMO of the carbonyl group. The excited states produced are called $\pi$-$\pi^*$ and n-$\pi^*$, respectively. They may react directly in their singlet state, or later, after intersystem crossing, in their triplet state. The second step of the reaction, if it is bimolecular, is between the photochemically excited molecule and a second molecule, which may or may not be the same compound, in its ground state.

For this kind of reaction, there will generally be two energetically profitable orbital interactions (Fig. 8.1): (1) the interaction between the singly occupied

*Molecular Orbitals and Organic Chemical Reactions: Student Edition*   Ian Fleming
© 2009 John Wiley & Sons, Ltd

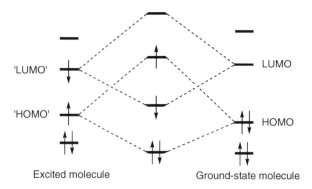

**Fig. 8.1** Frontier orbital interactions between a photochemically excited molecule and a ground-state molecule

$\pi^*$ orbital of the excited molecule, labelled 'LUMO' and the LUMO of the molecule which is in its ground state; and (2) the interaction of the singly occupied n or $\pi$ orbital of the excited molecule, labelled 'HOMO' and the HOMO of the molecule which is in its ground state. Both interactions will usually be strong, because the interacting orbitals are likely to be close in energy. Partly for this reason, this step of a photochemical reaction is often very fast. As they are so strong, the perturbations are first order, and the mathematical treatment of them would not take the form of the third term of Equation 3.4.

Figure 8.1 is shown for the singlet excited state, with the electrons having opposite spins in the singly occupied orbitals, but the interaction is not significantly different for the triplet state, when they have the same spin—it creates a similar set of four new orbitals. The triplet state cannot give the ground-state product until another intersystem crossing has taken place, and this delays the final bond-forming steps. Singlet photochemistry and triplet photochemistry are often different in their final outcome, and it is important to find out which pathway is being followed. However, it does not affect many of the arguments in this chapter, which are about the first step. What it does significantly affect is the lifetime and pathways available for the intermediate created by the orbital interactions.

In ground-state reactions, the first-order interactions of occupied orbitals with occupied orbitals are antibonding in their overall effect, and there is therefore a large repulsion between the two components of a bimolecular reaction. The bonding interactions of occupied orbitals with unoccupied orbitals are merely second-order effects lowering the energy of the transition structure. In photochemical reactions, however, the strong interactions shown in Fig. 8.1 can create an intermediate in which the *total* energy is lower than when the two components of the reaction were not interacting. This lower-energy intermediate can be identified as the excimer or exciplex, now well established in some photochemical reactions. The two molecules have become stuck together. They are still in an excited state, many steps may have to be taken before they can settle down into the

ground state of the product or products, but we shall not always be able to deal
with these later steps in any simple way.

It is a useful convention to label the orbitals which were the HOMO and LUMO
when the excited molecule was in its ground state as the 'HOMO', and the
'LUMO'. (The single quotation marks remind us that these orbitals are no longer
the actual HOMO and LUMO at the time of the reaction, but were the HOMO and
LUMO in the ground state, before the excitation took place.) With $\pi$-$\pi$* reactions,
the acronym 'HOMO' used in this generalisation needs some qualification, since
the important high-lying orbital from which the electron was promoted may not
precisely be the occupied orbital with the highest energy. Any lone pairs that may
be present, but playing no direct part, are often the occupied orbitals with the
highest energy, and when they are, it is not uncommon to see the $\pi$ orbital that is
involved referred to as the HOMO-1 or HOMO-2, or whatever it happens to be.

Bearing these qualifications in mind, the important frontier orbitals in a photochemical reaction are HOMO/'HOMO' and LUMO/'LUMO'. We now see why
so many photochemical reactions are complementary to the corresponding thermal reactions. Photochemical reactions often seem to do the opposite of what you
would expect of the equivalent thermal reaction, when there is one. In the latter it
is the HOMO/LUMO interactions which predominate in bond-making processes,
and in the former it is HOMO/'HOMO' and LUMO/'LUMO'.

## 8.2   Photochemical Ionic Reactions

### 8.2.1   Aromatic Nucleophilic Substitution

In certain cases, light promotes substitution reactions in aromatic compounds.[34]
One of the fascinating features of these reactions is an almost complete change in
regioselectivity from that observed in the ground-state reactions. When the
nitrocatechol ether **8.1** is irradiated in alkali or in methylamine, the nucleophilic
substitution takes place *meta* to the nitro group. The nucleophilic substitution of
*p*-nitroanisole **8.2**, however, takes place *para* to the X-substituent.

These examples have a nucleophile attacking a photoexcited aromatic ring at a site where nucleophiles do not attack the aromatic ring in the ground state. We can now easily see why this should happen. Benzene **8.3** without activating substituents can also be attacked by nucleophiles, provided that it is in an excited state; this time the result is an addition reaction giving a 1-substituted cyclohexa-2,5-diene. In the excited-state reaction, the HOMO of the nucleophile, a lone pair on the oxygen or nitrogen atoms or the p orbital at the 2-position of pyrrole, can interact productively with what were, before excitation, the HOMOs of the benzene ring (ignoring any unconjugated lone-pair orbitals). These orbitals, as we saw in Chapter 4, are those which, in aromatic *electrophilic* substitution, lead to *meta* attack on Z-substituted benzenes, and to *ortho* and *para* attack on X-substituted benzenes. It is the same here for *nucleophilic* substitution. The reaction with the unsubstituted benzene is another consequence of the ability of the HOMO of the nucleophile to interact productively with a singly occupied *bonding* orbital of the benzene ring which becomes available when the latter is in an excited state.

### 8.2.2   Aromatic Electrophilic Substitution

A nearly complementary pattern of reactivity has been found for photochemical *electrophilic* substitution. Proton exchange in the photolysis of toluene **8.4** takes place most rapidly at the *meta* position. In anisole **8.5**, the corresponding reaction is predominantly *ortho* and *meta*. Nitrobenzene **8.6**, however, exchanges protons most rapidly at the *para* position.

Again, because of photoexcitation, the important frontier orbital of the electrophile (the LUMO) is able to interact productively with the 'LUMO' of the benzene ring which was not productive in the ground state of any drop in energy. Certainly $\psi_5^*$ has an electron distribution ideal for explaining *ortho/meta* attack in anisole, and, as we saw in Chapter 4, the LUMOs of nitrobenzene do lead to reactivity at the *para* position. Now that photoexcitation has placed an electron in these orbitals, an electrophile can take advantage of this electron distribution, whereas, in the ground state, only a nucleophile could.

### 8.2.3 Aromatic Side-Chain Reactivity

A parallel series of changes in reactivity is found in the side-chains of aromatic compounds. Whereas in thermal reactions, the hydrolysis of phenolic esters and ethers is faster if there is a *para* Z-substituent, the corresponding photochemical reactions are faster with a *meta* Z-substituent, as in the *m*-nitrophenyl phosphate **8.7**. The breaking of the P—O bond in the first excited singlet state is the rate-determining step. It is evidently faster for the phosphate **8.7** giving the phenate ion **8.8** than it is for the phosphate **8.9** giving the phenate ion **8.10**. In thermal reactions, the *p*-nitro group is conjugated to the oxyanion, and so stabilises the phenate ion **8.10**, making it into a better nucleofugal group. In the photochemically excited state, either the *m*-nitrophenate anion **8.8** must have become better stabilised, or the *p*-nitrophenate anion **8.10** must have become worse.

To try to explain this change, we can use, as a crude model, the orbital of a carbanion in place of the oxyanion and of a carbocation in place of the nitro group. When these p orbitals are attached to a benzene ring in the *para* positions, they are equivalent to *p*-xylylene. In the *meta* arrangement, which can equally sensibly be drawn as a diradical, the *m*-xylylene does not have a structure with all the electrons paired in bonds, illustrating the lack of ground state conjugation between the two substituents. The $\pi$ molecular orbitals for these two systems are shown in Fig. 8.2, where we can see that the orbitals, $\psi_{1-4}$ in the *p*-xylylene case on the right are all bonding, below the $\alpha$ level, but that the highest orbitals in the *m*-xylylene on the left are the degenerate pair $\psi_4$ and $\psi_5$, which are nonbonding. In the ground state, with $\psi_{1-4}$ filled in both cases, the overall $\pi$ energy is $0.5\beta$ lower for *p*-xylylene than for *m*-xylylene, but in the excited state, *m*-xylylene is $0.1\beta$ lower in $\pi$ energy than *p*-xylylene with this crude model. Effectively, the nitro group and the oxyanion are better conjugated in the first excited state when they are *meta* than when they are *para*.

There is a parallel pattern for the photochemical ionisation of benzyl acetates **8.11** and **8.13** giving benzyl cations **8.12** and **8.14**, in which a *m*-methoxy group is effective in promoting the reaction, but a *p*-methoxy group is not (in fact, the *para* isomer **8.13** cleaves homolytically rather than heterolytically). The explanation is

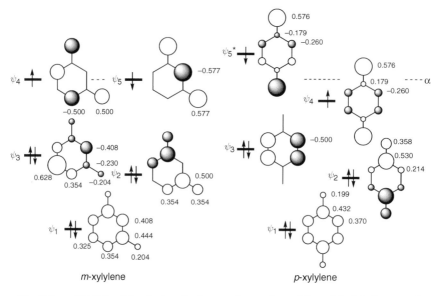

**Fig. 8.2** π Orbitals of the first excited state of *m*- and *p*-xylylene. Models for *m*- and *p*-nitrophenate ions and for *m*- and *p*-methoxybenzyl cations

the same, except that the roles of the cationic and anionic sites in the xylylene models have been exchanged.

## 8.3 Photochemical Pericyclic Reactions and Related Stepwise Reactions

### 8.3.1 The Photochemical Woodward-Hoffmann Rule

The most striking consequence of the change in what constitutes the frontier orbitals is the complete reversal of the Woodward-Hoffmann rule for thermal reactions quoted earlier [see (Section 6.3.2) page 201]. Thus, for a photochemical reaction, $[_\pi 2_s + _\pi 2_s]$ cycloadditions are allowed and $[_\pi 4_s + _\pi 2_s]$ cycloadditions forbidden, as we saw in Fig 6.20. The photochemical $[_\pi 2_s + _\pi 2_s]$ reaction does not meet a symmetry-imposed barrier like that for the ground-state reaction. More simply, but less rigorously, the frontier orbitals in Fig. 8.3a show that the HOMO/'HOMO' and the LUMO/'LUMO' interactions for a photochemical [2 + 2] cycloaddition are bonding, and in Fig. 8.3b the same orbitals for a Diels-Alder reaction are antibonding

(a) Photochemically allowed [$_\pi 2_s + _\pi 2_s$] cycloaddition   (b) Photochemically forbidden [$_\pi 2_s + _\pi 4_s$] cycloaddition

**Fig. 8.3**   Frontier orbitals for photocycloadditions

at one end. This is borne out in practice: there are very few photochemical Diels-Alder reactions, and a great many [2 + 2] cycloadditions, some of which have been proved to be stereospecifically suprafacial on both components.

We saw earlier [see (Section 8.1) page 300] that orbital interactions like those in Fig. 8.3a are strong enough for the energy to fall as the two components of a photochemically initiated reaction approach one another. In contrast, we also saw [see (Section 6.4.3.2) page 220], in the context of a thermal pericyclic [2 + 2] cycloaddition, that the energy rises dramatically as two alkenes approach each other, leading to the symmetry-imposed barrier for the concerted cycloaddition. The similarity between the attractive 'HOMO'/HOMO interactions in photochemical reactions and the repulsive HOMO/HOMO interactions in thermal reactions, means that the bottom of the one curve and the top of the other will have similar geometries and energies, and so we can follow the overall change in energy in the photochemical [2 + 2] reaction as one in which the two curves meet, allowing the exciplex or photoexcited molecule to cross over to the ground-state curve, following the dashed line in Fig. 8.4. The downward-sloping curve in photochemical reactions in general, not just pericyclic reactions, has been pictured as a funnel reaching down to the energy surfaces of the ground state, providing the electronic pathway that eventually allows an excited state molecule or pair of molecules to give a ground-state product.

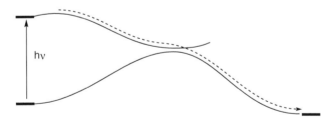

**Fig. 8.4**   Energy change in a photochemical reaction giving a ground-state product

Several cases of photochemical reactions, for which the thermal equivalents were forbidden, are shown below. In some cases the reactions simply did not occur thermally, like the [2 + 2] and [4 + 4] cycloadditions, and the 1,3- and 1,7-suprafacial sigmatropic rearrangements. In others, the photochemical reactions show different stereochemistry, as in the antarafacial cheletropic extrusion of sulfur dioxide, and in the electrocyclic reactions, where the 4-electron processes are now disrotatory and the 6-electron processes conrotatory. In each case, the observed reaction fits the opposite of the Woodward-Hoffmann rule for thermal reactions.

---

A pericyclic change in the first excited state is symmetry allowed when the total number of $(4q+2)_s$ and $(4r)_a$ components is even.

---

All-suprafacial cycloadditions:

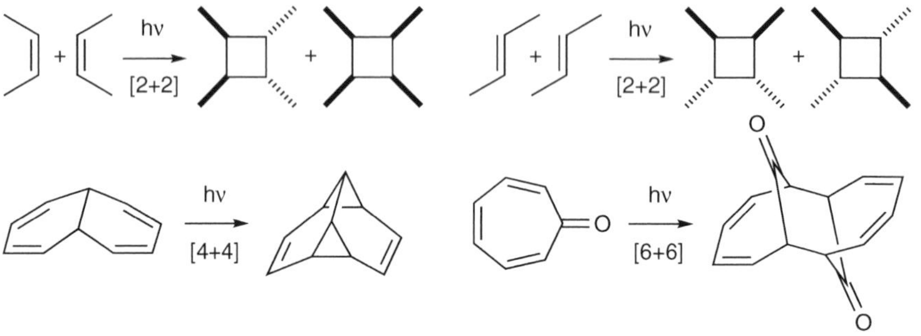

Cheletropic reactions:

Electrocyclic reactions:

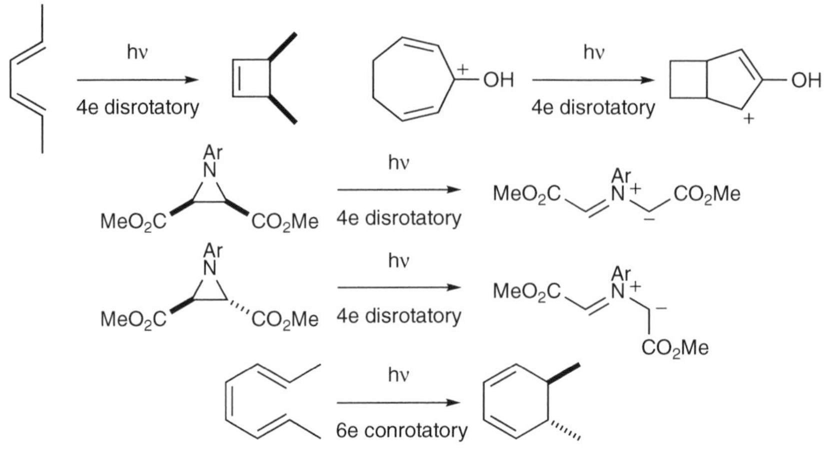

Sigmatropic rearrangements:

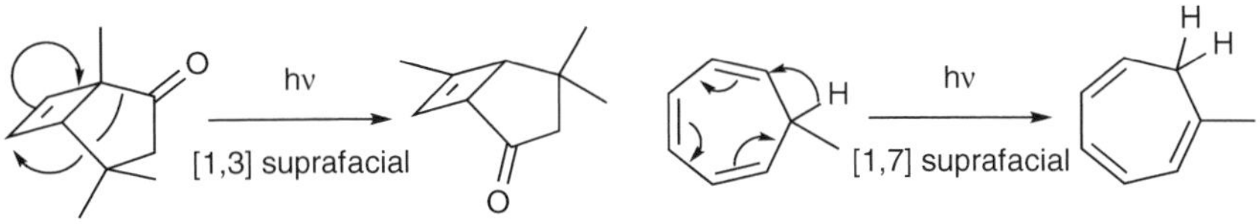

In photochemical reactions, it is even harder to *prove* the pericyclic nature of a process than it was in thermal reactions. Thus we should remember that some of the reactions in the examples above may not actually be pericyclic, and many of the reactions discussed below are certainly not straightforwardly pericyclic, since they take place from the excited triplet state. The large amount of energy imparted to a molecule when it absorbs a quantum of light might be enough to set off a large number of reactions that would not be feasible when a molecule is merely heated in the normal temperature range. Nevertheless, the contrast in the patterns of reactivity above with the corresponding thermal reactions in Chapter 6 is striking, and lends some support to the idea that some of these reactions are controlled by the orbital symmetry constraints expressed in the photochemical Woodward-Hoffmann rule. However uncertain their pericyclic status may be, at least they do not meet a symmetry-imposed barrier.

### 8.3.2 Regioselectivity of Photocycloadditions[35]

We have just seen that many photochemical reactions are complementary to the corresponding ground-state reactions. As with thermal pericyclic reactions, the frontier orbitals can explain some of the finer points of these reactions, most notably the regioselectivity of photocycloadditions. It is not important when predicting regiochemistry whether the cycloadditions are pericyclic or stepwise—regardless of whether both bonds are formed at once, or whether they are formed one at a time, the orientation should be determined by the large-large interaction in the appropriate frontier orbitals, and it is largely immaterial whether the second bond is forming at the same time or later.

*8.3.2.1 The Paterno-Büchi Reaction.* One well-known class of photocycloadditions is the Paterno-Büchi reaction in which aldehydes or ketones combine with alkenes to give oxetanes. The excited state of the ketone is n-π*, and it is the orbitals of this state which interact with the ground-state orbitals of the alkene. The orientation usually observed for C- and X-substituted alkenes is shown for benzophenone **8.15** and 2-methylpropene **8.16**.

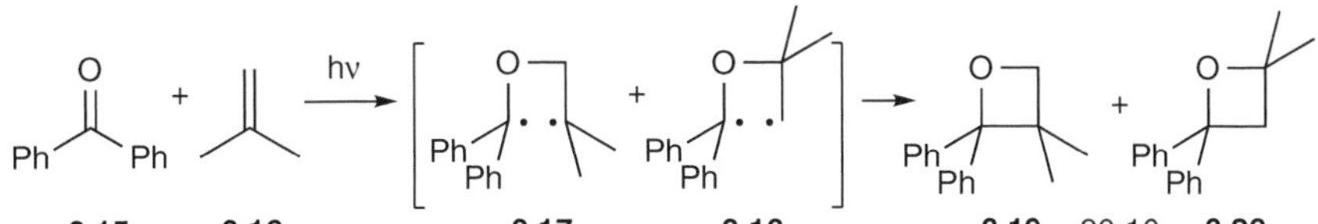

The more usual explanation looks at the energy of the intermediates produced—the lower energy diradical **8.17** usually gives the major product **8.19**. This argument applies safely to the triplet-state reaction, where an intermediate is likely to be involved. Singlet-state reactions may be pericyclic and may or may not involve the diradical. We can also explain the orientation if we assume that the major interaction is from the singly occupied n orbital of the ketone with the HOMO of the alkene. C- and X-substituted alkenes have high-energy HOMOs, which makes it reasonable that there should be a strong interaction with a nonbonding orbital. The carbon atom with the larger coefficient in the HOMO of the alkene is the one to which the oxygen atom becomes bonded.

However, the photocycloaddition of ketones to Z-substituted alkenes does not fit the explanation based on the relative stability of the diradicals. Irradiation of a solution of acrylonitrile **8.22** in acetone **8.21** gives the adduct **8.23**, together with dimers of acrylonitrile. This regioselectivity is consistent with a frontier orbital argument.

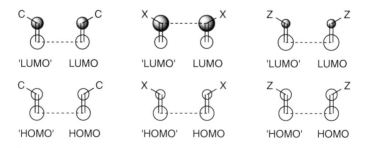

A Z-substituted alkene has a lower-energy HOMO than C- and X-substituted alkenes, and it has a correspondingly lower-energy LUMO. Thus the interaction of the $\pi^*$ orbital of the ketone **8.21** with the LUMO of the alkene **8.22** ought to be more important for a Z-substituted alkene than it was for C- or X-substituted alkene. In this interaction, the two larger lobes are on the carbonyl carbon **8.24** and on the $\beta$ carbon **8.25**, and it is these two which become bonded. Furthermore, this reaction is a singlet-state reaction, and, with $\beta$-substituted acrylonitriles, it is stereospecific, with retention of configuration on the alkene component.

**8.3.2.2 Photodimerisation of Alkenes.** When alkenes are irradiated, they often dimerise to give cyclobutanes. In a dimerisation reaction, the set of orbitals on the left of each pair in Fig. 8.1 and the set on the right will have identical

**Fig. 8.5** Frontier orbitals in the regioselectivity of photodimerisation of alkenes

energies before excitation. The bonding interactions should be very strong, and they should lead to the formation of what are known as head-to-head (HH) dimers as shown in Fig. 8.5. If we were to look only at the simplest examples of each kind of alkene, C-, X-, and Z-substituted **8.26**, **8.27** and **8.28**, respectively, we would find that this analysis seemed to be supported.

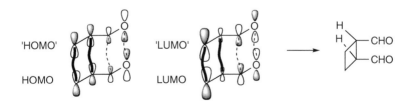

A 'sensitised' photolysis is one in which the light is absorbed by one compound, such as acetone or benzophenone, which then interacts with the substrate (butadiene, phenyl vinyl ether and acrylonitrile) in such a way that the latter is promoted to an excited state and the former reverts to the ground state. The sensitisers generally used fulfil this function from their *triplet states*, because they are molecules which rapidly undergo intersystem crossing from their excited singlet to excited triplet states. The triplet states then live long enough to collide with the substrate, which is similarly created in its triplet state, and so almost all sensitised photolyses are triplet-state reactions.

For a singlet-state reaction, where both bonds might be being formed at the same time, we might make a further but tentative prediction. Since the orbitals which are interacting are identical on each component, the *endo*-HH adduct should be preferred over the *exo*-HH adduct, because the secondary interactions (Fig. 8.6, dashed lines) will always be bonding. In two cases, where the singlet-and triplet-state reactions have been carefully looked at and separated, this proves to be true. Thus coumarin **8.29**, in the excited singlet state, dimerises to give only the syn-HH dimer **8.30**, but in the triplet state it gives both *syn*-HH **8.30** and anti-HH isomers **8.31**, with only a trace of head-to-tail (HT) products. Acenaphthylene also gives the syn dimer from the singlet-state reaction and a mixture of the syn and anti dimers from the triplet-state reaction.

**Fig. 8.6**   Secondary orbital interactions (dashed lines) in photocycloadditions

8.29                                    8.30                         8.31

The whole truth about regio- and stereochemistry, however, is not nearly as simple as this. The experimental evidence has not been collected systematically, and it is not always known whether singlet or triplet states are involved. Scores of papers have been published in which reasonably reliable structures have been assigned to the cyclobutane dimers produced by irradiation of a great variety of unsymmetrical alkenes in solution. (The stereo- and regioselectivity in *solid-state* photodimerisation reactions have also received a lot of attention, but they are determined by the alignment of the monomers in the crystal lattice, not by orbital effects.) For irradiation in solution, head-to-head dimers are the major products in a majority of these papers, but head-to-tail dimers are the major products in a substantial minority. [2 + 2] Cycloadditions are not the only ones for which the regiochemistry is difficult to predict—almost all [4 + 4] photo-dimerisations of 9-substituted anthracenes give the head-to-head dimer, whereas the photodimerisations of 2-pyridones give the head-to-tail, centrosymmetric dimers.

The most obvious factors which might lead molecules to adopt the head-to-tail course are dipole-dipole repulsions and steric effects. That both of these factors are often overridden is evidence for the importance of other effects, which may or may not be best accounted for by orbital interactions or by the formation of the most stable diradical. The single biggest complication is with those reactions which do pass through triplet diradical intermediates. With unpaired spins these have a long lifetime before intersystem crossing allows them to form the second bond to give the cyclobutane product. In this time, the diradical could break apart, and give back starting materials. The regiochemistry may therefore depend not only upon which bond is formed most rapidly, but also upon how efficiently each of the possible diradicals undergoes intersystem crossing in order to close to give a cyclobutane.

### 8.3.2.3 The Photochemical Cross-Coupling of Alkenes.[36]  When both alkenes have Z-substituents, the regiochemistry ought to resemble that shown by the dimerisation of alkenes with Z-substituents, but irradiation of cyclopentenone 8.32 in the presence of methyl acrylate 8.33 gives four 1:1 *cis*-fused adducts, two stereoisomeric head-to-head products 8.34 and two stereoisomeric head-to-tail products 8.35. This is just one example of poor regioselectivity that is often observed, in spite of the orbital predictions in favour of head-to-head products (Fig 8.5). The reactions are known to take place with the cyclopentenone absorbing light in a $\pi$-$\pi$* transition, undergoing intersystem crossing to a

triplet state, and then reacting with the methyl acrylate in its ground state. It must therefore involve one or more of the four possible triplet diradicals. The long lifetime of the intermediate diradicals appears to explain many anomalous features of photochemical cross-couplings like this. Without going into the details,[37] it appears that the bond that is formed most easily is between the $\beta$ position of the cyclopentenone and the $\alpha$ position of the ester, the opposite of the frontier orbital prediction in Fig. 8.5. The large amount of head-to-head product **8.34** is actually the result of relatively efficient ring closure of the least stabilised diradical.

**8.32**   **8.33**                     **8.34**  46.5%              **8.35**  53.5%

When one of the alkenes undergoing a cross-coupling has an X-substituent and the other a Z-substituent, the frontier orbitals in Fig. 8.7 do not make unambiguous predictions. The energy separations for the HOMOs and LUMOs (Fig. 6.22) might suggest that the HOMO/'HOMO' interaction ($\Delta E$ 1.9 eV) is more effective than the LUMO/'LUMO' interaction ($\Delta E$ 3.0 eV). However, the HOMO of a Z-substituted alkene is not highly polarised, with some calculations giving the $\alpha$ position the larger coefficient. Head-to-head and head-to-tail regiochemistry are delicately balanced whatever the truth, and it may be that regiochemistry is determined more by the LUMO/'LUMO' interaction. In practice there is almost always a high level of selectivity in favour of the head-to-tail regioisomer. Clearly the frontier orbital analysis is insecure.

'HOMO'   HOMO                    'LUMO'   LUMO

**Fig. 8.7**   Frontier orbitals for the photocycloaddition of an X-substituted alkene with a Z-substituted alkene

The best studied reaction in this class is that between cyclopentenone **8.32** and ethyl vinyl ether **8.36**, which again gives four *cis*-fused 1:1 adducts, two stereoisomeric head-to-head products **8.39** and two stereoisomeric head-to-tail products **8.40**, 3:1 in favour of the head-to-tail isomers. Again, without going into details, only two diradicals **8.37** and **8.38** appeared to have been formed, and in equal amounts. It again seems that the regiochemistry of the cycloaddition is determined by the more efficient closure of the diradical **8.37**, than by any selective interaction in the initial bond-making step. The efficiency in the ring closing may stem from faster intersystem crossing or from a less favourable fragmentation to the starting materials.

The cycloadditions of several cyclopentenones to 2-methylpropene show similar patterns, and the corresponding cycloadditions of cyclohexenones also give largely head-to-tail products, except that the major product has a *trans* ring fusion. One reason for this may be that the intermediate triplet derived from the cyclohexenone will twist, because the two radicals will no longer be in overlapping orbitals. If the twist is maintained through the sequence it could lead to a *trans* ring junction. Alternatively, it is possible that a *trans*-cyclohexenone is an intermediate.

Houk has pointed out that in the twisted triplet from photoexcitation of an $\alpha,\beta$-unsaturated ketone, the $\beta$ position will resemble a simple alkyl radical, with a high-energy SOMO, while the $\alpha$ position will be an alkyl radical conjugated to an acyl group, with a low-energy SOMO. These radicals should behave like ground-state radicals (Chapter 7), the $\beta$ radical will be relatively nucleophilic and the $\alpha$ radical will be relatively electrophilic, and this difference may account for the regioselectivity if the cycloaddition is determined by the initial coupling. The nucleophilic radical at the $\beta$ position will interact most favourably with the $\beta$ position of a Z-substituted alkene **8.41**, leading to the head-to-head adduct found more often than not in that series. Similarly, the electrophilic radical at the $\alpha$ position will interact most favourably with the $\beta$ position of an X-substituted alkene **8.42**, leading to the head-to-tail adduct usually seen in that series. It remains to be seen whether this orbital control is actually a factor in the regioselectivity of photocycloadditions. The current position based on calculations for the simplest system, the gas phase reaction of acrolein and ethylene, is that selectivity in both the initial attack and in the relative rates of ring closure are likely to be factors.

One solution to the regiochemical problem is to tie the partners together, and make the reaction intramolecular. Thus both cyclohexenones *cis*- and *trans*-**8.43** give a single regioisomer **8.45**, but the reaction is not stereospecific. The C—C coupling determined by the nearer ends of the components forming the six-membered ring gives a triplet diradical **8.44** with a long enough lifetime to lose the configuration at the carbon atom carrying the methyl group.

The loss of stereospecificity was avoided by converting the $\alpha,\beta$-unsaturated ketone into the corresponding eniminium ion. This has no possibility of giving an n-$\pi$* transition, and will have a slow rate of intersystem crossing. The reaction can then be expected to stay in the singlet manifold and give rise to a concerted cycloaddition. It is indeed stereospecific, with the geometry of the double bond, *trans* or *cis*, retained in the product **8.45**.

There are some other singlet state cross-couplings that are straightforwardly explained as more or less concerted [2 + 2] cycloadditions. They are often but not always regioselective in the head-to-head sense, usually stereospecific, retaining the double bond geometry, and sometimes endo selective. Examples are the counter-thermodynamic combination of *trans*-stilbene and methyl crotonate **8.46** giving the more hindered cyclobutane **8.47** as the major adduct, and the formation from 9-cyanophenanthrene with *trans*-methylstyrene **8.48** of the adduct **8.49**, implying the maximum amount of $\pi$ stacking. $\pi$ Stacking between a benzene ring in an excited state and one in the ground state is bonding, whereas it was not, unless the stack was displaced [see (Section 2.4.3.4) pages 94–95], between two ground states.

### 8.3.2.4 Photocycloadditions with Cumulated Double Bonds.
Wiesner discovered that the regioselectivity in intermolecular cycloadditions of allene to $\alpha,\beta$-unsaturated ketones **8.50** gave the cyclobutane **8.51** with the central carbon of the allene bonded to the $\alpha$ position. The addition also took place with high stereoselectivity for attack on the lower face of the double bond in **8.50**, surprisingly, because the lower face is the more hindered. The approach

leading to the ketone **8.51** is more crowded than the approach leading to the alternative **8.52**.

**8.50**          **8.51**          **8.52**

The regiochemistry is not easily explained using the frontier orbitals, since the $\alpha,\beta$-unsaturated ketone is likely to be in the n-$\pi^*$ excited state. The 'HOMO' is then the singly occupied orbital on the oxygen atom, which is not involved in the reaction. The 'LUMO' is $\psi_3^*$, with a large coefficient on the $\beta$ carbon, but the LUMO of an allene [see (Section 6.5.4.3) page 256] is not significantly polarised. Equally, but less likely, if it were a $\pi$-$\pi^*$ excited state, although the HOMO of the allene is polarised, the 'HOMO' of the enone, $\psi_2$, is only a little polarised. With barely decisive frontier orbitals, it may be that the regiochemistry is a consequence only of steric effects. Somewhat unusually, the regiochemistry in these photochemical cycloadditions is the same as that of the thermal cycloaddition of an allene with a Z-substituted alkene (see page 256). However, the stereochemistry in the reaction with the octalone **8.50** is explicable on the same basis as the stereochemistry of the dissolving-metal reduction of similar $\alpha,\beta$-unsaturated ketones [see (Section 7.6.2.2) page 294]. The 'LUMO', $\psi_3^*$, will have a high electron population on C-5 (steroid numbering), which pyramidalises towards a tetrahedral geometry with the larger lobe in the atomic orbital on the lower surface, making the AB ring system more like that of a *trans*-decalin. The allene bonds more strongly to the larger lobe on C-5, and the bond to C-4 follows because of the suprafacial nature of the cycloaddition.

The regiochemistry of the reaction between an $\alpha,\beta$-unsaturated ketone and a ketene is the opposite, even in an intramolecular reaction **8.53** → **8.54**, which involves substantial twisting and ring strain. This time it is in the sense easily explicable by the frontier orbitals. The 'LUMO' of the unsaturated ketone and the LUMO of the ketene show that the initial bonding will be between the $\beta$ carbon of the enone and the carbonyl carbon of the ketene, and there is an orthogonal orbital on the other carbon atom of the ketene able to complete the cycloaddition, just as we saw earlier (see pages 212 and 253) for thermal reactions.

**8.53**          **8.54**

**8.3.2.5  Photocycloadditions to Aromatic Rings.**[38]   The benzene ring under-goes several remarkable photocycloadditions, sometimes behaving as a $_\pi 2$ com-ponent, with bonds forming to the *ortho* carbons **8.55**, sometimes as a $_\pi 4$ component, with bonds forming to the *para* carbons **8.57**, and more often than not in a more complicated reorganisation of the molecule, with bonds forming to the *meta* carbons **8.58**.

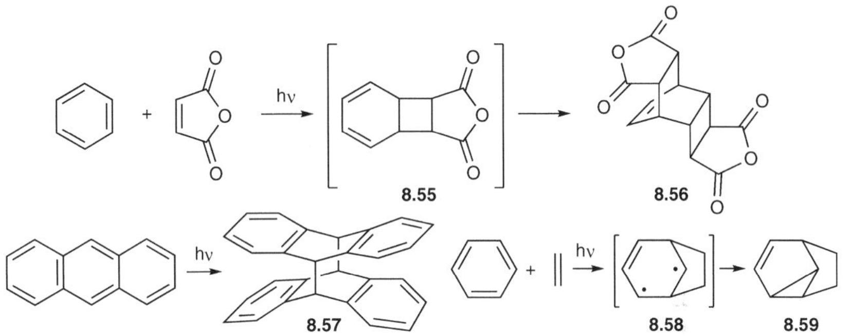

At first sight, the formation of the [2 + 2] adducts **8.55**, the intermediate in the formation of the compound actually isolated **8.56**, looks like a photochemically allowed pericyclic reaction—it would be a $[_\pi 2_s + _\pi 2_s]$ cycloaddition, which is common for one alkene with another. Surprisingly, a correlation diagram shows that this is not the case when one of the 'alkenes' is a benzene ring, and it is the *meta* adduct **8.58** that is the symmetry-allowed product. Perhaps for this reason the great majority of cycloadditions between an alkene and a benzene ring give *meta* adducts. The diradical **8.58** is a possible intermediate in the formation of the product actually isolated **8.59**, and invoking it is certainly a help in understanding the reorganisation of the bonds leading to the final product.

Let us look at the correlation diagrams for *ortho* and *meta* cycloadditions. Neither the ground state nor the excited states of benzene can be described with the same simplicity as the ground state and excited states of alkenes and linear polyenes, because of the degeneracy of the HOMOs and the LUMOs in benzene. The lowest excited singlet state, $^1B_{2u}$, is a combination of the con-figuration in which $\psi_2$ has lost an electron and $\psi_5^*$ has gained one, and the equally probable configuration in which $\psi_3$ has lost an electron and $\psi_4^*$ has gained one. The promotion of an electron from the ground state to the first excited state is actually forbidden by the symmetries of the four orbitals involved $\psi_2$, $\psi_3$ $\psi_4^*$ and $\psi_5^*$, and takes place only because of vibrational motions disturbing the strict symmetry. It is responsible for the low intensity absorption at 254 nm with its characteristic vibrational fine structure. We shall approximate the orbitals of a photoexcited benzene using only one of these configurations, namely the $\psi_2\psi_5^*$. The starting materials for a cycloaddition of

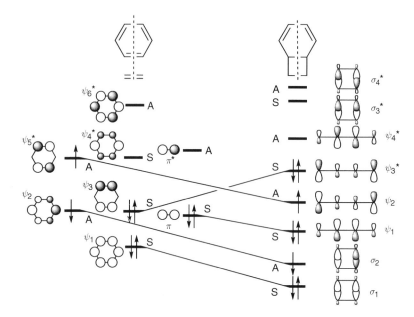

**Fig. 8.8** Correlation diagram for the $[2 + 2]$ cycloaddition of ethylene to benzene

photoexcited benzene to an alkene are shown on the left of Fig. 8.8, and the orbitals of the product of the $[2 + 2]$ cycloaddition are shown on the right. The only element of symmetry which would be preserved in a $[_\pi2_s + _\pi2_s]$ reaction is the plane perpendicular to the page bisecting both molecules. In contrast to the $[_\pi2_s + _\pi2_s]$ cycloaddition of two alkenes (see page 219), the first excited state on the left does not correlate with the first excited state on the right, but rather with a much higher-energy excited state corresponding to the promotion of an electron from the $\sigma_2$ orbital to $\psi_2$ of the diene unit and with $\psi_3*$ of the diene filled. Had we used the other configuration, with the electron promoted from $\psi_3$ to $\psi_4*$, we would have come to a similar conclusion, although a different set of orbitals in the product would have been occupied, reflecting the incompleteness of the picture we are using.

In contrast to the correlation diagram for the *ortho* cycloaddition, the equivalent correlation diagram for the *meta* cycloaddition in Fig. 8.9 shows that the first excited state on the left correlates directly with the ground state for the diradical **8.58**. It may be that the diradical is not actually an intermediate, but the correlation diagram can only be constructed by using it. The final product may be formed directly in a fully concerted cycloaddition if the diradical in the transition structure is already starting to form the second bond. Photoreactions forming this type of product are common, and they show the necessary conditions for a concerted pericyclic reaction: they are singlet-state reactions, stereospecifically suprafacial on both components.

It appears to be a general feature of photocycloadditions that they are more likely to be rule-obeying if the interacting orbitals of the two components are close

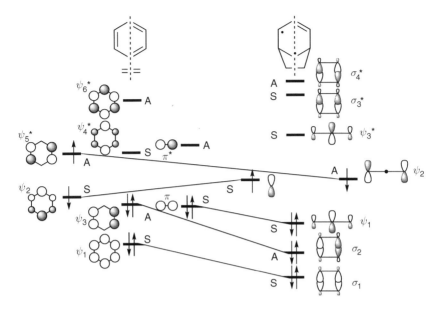

**Fig. 8.9** Correlation diagram for the *meta* cycloaddition of ethylene
to benzene

in energy. In the particular case of cycloadditions to benzene, Bryce-Smith and Gilbert have pointed out that the formation of the *meta* adduct is favoured when the alkene is not strongly polarised, and they narrow this down, with some exceptions, to those alkenes having ionisation potentials between 9.6 eV and 8.65 eV. This range straddles the HOMO energy of benzene at 9.24 eV, supporting the idea that the concerted cycloaddition is best achieved when the HOMO/ 'HOMO' interaction is between orbitals close in energy. More polar alkenes, like maleic anhydride, with high ionisation potentials, and correspondingly low-energy LUMOs, are more apt to react in the [2 + 2] mode, across *ortho* positions, as in the formation of the adduct **8.55**, perhaps because the interacting orbitals do not match well for the concerted pericyclic process.

Houk has suggested that the LUMO of a Z-substituted alkene is low enough in energy to encourage charge transfer from the 'LUMO' of the photoexcited benzene to the LUMO of the alkene, and the subsequent reactions of the charge-transfer complex will have different selectivities from the direct concerted reaction. Alkenes with electron-donating substituents are also apt to give *ortho* adducts rather than *meta*, as in the reactions of tetramethylethylene. In this case the charge transfer will be from the high-energy HOMO of the alkene to the 'HOMO' of the photoexcited benzene, and again the subsequent reactions are those of the charge-transfer complex, with a kinetic preference for *ortho* bonding.

There are many more detailed features of cycloadditions to the benzene ring which can be explained by orbital involvement—regioselectivity, endo-exo selectivity, and chemoselectivity especially—but the main points are those described above.

*8.3.2.6 Photochemical Di-π-Methane Rearrangements of Aromatic Systems.* The photochemical reaction of the aromatic alkene embedded in the bridged structure **8.60** does not result in cycloaddition of the double bond to the benzene ring, presumably because of the ring strain that would be created, whether it was *meta* or *ortho*. Instead the excited state undergoes a rearrangement either before or after intersystem crossing. Best results are often obtained by triplet sensitisation, which excludes the singlet-state pathway. This type of rearrangement belongs to a class called di-π-methane rearrangements. The triplet-state reaction is likely to be stepwise, and to involve one or more of the diradical intermediates **8.61** and **8.62**. The step from the intermediate **8.61** to **8.62** may be concerted with the formation of **8.61**, but the last step giving the cyclopropane **8.63** is likely to be separate, since it must wait for or be concerted with the intersystem crossing during which the spins become paired.

**8.60**       **8.61**       **8.62**       **8.63**

Overall the reaction corresponds to the cycloaddition of one of the π bonds to one of the allylic σ bonds, and it is formally possible that this is a concerted process giving the cyclopropane **8.63** directly in a single symmetry-allowed $[_\sigma 2_a + _\pi 2_a]$ step **8.64**. The stereochemistry of most di-π-methane rearrangements, largely investigated by Zimmerman, conforms to this pattern, and perhaps some of the singlet-state reactions are concerted. Nevertheless, they are more usually thought of as stepwise, and this certainly helps to explain the regiochemistry of the reactions observed when the aromatic ring has electron-donating and electron-withdrawing substituents.

$[_\sigma 2_a + _\pi 2_a]$

**8.64**

The observations are that a donor substituent, as in the benzonorbornene **8.65** forms the first bond to the *meta* position, which is best pictured as giving the diradical **8.66**, and hence the tricyclic product **8.67**. However, an electron-withdrawing group in the benzonorbornene **8.68** forms the first bond to the *para* position, which is best pictured as giving the diradical **8.69**, and hence the tricyclic product **8.70**. Both of these reactions are highly regioselective, yet the diradical **8.66** at least is probably the less stable of the two possible radicals in that case.

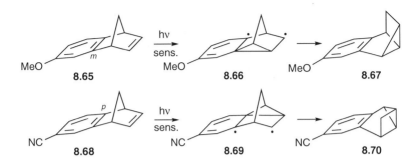

We can however find an explanation for this striking regioselectivity by looking at the antibonding orbitals $\psi_4^*$ and $\psi_5^*$, a weighted sum of which will be populated as the 'LUMO', a device similar to the weighted sum of the HOMOs for aromatic electrophilic substitution in the ground state [see (Section 4.3.4.1) page 129]. Although these are degenerate in benzene, the substituent lifts the degeneracy, and its nature determines which of the orbitals $\psi_4^*$ or $\psi_5^*$ is raised or lowered in energy, affecting the weighting of the sum of the pair. With an X-substituted benzene, the LUMO will be $\psi_5^*$ **8.71**, the energy of which is unaffected by the substituent, because it is at the node. This orbital will contribute disproportionately to the 'LUMO'. In this orbital the coefficient at the *para* position is zero and so bonding does not take place there. With a Z-substituent **8.68**, the frontier orbitals are the LUMO of the alkene and $\psi_4^*$ **8.72**, of the aromatic ring, because the energy of this orbital is lowered by the presence of the Z-substituent. In this orbital the coefficient at the *para* position is larger than at the *meta*, and bond formation takes place there. The only difference in the triplet state is that the spins of two of the singly occupied orbitals are inverted, which does not affect the argument.

### 8.3.3  Other Kinds of Selectivity in Pericyclic Photochemical Reactions

*8.3.3.1  Electrocyclic Reactions.*   Cycloheptatriene undergoes a photochemical, symmetry-allowed, disrotatory ring closure to give a bicyclo[3.2.0]heptadiene. A substituent placed at each of the three trigonal positions, as in 1-, 2- and 3-methoxycycloheptatriene, and in the corresponding ethoxycarbonyl cycloheptatrienes, affects which pair of double bonds is involved, and the regioselectivity in each case (Fig. 8.10) is different for an electron-donating and an electron-withdrawing substituent.

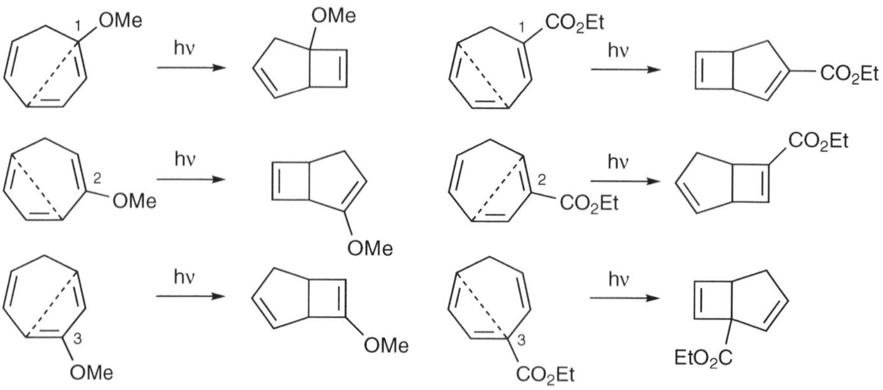

**Fig. 8.10**    Regioselectivity of electrocyclic closings in cycloheptatrienes

Clearly there is a strong electronic effect determining the regioselectivity. An explanation has been put forward by Houk,[39] using the change in conformation that must take place on excitation. He argues that one of the two trigonal carbon atoms at the terminus of the triene unit must twist out of conjugation with the rest of the conjugated system to give an excited state with a different conformation from the more nearly planar ground state. This is a consequence of promoting an electron into the LUMO, which has a node between C-1 and C-2. With one electron in the 'HOMO', which has a bonding interaction between these two atoms, and one in the 'LUMO', the $\pi$ bonding between these atoms is removed. Calculations suggest that there is a preference for partial positive charge to reside at the isolated p orbital and for the partial negative charge to be in the conjugated system **8.73**. The change in the charge distribution on photoexcitation has been called 'sudden polarisation'. The $\sigma$ bond of the cyclobutene ring then forms from the isolated p orbital, which is moving into the right position for bonding to C-4.

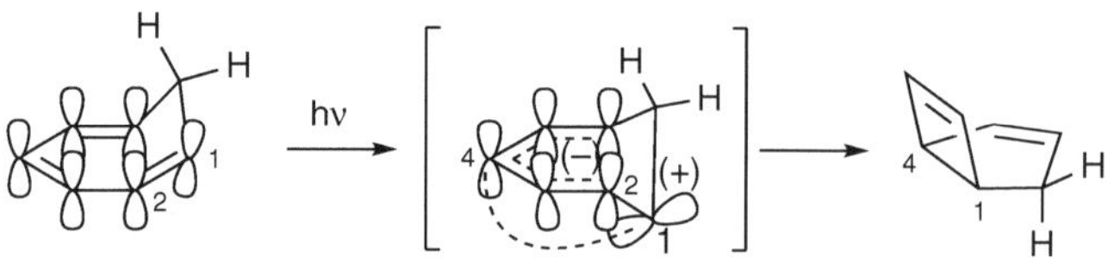

**8.73**

Positive charge at C-1 will be stabilised by a donor substituent at that site **8.74**. If the donor substituent is at C-2 or C-3, it will be best accommodated if it is at the nodes in $\psi_3$ of the pentadienyl anion **8.75** and **8.76**, where it will least destabilise that orbital. If an electron-withdrawing group is present it will best stabilise the negative charge if it is placed at the terminus or in the middle of the pentadienyl anion **8.77**–**8.79**. The pattern of charge distribution in the structures **8.74**–**8.79** matches the observed regioselectivity in all six cases.

**8.74**    **8.75**    **8.76**    **8.77**    **8.78**    **8.79**

***8.3.3.2   Sigmatropic   Rearrangements.*** Cycloheptatriene also undergoes a photochemical, symmetry-allowed, suprafacial [1,7]-shift of hydrogen, and, less readily, of other groups. In an unsymmetrical cycloheptatriene, there are two directions in which it can shift, and a little is known about this preference. A donor substituent at C-1 promotes the migration of the hydrogen atom towards C-1 **8.80**, whereas an electron-withdrawing group on C-1 **8.81** promotes the migration in the opposite direction.

**8.80**                                            **8.81**

The explanation for this electronically controlled selectivity is satisfyingly the same as that used in the previous section: in the excited state, either C-1 or C-6 experiences the sudden polarisation, and twists in whichever direction places either the positive charge on the carbon atom with the donor substituent **8.80** or the negative charge at the terminus of the pentadienyl anion system **8.81**. The shift of the hydrogen atom is then like the [1,2]-hydride shift in a cationic rearrangement, it takes place towards the carbon carrying the partial positive charge, C-1 in the former case and C-6 in the latter.

The two pathways, electrocyclic closure to the bicyclo[3.2.0]heptadiene and the [1,7]-shift of hydrogen, are in competition. This makes the experimental work more difficult, because the four possible isomers of the cycloheptatriene interconvert during the photolysis. Nevertheless the pattern of reactivity is clear, and the observation is that the better the donor substituent, the more it favours the formation of the bicyclo[3.2.0]heptadiene at the expense of the [1,7]-shift. This too is reasonable, since the hydride migration towards the donor-substituted atom **8.80** leads to a less stable cation, and ought to be slow.

# 8.4   Photochemically Induced Radical Reactions

As so much energy can be introduced at once, one of the most common ways of setting up a pair of radicals is by photochemical excitation. The excited state has a chemistry of its own, as we have seen in the earlier part of this chapter, but one outcome for the excited state is for a bond to break homolytically, with the two radicals (or the diradical if it is an internal bond that breaks) either produced in, or able rapidly to revert to, their electronic ground states. When this happens, the radicals usually show the same patterns of reactivity and selectivity as the corresponding radicals generated by pyrolysis of a strained or weak single bond.

A difference between the thermal and photochemical methods of generation is in sensitised photolysis, where the two radicals, or the diradical if they remain in the same molecule, are produced with unpaired spins, giving them a longer lifetime. An example is provided by the azo compound **8.82**, which gives octa-1,7-diene both on pyrolysis and on sensitised photolysis. The subtle difference is revealed by the deuterium labelling. In thermolysis, the diradical **8.83** is set up with the spins paired and with good overlap for easy cleavage of the $\sigma$ bond—the product is largely the $E,E$-diene **8.84**, in which the lifetime of the diradical **8.83** was too short to lose the configuration. In the sensitised photolysis, however, the triplet diradical **8.85** lived long enough for the molecule to change shape. The result is that the fragmentation of an intermediate, which might look something like **8.86**, gives more of the $E,Z$-diene **8.87** than of the $E,E$-diene **8.84**.

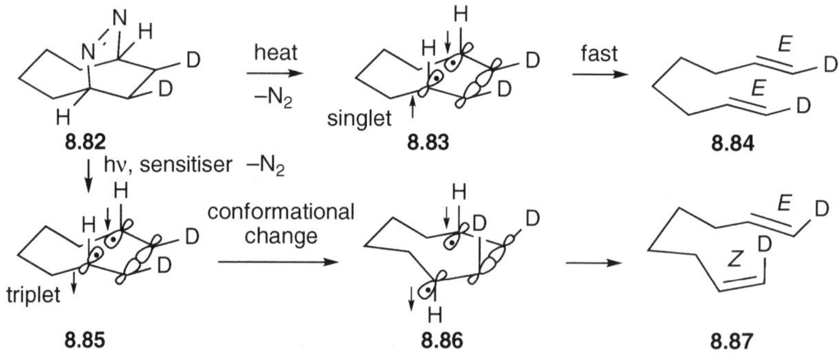

The photolysis of phenyl benzoate **8.88** is an example of the photo-Fries rearrangement, in which the acyl radical is created close to a phenoxy radical **8.89**. The C—O bond is weakened by the presence of an electron in the $\pi^*$ orbital of the ester conjugated system. The radical coupling then shows the usual selectivity for C—C bond formation, with the *para* selectivity in line with the coefficients of the SOMO of the phenoxy radical (see page 291).

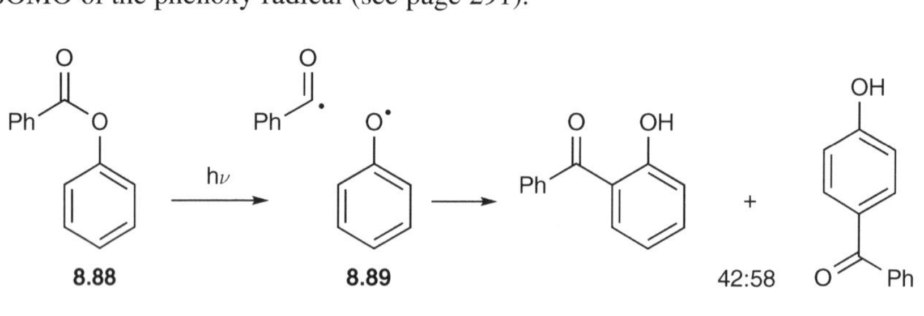

The formation of benzpinacol **8.92** by the action of light on benzophenone is one of the oldest known organic photochemical reactions. It is now known that the reaction takes place from the n-$\pi$* triplet state **8.90**. The electron left behind in the p orbital on oxygen will be low in energy, and therefore close to the HOMO of the C—H bond of the solvent, which is conjugated to the oxygen lone pair, and raised in energy relative to the other C—H bonds. The other unpaired electron will be in the carbonyl $\pi$* orbital, which is comparatively low in energy for an antibonding orbital, and therefore well separated from the rather high energy of the $\sigma$* orbital of the C—H bond. The interaction productive of hydrogen atom transfer is therefore to the oxygen **8.90** (arrows). After the hydrogen abstraction has taken place, the radical produced **8.91** will be in its electronic ground state, and this will dimerise with the usual regioselectivity of a benzyl radical.

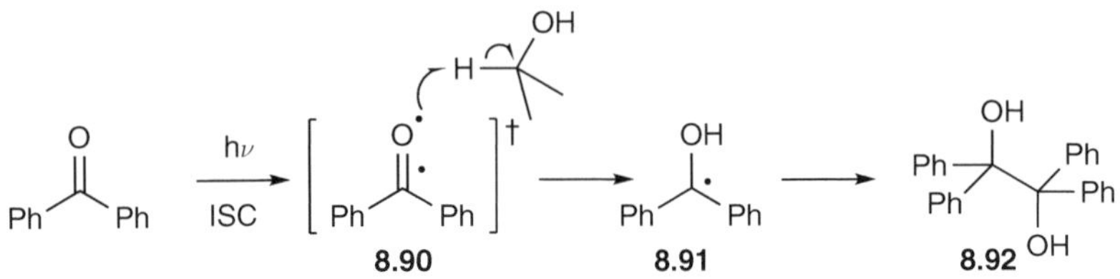

## 8.5 Chemiluminescence

One final reaction only marginally belongs in this chapter. It is the opposite of a photochemical reaction—instead of the organic molecule absorbing light and then undergoing chemical reaction some organic molecules undergo a chemical reaction in which light is emitted. When dioxetanes are heated, they release a large amount of energy, enough for a photon to be emitted.[40] The dioxetane **8.93** easily cleaves on warming, in one or two steps, to give the radical cation **8.94** and the radical anion **8.95**. This step is helped by the presence of the nitrogen atom, which is an X-substituent stabilising the electron deficiency in the radical cation. The radical anion **8.95** transfers an electron to the radical cation **8.94** to give acridone **8.96** and adamantanone **8.97**. The remarkable feature of this reaction is that the acridone is produced in a photochemically excited state.

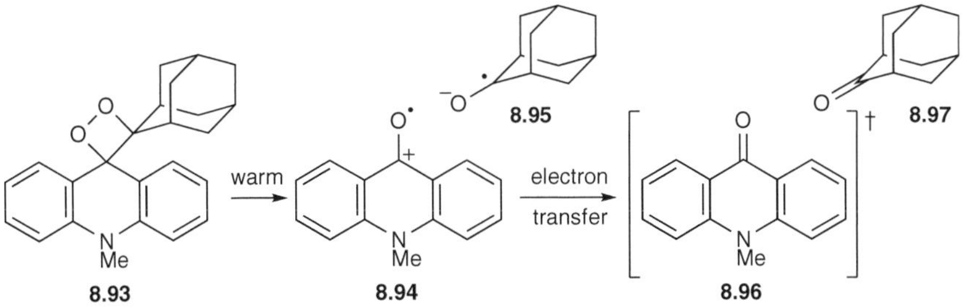

The electron in the high-energy SOMO of the radical anion **8.95** (effectively the $\pi$* orbital of the carbonyl group) is evidently transferred to the relatively low-energy LUMO of the radical cation (Fig. 8.11), giving the acridone product **8.96** with the

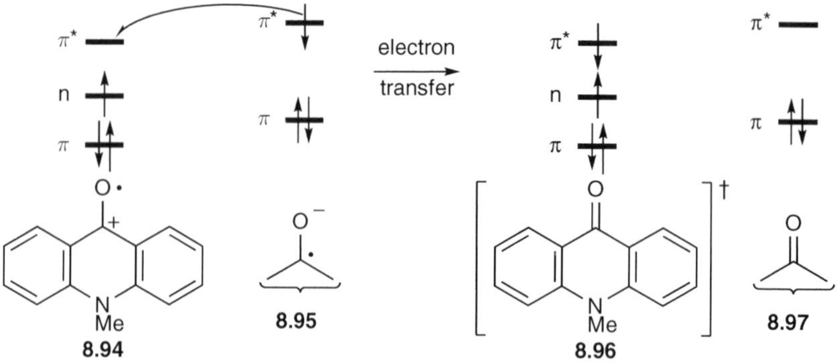

**Fig. 8.11**   Orbitals for the electron transfer from the radical anion **8.95** to the radical cation **8.94**

orbital occupancy of the n-π* excited state, which simply emits a photon when the excited state decays to the ground state.

Chemiluminescence can also occur with electron transfer in the opposite direction, as in the oxidation of sodium 9,10-diphenylanthracenide **8.98** by dibenzoyl peroxide. An electron is transferred from the anthracenide to the oxidising agent to give a benzoyloxy radical, a benzoate ion and the excited state of anthracene **8.99**, which emits a photon. The $\sigma$ and $\sigma^*$ orbitals of an O—O bond are low in energy, a characteristic of an oxidising agent, and the electron transferred to it comes most easily from what was the HOMO of the anthracene rather than from the SOMO of the anthracenide, leading the anthracene to be created in its first excited state. It is also likely that another electron transfer takes place from the anthracenide to the benzoyl radical, the reaction being characteristic of an oxidising agent with a hydrocarbon radical anion.

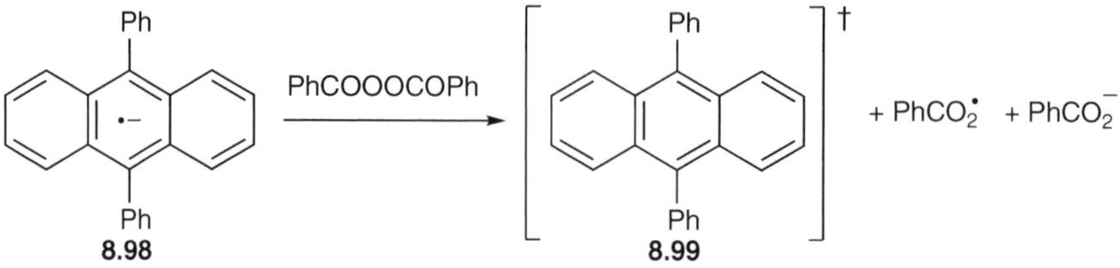

## 8.6   Exercises

1. Explain the regioselectivity in this reaction:

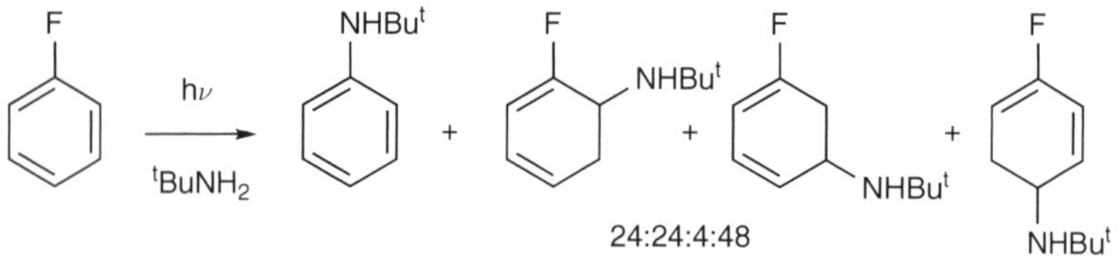

24:24:4:48

2. In contrast to the different regiochemistry seen in the photochemical ring
closures of the benznorbornadienes **8.65** and **8.68**, both benznorbornadienes
**8.100** show the same regioselectivity, probably by way of the diradical **8.101**.
Explain this observation.

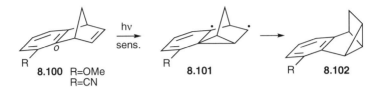

8.100  R=OMe
      R=CN

8.101

8.102

# References

## Note

The reference version of this book (Ref. 3) has more than 1850 references to the sources for everything in this book, as well as for those subjects covered there but not included here. The following is a selection of those references that might be the most immediately useful.

1. For a defence of valence bond theory, see L. Pauling, *J. Chem. Educ.*, 1992, **69**, 519 and some letters in the same issue on pp. 600–602. For some other champions of valence bond theory, and for its development to explain reactivity, see A. Pross and S. S. Shaik, *Acc. Chem. Res.*, 1983, **16**, 362.
2. D. Young, *Computational Chemistry: A Practical Guide for Applying Techniques to Real World Problems*, John Wiley & Sons, Ltd, New York, 2001.
3. I. Fleming, *Molecular Orbitals and Organic Chemical Reactions*, Reference Edition, John Wiley & Sons, Ltd, Chichester, in press.
4. C. A. Coulson and H. C. Longuet-Higgins, *Proc. R. Soc. London, Ser. A*, 1947, **192**, 16.
5. K. Fukui, *Acc. Chem. Res.*, 1971, **4**, 57; K. Fukui, Angew. *Chem., Int. Ed. Engl.*, 1982, **21**, 801.
6. I. Fleming, *Frontier Orbitals and Organic Chemical Reactions*, John Wiley & Sons, Ltd, Chichester, 1976.
7. G. C. Pimentel and R. D. Spratley, *Chemical Bonding Clarified through Quantum Mechanics*, Holden-Day, San Francisco, 1969; R. McWeeny, *Coulson's Valence*, 3rd Edn, Oxford University Press, Oxford, 1979; A. Szabo and N. S. Ostlund, *Modern Quantum Chemistry*, Dover, New York, 1996; N. J. B. Green, *Quantum Mechanics 1: Foundations*, Oxford University Press, Oxford, 1997; D. A. McQuarrie and J. D. Simon, *Physical Chemistry: A Molecular Approach*, University Science Books, Sausalito, 1997; V. M. S. Gil, *Orbitals in Chemistry*, Cambridge University Press, Cambridge, 2000; A. Vincent, *Molecular Symmetry and Group Theory*, 2nd Edn, John Wiley & Sons, Ltd, Chichester, 2001; A. Rauk, *Orbital Interaction Theory of Organic Chemistry*, 2nd Edn, John Wiley & Sons, Ltd, New York, 2001; D. O. Hayward, *Quantum Mechanics for Chemists*, Royal Society of Chemistry, Cambridge, 2002; J. E. House, *Fundamentals of Quantum Chemistry*, 2nd Edn, Elsevier, Amsterdam, 2004; N. T. Anh, *Frontier Orbitals; A Practical Manual*, John Wiley & Sons, Ltd, Chichester, 2007; J. Keeler and P. Wothers, *Chemical Structure and Reactivity*, Oxford University Press, Oxford, 2008.

*Molecular Orbitals and Organic Chemical Reactions: Student Edition*   Ian Fleming
© 2009 John Wiley & Sons, Ltd

8. R. F. Hout, W. J. Pietro and W. J. Hehre, *A Pictorial Approach to Molecular Structure and Reactivity*, John Wiley & Sons, Ltd, New York, 1984; T. A. Albright, J. K. Burdett and M.-H. Whangbo, *Orbital Interactions in Chemistry*, John Wiley & Sons, Ltd, New York, 1985; W. J. Hehre, L. Radom, P. Von R. Schleyer and J. A. Pople, *Ab Initio Molecular Orbital Theory*, John Wiley & Sons, Ltd, New York, 1986; R. G. Parr and W. Yang, *Density Functional Theory of Atoms and Molecules*, Oxford University Press, Oxford, 1989; Y. Jean and F. Volatron, translated and edited by J. Burdett, *An Introduction to Molecular Orbitals*, Oxford University Press, Oxford, 1993; J. K. Burdett, *Chemical Bonds: a Dialog*, John Wiley & Sons, Ltd, Chichester, 1997; A. Hinchliffe, *Modelling Molecular Structures*, 2nd Edn, John Wiley & Sons, Ltd, Chichester, 2000; W. Koch and M. C. Holthausen, *A Chemist's Guide to Density Functional Theory*, Wiley-VCH, New York, 2001; R. J. Gillespie and P. L. A. Popelier, *Chemical Bonding and Molecular Geometry*, Oxford University Press, New York, 2001; C. J. Cramer, *Essentials of Computational Chemistry*, John Wiley & Sons, Ltd, New York, 2002; P.W. Atkins and R. S. Friedman, *Molecular Quantum Mechanics*, 4th Edn, Oxford University Press, Oxford, 2005; F. Weinhold and C. Landis, *Valency and Bonding*, Cambridge University Press, Cambridge, 2005; F. Jensen, *Introduction to Computational Chemistry*, John Wiley & Sons, Ltd, New York, 2007.

9. Captodative Stabilisation: R. Sustmann and H.-G. Korth, *Adv. Phys. Org. Chem.*, 1989, **26**, 131.

10. A. J. Kirby, *The Anomeric Effect and Related Stereoelectronic Effects at Oxygen*, Springer-Verlag, Berlin, 1983.

11. R. G. Pearson, *Chemical Hardness*, John Wiley & Sons, Ltd, New York, 1997.

12. Tables 3.1–3.5: R. G. Parr and R. G. Pearson, *J. Am. Chem. Soc.*, 1983, **105**, 7512; R. G. Pearson, *J. Am. Chem. Soc.*, 1985, **107**, 6801; R. G. Pearson, *J. Am. Chem. Soc.*, 1988, **110**, 7684; R. G. Pearson, *J. Org. Chem.*, 1989, **54**, 1423.

13. The Hammond postulate: G. S. Hammond, *J. Am. Chem. Soc.*, 1955, **77**, 334.

14. The Salem-Klopman equation: G. Klopman, *J. Am. Chem. Soc.*, 1968, **90**, 223; L. Salem, *J. Am. Chem. Soc.*, 1968, **90**, 543 and 553.

15. G. Hagen and H. Mayr, *J. Am. Chem. Soc.*, 1991, **113**, 4954; H. Mayr, R. Schneider, B. Irrgang and C. Schade, *J. Am. Chem. Soc.*, 1990, **112**, 4454.

16. R. Loos, S. Kobayashi and H. Mayr, *J. Am. Chem. Soc.*, 2003, **125**, 14126; A. A. Tishkov and H. Mayr, *Angew. Chem., Int. Ed. Engl.*, 2005, **44**, 142; H. F. Schaller, U. Schmidhammer, E. Riedle and H. Mayr, *Chem. Eur. J.*, 2008, **14**, 3866.

17. For an account of the many ideas in this area, see *Chem. Rev.*, 1999, 1067.

18. S. D. Kahn, C. F. Pau, A. R. Chamberlin and W. J. Hehre, *J. Am. Chem. Soc.*, 1987, **109**, 650.

19. A. S. Cieplak, *Chem. Rev.*, 1999, **99**, 1265; A. S. Cieplak, Organic Addition and Elimination Reactions; Transformation Paths of Carbonyl Derivatives, Ch. 6 in *Structure Correlation*, Eds H.-B. Bürgi and J. D. Dunitz, VCH, Weinheim, 1994, Vol. 1, p. 205.

20. R. B. Woodward and R. Hoffmann, *Angew. Chem., Int. Ed. Engl.*, 1969, **8**, 781, later republished as *The Conservation of Orbital Symmetry*, Verlag Chemie, Weinheim, 1970.

21. For much of the material in this chapter, and more examples, see I. Fleming, *Pericyclic Reactions*, Oxford University Press, Oxford, 1998.

22. K. N. Houk, *J. Am. Chem. Soc.*, 1973, **95**, 4092.

23. P. Caramella, P. Quadrelli and L. Toma, *J. Am. Chem. Soc.*, 2002, **124**, 1130; A. G. Leach and K.N. Houk, *Chemtracts: Org. Chem.*, 2002, **15**, 611.

24. K. N. Houk, J. Sims, R. E. Duke, Jr, R. W. Strozier and J. K. George, *J. Am. Chem. Soc.*, 1973, **95**, 7287; K. N. Houk, J. Sims, C. R. Watts and L. J. Luskus, *J. Am. Chem. Soc.*, 1973, **95**, 7301.

25. B. K. Carpenter, *Tetrahedron*, 1978, **34**, 1877; C. F. Wilcox, Jr and B. K. Carpenter, *J. Am. Chem. Soc.*, 1979, **101**, 3897.

26. N. G. Rondan and K. N. Houk, *J. Am. Chem. Soc.*, 1985, **107**, 2099.

27. A. M. Boersma, J. W. de Haan, H. Kloosterziel and L. J. M. van de Ven, *J. Chem. Soc., Chem. Commun.*, 1970, 1168; F.-G. Klärner and M. Wette, *Chem. Ber.*, 1978, **111**, 282; F. Toda, K. Tanaka, T. Tamashima and M. Kato, *Angew. Chem., Int. Ed. Engl.*, 1998, **37**, 2724.

28. J.-C. Paladini and J. Chuche, *Bull. Soc. Chim. Fr.*, 1974, 197; J. A. Berson, P. B. Dervan and J. A. Jenkins, *J. Am. Chem. Soc.*, 1972, **94**, 7598; W. D. Huntsman and H. J. Wristers, *J. Am. Chem. Soc.*, 1967, **89**, 342; J. B. Lambert, D. M. Fabricius and J. J. Napoli, *J. Am. Chem. Soc.*, 1979, **101**, 1793.

29. P. Schiess and R. Dinkel, *Tetrahedron Lett.*, 1975, **16**, 2503; K. S. Replogle and B. K. Carpenter, *J. Am. Chem. Soc.*, 1984, **106**, 5751; N. Choony, P. G. Sammes, G. Smith and R. W. Ward, *Chem. Commun.*, 2001, 2062.

30. H. E. Zimmerman and D. S. Crumrine, *J. Am. Chem. Soc.*, 1968, **90**, 5612; T. M. Brennan and R. K. Hill, *J. Am. Chem. Soc.*, 1968, **90**, 5614; Y. Yamamoto, J. Oda and Y. Inouye, *Tetrahedron Lett.*, 1979, 2411 reports the reaction without one of the methyl groups; I. Fleming and N. K. Terrett, J. Chem. Soc., Perkin Trans. 1, 1998, 2645; S. M. Sheehan, G. Lalic, J. S. Chen and M. D. Shair, *Angew. Chem., Int. Ed. Engl.*, 2000, **39**, 2714; T. P. Yoon, V. M. Dong and D. W. C. MacMillan, *J. Am. Chem. Soc.*, 1999, **121**, 9726; H. Hu, D. Smith, R. E. Cramer and M. A. Tius, *J. Am. Chem. Soc.*, 1999, **121**, 9895; M. P. Doyle, W. Hu and D. J. Timmons, *Org. Lett.*, 2001, **3**, 3741.

31. For a more thorough frontier orbital treatment of radical reactions, see D. Lefort, J. Fossey and J. Sorba, *Free Radicals in Organic Chemistry*, John Wiley & Sons, Ltd, Chichester, 1995.

32. G. Moad and D. H. Solomon, *The Chemistry of Free Radical Polymerization*, Pergamon, Oxford, 1995.

33. D. P. Curran, N. A. Porter and B. Giese, *Stereochemistry of Radical Reactions*, VCH, Weinheim, 1996.

34. E. Havinga and M. E. Kronenberg, *Pure Appl. Chem.*, 1968, **16**, 137; J. Cornelisse, *Pure Appl. Chem.*, 1975, **41**, 433; J. Cornelisse and E. Havinga, *Chem. Rev.*, 1975, **75**, 353; J. Cornelisse, G. P. de Gunst and E. Havinga, *Adv. Phys. Org. Chem.*, 1975, **11**, 225.

35. For a compilation of photochemical [2 + 2] cycloadditions to enones, see M. T. Crimmins and T. L. Reinhold, *Org. React.*, 1993, **44**, 297.

36. D. I. Schuster, G. Lem and N. A. Kaprinidis, *Chem. Rev.*, 1993, **93**, 3.

37. D. Andrew, D. J. Hastings and A. C. Weedon, *J. Am. Chem. Soc.*, 1994, **116**, 10 870.

38. D. Bryce-Smith, A. Gilbert and J. Mattay, *Tetrahedron*, 1977, **33**, 2459; D. Bryce-Smith, A. Gilbert and J. Mattay, *Tetrahedron*, 1986, **42**, 6011; P. A. Wender, L. Siggel and J. M. Nuss, *Org. Photochem.*, 1989, **10**, 357; J. Cornelisse, *Chem. Rev.*, 1993, **93**, 615.

39. T. Tezuka, O. Kikuchi, K. N. Houk, M.N. Paddon-Row, C. M. Santiago, N. G. Rondan, J. C. Williams, Jr and R. W. Gandour, *J. Am. Chem. Soc.*, 1981, **103**, 1367.

40. F. McCapra, I. Beheshti, A. Burford, R. A. Hann and K. A. Zaklika, *J. Chem. Soc., Chem. Commun.*, 1977, 944; F. McCapra, *J. Chem. Soc., Chem. Commun.*, 1977, 946; F. McCapra, *Tetrahedron Lett.*, 1993, **34**, 6941.

# Index

Made in the USA
San Bernardino, CA
29 September 2014